# ELEMENTS OF ALGEBRA

BY

## ARTHUR SCHULTZE, Ph.D.

FORMERLY ASSISTANT PROFESSOR OF MATHEMATICS, NEW YORK UNIVERSITY
HEAD OF THE MATHEMATICAL DEPARTMENT, HIGH
SCHOOL OF COMMERCE, NEW YORK CITY

D0861094

New York
THE MACMILLAN COMPANY
1917

*All rights reserved*

COPYRIGHT, 1910,

BY THE MACMILLAN COMPANY.

Set up and electrotyped. Published May, 1910. Reprinted
September, 1910 ; January, 1911; July, 1912 ; February, 1913;
January, 1915; May, September, 1916; August, 1917.

Norwood Press
. J. S. Cushing Co. — Berwick & Smith Co.
Norwood, Mass., U.S.A.

# PREFACE

In this book the attempt is made to shorten the usual course in algebra, while still giving to the student complete familiarity with all the essentials of the subject. While in many respects similar to the author's "Elementary Algebra," this book, owing to its peculiar aim, has certain distinctive features, chief among which are the following:

1. *All unnecessary methods and "cases" are omitted.* These omissions serve not only practical but distinctly pedagogic ends. Until recently the tendency was to multiply "cases" as far as possible, in order to make every example a special case of a memorized method. Such a large number of methods, however, not only taxes a student's memory unduly but invariably leads to mechanical modes of study. The entire study of algebra becomes a mechanical application of memorized rules, while the cultivation of the student's reasoning power and ingenuity is neglected. Typical in this respect is the treatment of factoring in many text-books In this book all methods which are of real value, and which are applied in advanced work are given, but "cases" that are taught only on account of tradition, short-cuts that solve only examples specially manufactured for this purpose, etc., are omitted.

2. *All parts of the theory which are beyond the comprehension of the student or which are logically unsound are omitted.* All practical teachers know how few students understand and appreciate the more difficult parts of the theory, and conse-

v

quently hardly ever emphasize the theoretical aspect of algebra. Moreover, a great deal of the theory offered in the average text-book is logically unsound; *e.g.* all proofs for the sign
of the product of two negative numbers, all elementary proofs
of the binomial theorem for fractional exponents, etc.

3. *The exercises are slightly simpler than in the larger book.*
The best way to introduce a beginner to a new topic is to offer
him a large number of simple exercises. For the more ambi-
tious student, however, there has been placed at the end of
the book a collection of exercises which contains an abundance
of more difficult work. With very few exceptions all the exercises in this book differ from those in the "Elementary Algebra"; hence either book may be used to supplement the other.

4. *Topics of practical importance, as quadratic equations and
graphs, are placed early in the course.* This arrangement will
enable students who can devote only a minimum of time to
algebra to study those subjects which are of such importance
for further work.

In regard to some other features of the book, the following
may be quoted from the author's "Elementary Algebra":

"Particular care has been bestowed upon those chapters
which in the customary courses offer the greatest difficulties to
the beginner, especially problems and factoring. The presen-
tation of problems as given in Chapter V will be found to be
quite a departure from the customary way of treating the subject, and it is hoped that this treatment will materially diminish the difficulty of this topic for young students.

"The book is designed to meet the requirements for admission to our best universities and colleges, in particular the
requirements of the College Entrance Examination Board.
This made it necessary to introduce the theory of proportions

and graphical methods into the first year's work, an innovation which seems to mark a distinct gain from the pedagogical point of view.

"By studying proportions during the first year's work, the student will be able to utilize this knowledge where it is most needed, viz. in geometry; while in the usual course proportions are studied a long time after their principal application.

"Graphical methods have not only a great practical value, but they unquestionably furnish a very good antidote against 'the tendency of school algebra to degenerate into a mechanical application of memorized rules.' This topic has been presented in a simple, elementary way, and it is hoped that some of the modes of representation given will be considered improvements upon the prevailing methods. The entire work in graphical methods has been so arranged that teachers who wish a shorter course may omit these chapters."

Applications taken from geometry, physics, and commercial life are numerous, but the true study of algebra has not been sacrificed in order to make an impressive display of sham applications. It is undoubtedly more interesting for a student to solve a problem that results in the height of Mt. McKinley than one that gives him the number of Henry's marbles. But on the other hand very few of such applied examples are genuine applications of algebra, — nobody would find the length of the Mississippi or the height of Mt. Etna by such a method, — and they usually involve difficult numerical calculations. Moreover, such examples, based upon statistical abstracts, are frequently arranged in sets that are algebraically uniform, and hence the student is more easily led to do the work by rote than when the arrangement is based principally upon the algebraic aspect of the problem.

It is true that problems relating to physics often offer a field for genuine applications of algebra. The average pupil's knowledge of physics, however, is so small that an extensive use of such problems involves as a rule the teaching of physics by the teacher of algebra.

Hence the field of genuine applications of elementary algebra suitable for secondary school work seems to have certain limitations, but within these limits the author has attempted to give as many simple applied examples as possible.

The author desires to acknowledge his indebtedness to Mr. William P. Manguse for the careful reading of the proofs and for many valuable suggestions.

ARTHUR SCHULTZE.

New York,
April, 1910.

# CONTENTS

## CHAPTER I

## CHAPTER II

## CHAPTER III

## CHAPTER IV

## CHAPTER V

## CHAPTER VI

## CHAPTER VII

## CHAPTER VIII

## CHAPTER IX

## CHAPTER X

## CHAPTER XI

## CHAPTER XII

## CHAPTER XIII

## CHAPTER XIV

## CHAPTER XV

# ELEMENTS OF ALGEBRA

# ELEMENTS OF ALGEBRA

## CHAPTER I

### INTRODUCTION

**1. Algebra** may be called an extension of arithmetic. Like arithmetic, it treats of numbers, but these numbers are frequently denoted by letters, as illustrated in the following problem.

### ALGEBRAIC SOLUTION OF PROBLEMS

**2. Problem.** The sum of two numbers is 42, and the greater is five times the smaller. Find the numbers.

Let $x =$ the smaller number.

Then $5x =$ the greater number,

and $6x =$ the sum of the two numbers.

Therefore, $6x = 42,$

$x = 7,$ the smaller number,

and $5x = 35,$ the greater number.

**3.** A **problem** is a question proposed for solution.

**4.** An **equation** is a statement expressing the equality of two quantities; as, $6x = 42.$

**5.** In algebra, problems are frequently solved by denoting numbers by letters and by expressing the problem in the form of an equation.

**6. Unknown numbers** are usually represented by the last letters of the alphabet; as, $x, y, z$, but sometimes other letters are employed.

B

1

## EXERCISE 1

Solve algebraically the following problems:

**1.** The sum of two numbers is 40, and the greater is four times the smaller. Find the numbers.

**2.** A man sold a horse and a carriage for $480, receiving twice as much for the horse as for the carriage. How much did he receive for the carriage?

**3.** A and B own a house worth $14,100, and A has invested twice as much capital as B. How much has each invested?

**4.** The population of South America is 9 times that of Australia, and both continents together have 50,000,000 inhabitants. Find the population of each.

**5.** The rise and fall of the tides in Seattle is twice that in Philadelphia, and their sum is 18 feet. Find the rise and fall of the tides in Philadelphia.

**6.** Divide $240 among A, B, and C so that A may receive 6 times as much as C, and B 8 times as much as C.

**7.** A pole 56 feet high was broken so that the part broken off was 6 times the length of the part left standing. Find the length of the two parts.

**8.** The sum of the sides of a triangle equals 40 inches. If two sides of the triangle are equal, and each is twice the remaining side, how long is each side?

**9.** The sum of the three angles of any triangle is 180°. If 2 angles of a triangle are equal, and the remaining angle is 4 times their sum, how many degrees are there in each?

**10.** The number of negroes in Africa is 10 times the number of Indians in America, and the sum of both is 165,000,000. How many are there of each?

**11.** Divide $280 among A, B, and C, so that B may receive twice as much as A, and C twice as much as B.

**12.** Divide $90 among A, B, and C, so that B may receive twice as much as A, and C as much as A and B together.

**13.** A line 20 inches long is divided into two parts, one of which is equal to 5 times the other. How long are the parts?

**14.** A travels twice as fast as B, and the sum of the distances traveled by the two is 57 miles. How many miles did each travel?

**15.** A, B, C, and D buy $2100 worth of goods. How much does A take, if B buys twice as much as A, C three times as much as B, and D six times as much as B?

## NEGATIVE NUMBERS

### EXERCISE 2

**1.** Subtract 9 from 16.

**2.** Can 9 be subtracted from 7?

**3.** In arithmetic why cannot 9 be subtracted from 7?

**4.** The temperature at noon is 16° and at 4 P.M. it is 9° What is the temperature at 4 P.M.? State this as an of subtraction.

**5.** The temperature at 4 P.M. is 7°, and at 10 P.M. it is 10° less. What is the temperature at 10 P.M.?

**6.** Do you know of any other way of expressing the last answer (3° below zero)?

**7.** What then is 7 − 10?

**8.** Can you think of any other practical examples which require the subtraction of a greater number from a smaller one?

---

**7.** Many practical examples require the subtraction of a greater number from a smaller one, and in order to express in a convenient form the results of these, and similar examples, it becomes necessary to enlarge our concept of number, so as to include numbers less than zero.

**8. Negative numbers** are numbers smaller than zero; they are denoted by a prefixed minus sign; as − 5 (read " minus 5 "). Numbers greater than zero, for the sake of distinction, are frequently called positive numbers, and are written either with a prefixed plus sign, or without any prefixed sign; as + 5 or 5.

The fact that a thermometer falling 10° from 7° indicates 3° below zero may now be expressed

$$7° - 10° = -3°.$$

Instead of saying a gain of $ 30, and a loss of $ 90 is equal to a loss of $ 60, we may write

$$\$30 - \$90 = -\$60.$$

**9.** The **absolute value** of a number is the number taken without regard to its sign.

The absolute value of − 5 is 5, of + 3 is 3.

**10.** It is convenient for many discussions to represent the positive numbers by a succession of equal distances laid off on a line from a point 0, and the negative numbers by a similar series in the opposite direction.

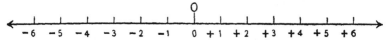

Thus, in the annexed diagram, the line from 0 to + 5 represents + 5, the line from 0 to − 4 represents − 4, etc. The addition of 3 is represented by a motion of three spaces toward the right, and the subtraction of 3 by a similar motion toward the left.

Thus, 5 added to − 1 equals 4, 5 subtracted from − 1 equals − 6, etc.

### EXERCISE 3

**1.** If in financial transactions we indicate a man's income by a positive sign, what does a negative sign indicate ?

**2.** State in what manner the positive and negative signs may be used to indicate north and south latitude, east and west longitude, motion upstream and downstream.

**3.** If north latitude is indicated by a positive sign, by what is south latitude represented ?

**4.** If south latitude is indicated by a positive sign, by what is north latitude represented ?

**5.** What is the meaning of the year $-20$ A.D.? Of an easterly motion of $-6$ yards per second ?

**6.** A merchant gains $\$200$, and loses $\$350$. (*a*) What is his total gain or loss ? (*b*) Find $200-350$.

**7.** If the temperature at 4 A.M. is $-8°$ and at 9 A.M. it is $7°$ higher, what is the temperature at 9 A.M.? What, therefore, is $-8+7$ ?

**8.** A vessel starts from a point in $25°$ north latitude, and sails $38°$ due south. (*a*) Find the latitude at the end of the journey. (*b*) Find $25-38$.

**9.** A vessel starts from a point in $15°$ south latitude, and sails $22°$ due south. (*a*) Find the latitude at the end of the journey. (*b*) Subtract $22$ from $-15$.

| | |
|---|---|
| **10.** From 30 subtract 40. | **18.** To $-6$ add 12. |
| **11.** From 4 subtract 7. | **19.** To $-2$ add 1. |
| **12.** From 7 subtract 9. | **20.** To $-1$ add 2. |
| **13.** From 19 subtract 34. | **21.** From 1 subtract 2. |
| **14.** From 0 subtract 14. | **22.** To $-8$ add 9. |
| **15.** From $-12$ subtract 20. | **23.** To $-7$ add 4. |
| **16.** From $-2$ subtract 5. | **24.** From $-1$ subtract 2. |
| **17.** From $-1$ subtract 1. | **25.** Add $-1$ and 2. |

**26.** Solve examples 16–25 by using a diagram similar to the one of § 10, and considering additions and subtractions as motions.

**27.** Which is the greater number:

(*a*) 1 or $-1$ ?    (*b*) $-2$ or $-4$ ?

**28.** By how much is $-7$ greater than $-12$ ?

**29.** Determine from the following table the range of temperature in each locality :

| | Highest | Lowest |
|---|---|---|
| Flagstaff, Ariz. . . . . | 93° | 20° |
| Chicago . . . . . . | 103° | −23° |
| Washington, D.C. . . . . | 104° | 15° |
| Springfield, Ill. . . . . | 107° | − 24° |
| New York City . . . . | 100° | − 6° |
| Key West . . . . . | 100° | 41° |
| Boston . . . . . . | 102° | −13° |

## NUMBERS REPRESENTED BY LETTERS

**11.** For many purposes of arithmetic it is advantageous to
express numbers by letters. One advantage was shown in § 2;
others will appear in later chapters (§ 30).

### EXERCISE 4

**1.** If the letter $t$ means 1000, what is the value of $5t$?

**2.** What is the value of $3b$, if $b = 3$? if $b = 4$?

**3.** What is the value of $a + b$, if $a = 5$, and $b = 7$? if $a = 6$,
and $b = -4$?

**4.** What is the value of $17c$, if $c = 5$? if $c = -2$?

**5.** If a boy has $9d$ marbles and wins $4d$ marbles, how
many marbles has he?

**6.** Is the last answer correct for any value of $d$?

**7.** A merchant had $20m$ dollars and lost $11m$ dollars.
How much has he left?

**8.** What is the sum of $8b$ and $6b$?

**9.** Find the numerical value of the last answer if $b = 15$.

**10.** If $c$ represents a certain number, what represents 9 times
that number?

**11.** From 26 *m* subtract 19 *m*.

**12.** What is the numerical value of the last answer if *m* = 2? if *m* = − 2?

**13.** From 22 *m* subtract 25 *m*, and find the numerical value of the answer if *m* = 12.

**14.** Add 13 *p*, 3 *p*, 6 *p*, and subtract 24 *p* from the sum.

**15.** From − 10 *q* subtract 20 *q*.     **17.** From 0 subtract 26 *x*.

**16.** Add − 10 *q* and + 20 *q*.     **18.** Add − 6 *x* and 8 *x*.

**19.** From − 22 *x* subtract 0.     **20.** From − 10 *p* subtract 10 *p*.

**21.** If *a* = 20, then 7 *a* = 140. What sign, therefore, is understood between 7 and *a* in the expression 7 *a*?

## FACTORS, POWERS, AND ROOTS

**12.** The **signs of addition, subtraction, multiplication, division,** and **equality** have the same meaning in algebra as they have in arithmetic.

**13.** If there is **no sign between two letters,** or a letter and a number, a sign of multiplication is understood.

5 × *a* is generally written 5 *a* ; *m* × *n* is written *mn*.

Between two figures, however, a sign of multiplication (either × or ·) has to be employed; as, 4 × 7, or 4 · 7.

4 × 7 cannot be written 47, for 47 means 40 + 7.

**14.** A **product** is the result obtained by multiplying together two or more quantities, each of which is a factor of the product.

Since 24 = 3 × 8, or 12 × 2, each of these numbers is a factor of 24.

Similarly, 7, *a*, *b*, and *c* are factors of 7 *abc*.

**15.** A **power** is the product of two or more equal factors; thus, *aaaaa* is called the "5th power of *a*," and written $a^5$; *aaaaaa*, or $a^6$, is "the 6th power of *a*," or *a* 6th.

The *second power* is also called the *square*, and the *third power* the *cube;* thus, $12^2$ (read "12 square") equals 144.

**16.** The **base** of a power is the number which is repeated as a factor.

The base of $a^3$ is $a$.

**17.** An **exponent** is the number which indicates how many times a base is to be used as a factor. It is placed a little above and to the right of the base.

The exponent of $m^6$ is 6; $n$ is the exponent of $a^n$.

### EXERCISE 5

**1.** Write and find the numerical value of the square of 7, the cube of 6, the fourth power of 3, and the fifth power of 2. Find the numerical values of the following powers:

| | | | |
|---|---|---|---|
| **2.** $7^2$. | **6.** $4^2$. | **10.** $(\frac{1}{2})^3$. | **14.** $25^1$. |
| **3.** $2^5$. | **7.** $2^4$. | **11.** $0^9$. | **15.** $.0001^2$. |
| **4.** $5^2$. | **8.** $10^6$. | **12.** $(4\frac{1}{2})^2$. | **16.** $1.1^2$. |
| **5.** $8^3$. | **9.** $1^{30}$. | **13.** $(1.5)^2$. | **17.** $2^2 + 3^2$. |

If $a = 3$, $b = 2$, $c = 1$, and $d = \frac{1}{2}$, find the numerical values of:

| | | | | |
|---|---|---|---|---|
| **18.** $a^3$. | **20.** $c^{10}$. | **22.** $a^2$. | **24.** $(2\,c)^3$. | **26.** $2\,ab$. |
| **19.** $b^2$. | **21.** $d^4$. | **23.** $(6\,d)^2$. | **25.** $ab$. | **27.** $(4\,bd)^2$. |

**28.** If $a^3 = 8$, what is the value of $a$?

**29.** If $m^2 = \frac{1}{16}$, what is the value of $m$?

**30.** If $4^a = 64$, what is the value of $a$?

---

**18.** In a product any factor is called the **coefficient** of the product of the other factors.

In $12\ mn^3p$, 12 is the coefficient of $mn^3p$, $12\ m$ is the coefficient of $n^3p$.

**19.** A **numerical coefficient** is a coefficient expressed entirely in figures.

In $-17\ xyx$, $-17$ is the numerical coefficient.

When a product contains no numerical coefficient, 1 is understood; thus $a = 1\ a$, $a^3b = 1\ a^3b$.

**20.** When several powers are multiplied, the beginner should remember that every exponent refers only to the number near which it is placed.

$3\,a^2$ means $3\,aa$, while $(3\,a)^2 = 3\,a \times 3\,a$.

$9\,aby^3 = 9\,abyyy$.

$2^4\,xy^3z^2 = 2 \cdot 2 \cdot 2 \cdot 2 \cdot xyyyzz$.

$7\,abc^3 = 7\,abccc$.

### EXERCISE 6

If $a = 4$, $b = 1$, $c = 2$, and $x = \frac{1}{2}$, find the numerical values of:

| | | | |
|---|---|---|---|
| 1. $4\,a$. | 6. $a^3$. | 11. $4\,ab^3$. | 16. $12\,ab^5c^3x$. |
| 2. $6\,x$. | 7. $b^{10}$. | 12. $4\,a^3b^2$. | 17. $\frac{1}{8}\,a^4x^4$. |
| 3. $abc$. | 8. $4\,ab^2$. | 13. $2\,bcx^2$. | 18. $\frac{1}{4}\,abc^2$. |
| 4. $abcx$. | 9. $5\,bc^2x$. | 14. $3\,abc^3$. | 19. $\dfrac{a^2}{c^3}$. |
| 5. $6\,bc$. | 10. $9\,a^3$. | 15. $2\,a^2bc^2$. | |

---

**21.** A **root** is one of the equal factors of a power. According to the number of equal factors, it is called a square root, a cube root, a fourth root, etc.

3 is the square root of 9, for $3^2 = 9$.

5 is the cube root of 125, for $5^3 = 125$.

$a$ is the fifth root of $a^5$, the $n$th root of $a^n$.

The $n$th root is indicated by the symbol $\sqrt[n]{\phantom{a}}$; thus $\sqrt[5]{a}$ is the fifth root of $a$, $\sqrt[3]{27}$ is the cube root of 27, $\sqrt[2]{a}$, or more simply $\sqrt{a}$, is the square root of $a$.

Using this symbol we may express the definition of root by $(\sqrt[n]{a})^n = a$.

**22.** The **index** of a root is the number which indicates what root is to be taken. It is written in the opening of the radical sign.

In $\sqrt[7]{a}$, 7 is the index of the root.

**23.** The **signs of aggregation** are: the *parenthesis*, ( ); the *bracket*, [ ]; the *brace*, { }; and the *vinculum*, ——.

They are used, as in arithmetic, to indicate that the expressions included are to be treated as a whole.

Each of the forms $10 \times (4+1)$, $10 \times [4+1]$, $10 \times \overline{4+1}$ indicates that 10 is to be multiplied by $4+1$ or by 5.

$(a-b)$ is sometimes read "quantity $a-b$."

### EXERCISE 7

If $a=2, b=3, c=1, d=0, x=9$, find the numerical value of:

| | | |
|---|---|---|
| 1. $\sqrt{x}$. | 7. $\sqrt{a^2}$. | 13. $4(a+b)$. |
| 2. $\sqrt{3\,b}$. | 8. $\sqrt[3]{b^3}$. | 14. $6(b+c)$. |
| 3. $\sqrt{2\,a}$. | 9. $4\sqrt{3\,bc}$. | 15. $(c+d)^2$. |
| 4. $\sqrt[3]{4\,a}$. | 10. $5\sqrt{16\,c}$. | 16. $[b-c]^3$. |
| 5. $\sqrt[3]{c}$. | 11. $a\sqrt{cx}$. | 17. $(b+c)\sqrt{x}$. |
| 6. $\sqrt{6\,ab}$. | 12. $bd\sqrt{c^2x^2}$. | |

## ALGEBRAIC EXPRESSIONS AND NUMERICAL SUBSTITUTIONS

**24.** An **algebraic expression** is a collection of algebraic symbols representing some number; *e.g.* $6\,a^2b - 7\sqrt{ac^2} + 9$.

**25.** A **monomial** or **term** is an expression whose parts are not separated by a sign $+$ or $-$; as $3\,ax^2$, $-9\sqrt{x}$, $\dfrac{-3\,ab}{5\,c^2}$.

$a(b+c+d)$ is a monomial, since the parts are $a$ and $(b+c+d)$.

**26.** A **polynomial** is an expression containing more than one term.

$4\,x+y$, $\dfrac{7\,x}{y} + \sqrt{z} - 3\,a^3b$, and $a^4 + b^4 + c^4 + d^4$ are polynomials.

**27.** A **binomial** is a polynomial of two terms.

$a^2 + b^2$, and $\tfrac{3}{4} - \sqrt[3]{a}$ are binomials.

**28.** A **trinomial** is a polynomial of three terms.

$a+b+c$, $a+9\,b+\sqrt{3}$ are trinomials.

**29.** In a polynomial each term is treated as if it were contained in a parenthesis, *i.e.* each term has to be computed before the different terms are added and subtracted. Otherwise all operations of addition, subtraction, multiplication, and division are to be performed in the order in which they are written from left to right.

*E.g.* $3 + 4 \cdot 5$ means $3 + 20$ or $23$.

**Ex. 1.** Find the value of $4 \cdot 2^3 + 5 \cdot 3^2 - \dfrac{9\sqrt{36}}{2}$.

$$4 \cdot 2^3 + 5 \cdot 3^2 - \frac{9\sqrt{36}}{2}$$
$$= 4 \cdot 8 + 5 \cdot 9 - \frac{9 \cdot 6}{2}$$
$$= 32 + 45 - 27$$
$$= 50.$$

**Ex. 2.** If $a = 5$, $b = 3$, $c = 2$, $d = 0$, find the numerical value of $6\,ab^2 - 9\,ab^2c + \frac{2}{5}\,a^3b - 19\,a^2bcd$.

$$6\,ab^2 - 9\,ab^2c + \tfrac{2}{5}\,a^3b - 19\,a^2bcd$$
$$= 6 \cdot 5 \cdot 3^2 - 9 \cdot 5 \cdot 3^2 \cdot 2 + \tfrac{2}{5} \cdot 5^3 \cdot 3 - 19 \cdot 5^2 \cdot 3 \cdot 2 \cdot 0$$
$$= 6 \cdot 5 \cdot 9 - 9 \cdot 5 \cdot 9 \cdot 2 + \tfrac{2}{5} \cdot 125 \cdot 3 - 0$$
$$= 270 - 810 + 150$$
$$= -390.$$

### EXERCISE 8 *

If $a=4$, $b=3$, $c=1$, $d=0$, $x=\frac{1}{2}$, find the numerical value of:

1. $a + 2\,b + 3\,c.$

2. $3\,a + 5\,b^2.$

3. $a^2 - b.$

4. $a^2 - 5\,c^2 + d^2.$

5. $5\,a^2 - 4\,bc + 2$

6. $2\,a^2 + 3\,ab + 4\,b^2 - 9\,ad.$

7. $6\,a^4 + 4\,ab^2 - bc^3 + 12\,ad^4.$

8. $27\,c^3 - 5\,ax - 50\,abcd.$

9. $5\,cb^2 + 6\,ac^3 - 17\,c^3 + 12\,x.$

10. $a^3 + b^3 + c^3 - d^3.$

11. $3\,ab^2 + 3\,a^2b - abc^2.$

12. $2\,a^2b - 3\,abc^2 + 7\,ab - 2\,x.$

13. $(a + b)c.$

14. $(a + b)^2 \cdot c + (a + b)c^2.$

15. $a^2 + x(2\,b + a^2).$

16. $4\,ab + \sqrt{a} - \sqrt{2\,x}.$

* For additional examples see page 258.

17. $\dfrac{a+b}{a-b}$.

18. $\dfrac{a^2 + 8\,b^2}{8}$.

19. $(a+b)(b+c)+(a+c)(b+d)-(a+d)(b-c)$.

20. $\dfrac{1}{a}+\dfrac{1}{c}-\dfrac{1}{b}$.

Find the numerical value of $8\,a^3 - 12\,a^2b + 6\,ab^2 - b^3$, if:

21. $a = 2,\ b = 1$.   26. $a = 3,\ b = 3$.

22. $a = 2,\ b = 2$.   27. $a = 4,\ b = 5$.

23. $a = 3,\ b = 2$.   28. $a = 4,\ b = 6$.

24. $a = 3,\ b = 4$.   29. $a = 3,\ b = 6$.

25. $a = 3,\ b = 5$.   30. $a = 4,\ b = 7$.

Express in algebraic symbols:

31. Six times $a$ plus 4 times $b$.

32. Six times the square of $a$ minus three times the cube of $b$.

33. Eight $x$ cube minus four $x$ square plus $y$ square.

34. Six $m$ cube plus three times the quantity $a$ minus $b$.

35. The quantity $a$ plus $b$ multiplied by the quantity $a^2$ minus $b^2$.

36. Twice $a^3$ diminished by 5 times the square root of the quantity $a$ minus $b$ square.

37. Read the expressions of Exs. 2–6 of the exercise.

38. What kind of expressions are Exs. 10–14 of this exercise?

---

**30.** The representation of numbers by letters makes it possible to state very briefly and accurately some of the principles of arithmetic, geometry, physics, and other sciences.

Ex. If the three sides of a triangle contain respectively $a$, $b$, and $c$ feet (or other units of length), and the area of the triangle is $S$ square feet (or squares of other units selected), then

$$S = \tfrac{1}{4}\sqrt{(a+b+c)(a+b-c)(a-b+c)(b-a+c)}.$$

*E.g.* the three sides of a triangle are respectively 13, 14, and 15 feet, then $a = 13$, $b = 14$, and $c = 15$; therefore

$$S = \tfrac{1}{4}\sqrt{(13+14+15)(13+14-15)(13-14+15)(14-13+15)}$$
$$= \tfrac{1}{4}\sqrt{42 \cdot 12 \cdot 14 \cdot 16}$$
$$= \tfrac{1}{4} \times 336$$

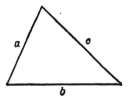

$= 84$, *i.e.* the area of the triangle equals 84 square feet.

### EXERCISE 9

**1.** The distance $s$ passed over by a body moving with the uniform velocity $v$ in the time $t$ is represented by the formula

$$s = vt.$$

Find the distance passed over by:

*a.* A snail in 100 seconds, if $v = .16$ centimeters per second.

*b.* A train in $4\tfrac{1}{2}$ hours, if $v = 30$ miles per hour.

*c.* An electric car in 40 seconds, if $v = 50$ meters per second

*d.* A carrier pigeon in 10 minutes, if $v = 5000$ feet per minute.

**2.** A body falling from a state of rest passes in $t$ seconds over a space $S = \tfrac{1}{2} gt^2$. (This formula does not take into account the resistance of the atmosphere.) Assuming $g = 32$ feet,

(*a*) How far does a body fall from a state of rest in 2 seconds?

(*b*) A stone dropped from the top of a tree reached the ground in $2\tfrac{1}{2}$ seconds. Find the height of the tree.

(*c*) How far does a body fall from a state of rest in $\tfrac{1}{100}$ of a second?

**3.** By using the formula

$$S = \tfrac{1}{4}\sqrt{(a+b+c)(a+b-c)(a-b+c)(b-a+c)},$$

find the area of a triangle whose sides are respectively

(*a*) 3, 4, and 5 feet.

(*b*) 5, 12, and 13 inches.

(*c*) 4, 13, and 15 feet.

**4.** If the radius of a circle is $R$ units of length (inches, meters, etc.), the area $S = 3.14 \cdot R^2$ square units (square inches, square meters, etc.). Find the area of a circle whose radius is

(*a*) 10 meters.     (*b*) 2 inches.     (*c*) 5 miles.

**5.** If $i$ represents the simple interest of $p$ dollars at $r\%$ in $n$ years, then $i = p \cdot n \cdot r \%$, or $\dfrac{pnr}{100}$.

Find by means of this formula:

(*a*) The interest on $800 for 4 years at $2\frac{1}{2}\%$.
(*b*) The interest on $500 for 2 years at 4%.

**6.** If the diameter of a sphere equals $d$ units of length, the surface $S = 3.14\, d^2$ (square units). (The number 3.14 is frequently denoted by the Greek letter $\pi$. This number cannot be expressed exactly, and the value given above is only an approximation.)

Find the surface of a sphere whose diameter equals:

(*a*) 8000 miles.     (*b*) 1 inch.     (*c*) 10 feet.

**7.** If the diameter of a sphere equals $d$ feet, then the volume

$$V = \frac{\pi d^3}{6} \text{ cubic feet.}$$

Find the volume of a sphere whose diameter equals:

(*a*) 10 feet.     (*b*) 3 feet.     (*c*) 8000 miles.

**8.** If $F$ denotes the number of degrees of temperature indicated on the Fahrenheit scale, the equivalent reading $C$ on the Centigrade scale may be found by the formula

$$C = \tfrac{5}{9}(F - 32).$$

Change the following readings to Centigrade readings:

(*a*) 122° F.     (*b*) 32° F.     (*c*) 5° F.

# CHAPTER II

## ADDITION, SUBTRACTION, AND PARENTHESES

### ADDITION OF MONOMIALS

**31.** While in arithmetic the word *sum* refers only to the result obtained by adding positive numbers, in algebra this word includes also the results obtained by adding negative, or positive and negative numbers.

In arithmetic we add a gain of $6 and a gain of $4, but we cannot add a gain of $6 and a loss of $4. In algebra, however, we call the aggregate value of a gain of 6 and a loss of 4 the sum of the two. Thus a gain of $2 is considered the sum of a gain of $6 and a loss of $4. Or in the symbols of algebra

$$(+\$6)+(-\$4)=+\$2.$$

Similarly, the fact that a loss of $6 and a gain of $4 equals a loss of $2 may be represented thus

$$(-\$6)+(+\$4)=(-\$2).$$

In a corresponding manner we have for a loss of $6 and a loss of $4

$$(-\$6)+(-\$4)=(-\$10).$$

Since similar operations with different units always produce analogous results, we *define* the sum of two numbers in such a way that these results become general, or that

$$6+(-4)=+2,$$
$$(-6)+(+4)=-2.$$
$$(-6)+(-4)=-10,$$

and
$$(+6)+(+4)=+10.$$

15

**32.** These considerations lead to the following principle:

*If two numbers have the same sign, add their absolute values ; if they have opposite signs, subtract their absolute values and (always) prefix the sign of the greater.*

**33.** The **average** of two numbers is one half their sum, the average of three numbers is one third their sum, and the average of *n* numbers is the sum of the numbers divided by *n*.

Thus, the average of 4 and 8 is 6.
The average of 2, 12, − 4 is 3⅓.
The average of 2, − 3, 4, 5, − 6, 10, is 2.

**EXERCISE 10**

Find the sum of:

| | | | |
|---|---|---|---|
| **1.** $-4$ | **4.** $+5$ | **7.** $+11$ | **10.** $+6$ |
| $+5$ | $+4$ | $-3$ | $-6$ |
| | | | |
| **2.** $-5$ | **5.** $-9$ | **8.** $-11$ | **11.** $-12$ |
| $+4$ | $+7$ | $-3$ | $-3$ |
| | | | |
| **3.** $-5$ | **6.** $-7$ | **9.** $-5$ | **12.** $0$ |
| $-4$ | $-8$ | $-17$ | $-5$ |
| | | | |
| **13.** $4$ | **14.** $-1$ | **15.** $-1$ | **16.** $-12$ |
| $5$ | $+2$ | $-2$ | $+15$ |
| $-6$ | $-3$ | $-3$ | $-3$ |

Find the values of:

**17.** $(-17)+(-14)$.

**18.** $15+(-9)$.

**19.** $0+(-12)$·

**20.** $1+(-2)$.

**21.** $(-2)+3+(-4)+5$.

**22.** $1+(-2)+(-3)+(-5)$.

In Exs. 23–26, find the numerical values of $a+b+c+d$, if:

**23.** $a=2,\ b=-3,\ c=-4,\ d=5$.

**24.** $a=5,\ b=5,\ \ \ c=-5,\ d=0$.

**25.** $a = 22$, $b = -23$, $c = 14$, $d = 1$.

**26.** $a = -13$, $b = 15$, $c = 0$, $d = 3$.

**27.** What number must be added to 9 to give 12?

**28.** What number must be added to 12 to give 9?

**29.** What number must be added to $-3$ to give 6?

**30.** What number must be added to $-3$ to give $-6$?

**31.** Add 2 yards, 7 yards, and 3 yards.

**32.** Add $2\,a$, $7\,a$, and $3\,a$.

**33.** Add $2\,a$, $7\,a$, and $-3\,a$.

Find the average of the following sets of numbers:

**34.** 3 and 25.          **37.** 2, 3, 4, $-7$, and 13.

**35.** 5 and $-13$.          **38.** $-3$, $-4$, $-5$, $-6$, $-7$, and 1.

**36.** 12, $-13$, and $-2$.

**39.** Find the average of the following temperatures: 4° F., $-8°$ F., 27° F., and 3° F.

**40.** Find the average temperature of New York by taking the average of the following monthly averages: 30°, 32°, 37°, 48°, 60°, 69°, 74°, 72°, 66°, 55°, 43°, 34°.

**41.** Find the average gain per year of a merchant, if his yearly gain or loss during 6 years was: $5000 gain, $3000 gain, $1000 loss, $7000 gain, $500 loss, and $4500 gain.

**42.** Find the average temperature of Irkutsk by taking the average of the following monthly temperatures: $-12°$, $-10°$, $-4°$, 1°, 6°, 10°, 12°, 10°, 6°, 0°, $-5°$, $-11°$ (Centigrade).

---

**34. Similar** or **like terms** are terms which have the same literal factors, affected by the same exponents.

$6\,ax^2y$ and $-7\,ax^2y$, or $-5\,a^2b$ and $\dfrac{a^2b}{7}$, or $16\sqrt{a+b}$ and $-2\sqrt{a+b}$, are similar terms.

**Dissimilar** or **unlike terms** are terms which are not similar.

$4\,a^2bc$ and $-4\,a^2bc^2$ are dissimilar terms.

c

**35. The sum of two similar terms** is another similar term.

The sum of $3\,x^2$ and $-\tfrac{1}{2}\,x^2$ is $\tfrac{5}{2}\,x^2$.

Dissimilar terms cannot be united into a single term. The sum of two such terms can only be indicated by connecting them with the $+$ sign.

The sum of $a$ and $a^2$ is $a + a^2$.

The sum of $a$ and $-b$ is $a + (-b)$, or $a - b$.

**36. Algebraic sum.** In algebra the word **sum** is used in a wider sense than in arithmetic. While in arithmetic $a - b$ denotes a difference only, in algebra it may be considered either the difference of $a$ and $b$ or the sum of $a$ and $-b$.

The sum of $-a$, $-2\,ab$, and $4\,ac^2$ is $-a - 2\,ab + 4\,ac^2$.

### EXERCISE 11

Add:

| 1. | 2. | 3. | 4. |
|---|---|---|---|
| $-2\,a$ | $2\,ab$ | $6\,c^2dn$ | $-2\,pqr^3$ |
| $+3\,a$ | $ab$ | $-7\,c^2dn$ | $-3\,pqr^3$ |
| $-4\,a$ | $-4\,ab$ | $c^2dn$ | $-4\,pqr^3$ |

| 5. | 6. | 7. | 8. |
|---|---|---|---|
| $b^3x$ | $12\,x^2y^3$ | $-19\,a^2bc$ | $-7\,a^3x$ |
| $-2\,b^3x$ | $13\,x^2y^3$ | $-20\,a^2bc$ | $-16\,a^3x$ |
| $-b^3x$ | $-7\,x^2y^3$ | $+7\,a^2bc$ | $8\,a^3x$ |

Find the sum of:

9. $2\,a^2, -3\,a^2, +4\,a^2, -5\,a^2, +6\,a^2.$

10. $-12\,mp^2, -13\,mp^2, -25\,mp^2, 7\,mp^2.$

11. $2(a+b), -3(a+b), -9(a+b), -12(a+b)$

12. $5(x+y^2), 6(x+y^2), -13(x+y^2), -24(x+y^2).$

13. $\sqrt{m+n}, 5\sqrt{m+n}, -12\sqrt{m+n}, -14\sqrt{m+n}.$

Simplify:

**14.** $2\,ab - 6\,ab + 12\,ab + 4\,ab - 50\,ab.$

**15.** $-17\,c^3 + 15\,c^3 + 18\,c^3 + 22\,c^3 + c^3.$

**16.** $-4\,p^3t - 4\,p^3t + p^3t + 3\,p^3t - 17\,p^3t.$

**17.** $-xyz + xyz - 12\,xyz + 13\,xyz + 15\,xyz.$

Add:

| **18.** $m$ | **19.** $+m$ | **20.** $-m^2$ | **21.** $a$ |
|---|---|---|---|
| $\underline{\quad n\quad}$ | $\underline{\,-n\,}$ | $\underline{\,-n^2\,}$ | $\underline{\,-a^3\,}$ |

| **22.** $b$ | **23.** $c^2$ | $\checkmark$ **24.** $b$ | **25.** $7$ |
|---|---|---|---|
| $\underline{-1}$ | $\underline{1}$ | $b^2$ | $\underline{-cx^2}$ |
| | | $\underline{\quad}$ | |

| **26.** $mn$ | **27.** $-xyz$ | $\checkmark$ **28.** $\sqrt{p+q}$ |
|---|---|---|
| $\underline{mn^2}$ | $\underline{xyz^3}$ | $\underline{-\sqrt{p-q}}$ |

Simplify the following by uniting like terms:

**29.** $3\,a - 7\,b + 5\,a + 2\,a - 3\,b - 10\,a + 11\,b.$

**30.** $2\,a^2 - 4\,a - 4 + 6\,a^2 - 7\,a^2 - 9\,a - 2\,a + 8.$

**31.** $2\,d + 3\,e + 4\,e - 5\,x - 7\,x - 12\,e - 4\,d + 5\,d.$

**32.** $3\,a^2 - 4\,ab + 7\,ab + 2\,b^2 - 4\,a^2 + 6\,ab - 4\,b^2.$

**33.** $\sqrt{x+y} - \sqrt{x+y^2} - 2\sqrt{x+y} + 2\sqrt{x+y^2} + 3\sqrt{x+y}.$

Add, without finding the value of each term:

**34.** $5 \times 173 + 6 \times 173 - 3 \times 173 - 7 \times 173.$

**35.** $4 \times 9^2 + 2 \times 9^2 - 10 \times 9^2 - 3 \times 9^2 + 8 \times 9^2.$

**36.** $10 \times 3^8 - 4 \times 3^8 - 2 \times 3^8 + 5 \times 3^8 - 6 \times 3^8 - 2 \times 3^8.$

## ADDITION OF POLYNOMIALS

**37. Polynomials are added** by uniting their like terms. It is convenient to arrange the expressions so that like terms may be in the same vertical column, and to add each column.

Thus, to add $26\,ab - 8\,abc - 15\,bc,\ -12\,ab + 15\,abc - 20\,c^2,$ $-5\,ab + 10\,bc - 6\,c^2,$ and $-7\,abc + 4\,bc + c^2,$ we proceed as follows:

$$
\begin{array}{l}
26\,ab - \phantom{1}8\,abc - 15\,bc \\
-12\,ab + 15\,abc \phantom{- 15\,bc} \quad -20\,c^2 \\
-\phantom{1}5\,ab \phantom{+ 15\,abc - 15\,bc} +10\,bc - \phantom{1}6\,c^2 \\
\phantom{-12\,ab} -\phantom{1}7\,abc + \phantom{1}4\,bc + \phantom{1}c^2 \\
\hline
\phantom{-1}9\,ab \phantom{+ 15\,abc} -\phantom{1} bc - 25\,c^2 \quad \text{Sum.}
\end{array}
$$

**38.** Numerical substitution offers a convenient method for **checking the sum** of an addition. To check the addition of $-3\,a + 4\,b + 5\,c$ and $+2\,a - 2\,b - c$ assign any convenient numerical values to $a$, $b$, and $c$, e.g. $a = 1$, $b = 2$, $c = 1$,

then
$$-3\,a + 4\,b + 5\,c = -3 + 8 + 5 = 10,$$
$$2\,a - 2\,b - \phantom{1}c = \phantom{1}2 - 4 - 1 = -3,$$
the sum
$$-\phantom{1}a + 2\,b + 4\,c = -1 + 4 + 4 = 7.$$

But $7 = 10 - 3$, therefore the answer is correct.

Note. While the check is almost certain to show any error, it is not an absolute test; e.g. the erroneous answer $-a + 6\,b - 4\,c$ would also equal 7.

**39.** In various operations with polynomials containing terms with different powers of the same letter, it is convenient to **arrange the terms** according to ascending or descending powers of that letter.

$7 + x + 5\,x^2 + 7\,x^3 + 5\,x^5$ is arranged according to ascending powers of $x$.   $5\,a^7 - 7\,a^6b + 4\,a^4bc - 8\,a^2bd^2 + 7\,ab^4d + 9\,b^5 + e^7$ is arranged according to descending powers of $a$.

**EXERCISE 12 \***

Add the following polynomials:

1. $2\,a - 3\,b + 6\,c,\ 3\,a - 4\,b - 7\,c,$ and $-4\,a + b - 2\,c.$

2. $9\,q + 7\,t - 5\,s,\ 2\,q + t - 3\,s,$ and $-12\,q - 5\,t + 2\,s.$

√3. $2\,x^2 - 4\,y^2 + 2\,z^2,\ -3\,x^2 - 2\,y^2 + z^2,\ 4\,x^2 - 3\,y^2 - 2\,z^2,$ and $5\,x^2 - 9\,z^2.$

\* For additional examples see page 259.

4. $14 d - 15 e + 2 f,\ 16 e + 17 f - 9 g,\ -18 f + 6 g + d$, and $g - 15 d$.

5. $-12 a^2 + 15 b^2 - 20 c^2,\ -12 b^2 - 5 a^2 - 5 c^2,\ 11 b^2 - 7 c^2$, and $- b^2 + a^2$.

6. $6 x^3 + 5 x^2 + 2 x - 7, 6 x^2 - 10 x + 10, - 7 x^3 + 3 x^2 - 5$, and $x^3 - 6 x^2 + 3 x + 5$.

7. $9 m^3 + 8 m^2 + 7 m + 6,\ 5 - 6 m - 7 m^2 - 8 m^3,\ 2 m^3 - 12$, and $- m^2 - m + 6$.

8. $- xy - 3 xz + yz,\ - 2 xy + 4 xz + 5 yz, 4 xy - xz - 6 yz$, and $- 2 xy + yz$.

9. $6 a^3 - 5 a^2 b + 7 ab^2 - 4 b^3,\ 5 a^3 + 4 ab^2 - 5 a^2 b + 7 b^3$, and $10 a^3 + 10 a^2 b - 11 ab^2$.

10. $a^4 + a^3 - a + 1, a^3 + a^2 - 1 - a^4, a^3 - a^2$, and $a^2 + a$.

11. $a - b, b - c, c - d, d - e, e - f$, and $f - a$.

12. $4 x^3 + 5 y^3 + 6 z^3,\ - 5 x^3 - 6 y^3 - 7 z^3, 2 x^3 + 2 y^3 + 2 z^3$, and $x^3 - y^3 - z^3$.

13. $4(a + b) - (b + c) - 3(c + a),\ 2(b + c) + (c + a)$, and $-4(a + b) - (b + c) - (c + a)$.

14. $a^4 - a - 1, a^2 + a + 1, a^2 - a$, and $- a^2 + a + 1$.

15. $6(a + b)^2 - 5(a + b) + 3, 7(a + b)^2 - 9(a + b) - 12$, and $- 12(a + b)^2 + 14(a + b) + 10$.

16. $7 x^4 + 5 x^3 y - 2 x^2 y^2 - 3 xy^3,\ - x^4 + x^2 y^2 + y^4$, and $- 6 x^4 - 4 x^3 y - 2 y^4$.

17. $5\sqrt{a} - \sqrt{b} + 2\sqrt{c},\ -\sqrt{a} + 2\sqrt{b} + 6\sqrt{c}, -4\sqrt{a} - 10\sqrt{c}$, and $5\sqrt{b} + \sqrt{c}$.

18. $a^3 + a^2 - a, a^2 + a - 1, - a^3 + a + 1$, and $1 + a^3 - a^2$.

19. $3 x^4 - 9 y^4, 3 x^3 y - 3 xy^3,\ 6 x^3 y + 6 xy^3 + 10 y^4$, and $- 3 x^4 - y^4 - x^3 y$.

20. $m^3 + 3 m^2 n + 3 mn^2 + n^3,\ m^3 - 3 m^2 n + 3 mn^2 - n^3$, and $2 m^3 - 2 n^3$.

**21.** $4\,m^5 + 3\,m^3 + 2\,m, \ 4\,m^4 - 2\,m + 2\,m^3, \ -m^5 + m,$ and $3\,m^3 - 7\,m.$

**22.** $a + b - c, \ b + c - d, \ c + d - e, \ d + e - a,$ and $e + a - b.$

**23.** $16\,m^2y - 12\,my^2 + 6\,y^3, \ -17\,y^3 + 4\,m^2y - m^3, \ 4\,m^2y - y^3,$ and $2\,m^3 - 5\,y^3.$

**24.** $3\,t^3 - 2\,t^2n + tn^2, 4\,t^3 + 3\,t^2n - 2\,tn^2 + n^3, 5\,t^3 - 4\,t^2n + 3\,tn^2 - 2\,n^3,$ and $-4\,t^3 + 3\,t^2n - 2\,tn^2 + n^3.$

**25.** $-1 - 2\,x - 3\,x^2, \ -4 - 5\,x - 6\,x^2, \ -7 - 8\,x - 9\,x^2,$ and $6 + 9\,x + 12\,x^2.$

**26.** $2 + a^4 - a^2 + 7\,a, \ a^3 - a^4 - 3\,a + 5, \ a^3 + 3\,a - 2,$ and $-5 + a^3.$

$\checkmark$ **27.** $3\,xyz + 2\,xy + x, \ 6\,xyz - 7\,x, \ 9\,xy - 12\,xyz,$ and $3\,xyz - 11\,xy + 12.$

## SUBTRACTION

### EXERCISE 13

**1.** If from the five negative units $-1, \ -1, \ -1, \ -1, \ -1,$ three negative units are taken, how many negative units remain? What is therefore the remainder when $-3$ is taken from $-5$?

**2.** Instead of subtracting in the preceding example, what number may be added to obtain the same result?

**3.** The sum total of the units $+1, \ +1, \ +1, \ +1, \ +1, \ -1,$ $-1,$ and $-1,$ is 2. What is the value of the sum if two negative units are taken away? If three negative units are taken away?

**4.** What is therefore the remainder when $-2$ is taken from 2? When $-3$ is taken from 2?

**5.** What other operations produce the same result as the subtraction of a negative number?

**6.** If you diminish a person's debts, does he thereby become richer or poorer?

**7.** State the other practical examples which show that the subtraction of a negative number is equal to the addition of a positive number.

_____

**40. Subtraction** is the inverse of addition. In addition, two numbers are given, and their algebraic sum is required. In subtraction, the algebraic sum and one of the two numbers is given, the other number is required. The algebraic sum is called the *minuend*, the given number the *subtrahend,* and the required number the *difference.*

Therefore any example in subtraction may be stated in a different form; *e.g.* from $-5$ take $-3$, may be stated : What number added to $-3$ will give $-5$? To subtract from $a$ the number $b$ means to find the number which added to $b$ gives $a$. Or in symbols,

$$a - b = x,$$

if

$$x + b = a.$$

**Ex. 1. From 5 subtract $-3$.**

The number which added to $-3$ gives 5 is evidently 8.
Hence, $\qquad 5 - (-3) = 8.$

**Ex. 2. From $-5$ subtract $-3$.**

The number which added to $-3$ gives $-5$ is $-2$.
Hence, $\qquad (-5) - (-3) = -2.$

**Ex. 3. From $-5$ subtract $+3$.**

This gives by the same method,
$$-5 - (+3) = -8.$$

**41.** The results of the preceding examples could be obtained by the following

**Principle.** *To subtract, change the sign of the subtrahend and add.*

NOTE. The student should perform *mentally* the operation of changing the sign of the subtrahend ; thus to subtract $-8\, a^2 b$ from $-6\, a^2 b$, change mentally the sign of $-8\, a^2 b$ and find the sum of $-6\, a^2 b$ and $+8\, a^2 b$.

**42. To subtract polynomials** we change the sign of each term of the subtrahend and add.

Ex. From $-6x^3 - 3x^2 + 7$ subtract $2x^3 - 3x^2 - 5x + 8$.

$$
\begin{array}{ll}
& \textit{Check,} \quad \text{If } x = 1 \\
-6x^3 - 3x^2 + 7 & \qquad = +2 \\
\underline{2x^3 - 3x^2 - 5x + 8} & \qquad = +2 \\
-8x^3 + 5x - 1 & \qquad = -4
\end{array}
$$

### EXERCISE 14

Subtract the lower number from the upper one:

1. $5$  
  $\underline{7}$

2. $-5$  
  $\underline{7}$

3. $-5$  
  $\underline{-7}$

4. $4$  
  $\underline{-2}$

5. $-5$  
  $\underline{6}$

6. $-9$  
  $\underline{5}$

7. $-9$  
  $\underline{8}$

8. $-9$  
  $\underline{-8}$

9. $9$  
  $\underline{-8}$

10. $-5$  
  $\underline{-5}$

11. $-12$  
  $\underline{12}$

12. $12$  
  $\underline{-19}$

13. $0$  
  $\underline{-7}$

14. $-7$  
  $\underline{0}$

15. $0$  
  $\underline{+9}$

16. $-7$  
  $\underline{-6}$

17. $-6$  
  $\underline{-7}$

18. $40$  
  $\underline{41}$

19. $2$  
  $\underline{20}$

20. $-2$  
  $\underline{-3}$

21. $17$  
  $\underline{-17}$

22. $14n^2$  
  $\underline{15n^2}$

23. $-14n^2$  
  $\underline{-15n^2}$

24. $16n^2m$  
  $\underline{-16n^2m}$

25. $-7x^3y^3$  
  $\underline{-7x^3y^3}$

26. $mpq^2$  
  $\underline{-3mpq^2}$

27. $-9p^2st$  
  $\underline{9p^2st}$

28. $abc$  
  $\underline{2abc}$

29. $3m^3x$  
  $\underline{13m^3x}$

30. $0$  
  $\underline{-7a^5}$

31. $a$  
  $\underline{b}$

32. $a$  
  $\underline{-b}$

33. $-a$  
  $\underline{+b}$

34. $-a$  
  $\underline{-b}$

35. $a^2$  
  $\underline{a}$

36. $a^2$  
  $\underline{-a}$

37. $1$  
  $\underline{-a^2}$

38. $ab$  
  $\underline{a}$

39. $a^2b$  
  $\underline{-a^2}$

40. $5\sqrt{a+b}$  
  $\underline{-6\sqrt{a+b}}$

**41.** Subtract the sum of $2\,m$ and $7\,m$ from $-10\,m$.

**42.** From $10\,a - 12\,b + 6\,c$ subtract $14\,a + 12\,b + 5\,c$, and check the answer.

**43.** From $x^3 - 7\,x^2y - 5\,xy^2 + y^3$ subtract $x^3 + 7\,x^2y - 5\,xy^2 - y^3$, and check the answer.

**44.** From $-a^3 + 2\,a^2 - 7\,a + 2$ subtract $a^3 + 2\,a^2 - 7\,a - 2$.

**45.** From $5\,a - 6\,b + 7\,c - 8\,d$ take $11\,a - 6\,b - 11\,c + 17\,d$.

**46.** From $2\,x^2 - 8\,xy + 2\,y^2$ take $2\,x^2 - 8\,xy - 2\,y^2$.

**47.** From $mn + np + qt - mt$ subtract $-mn + np + mt$.

**48.** From $a^3 - 1$ subtract $a^2 + a + 1$.

**49.** From $6\,a + 9\,b - c$ subtract $10\,b - c + d$.

**50.** From $1 + b$ take $1 + b^2 + b - b^3$.

**51.** From $6\,x^4 + 3\,x + 12$ take $3\,x^3 + 4\,x + 11$.

**52.** From $2\,a + b$ take $a + b + c$.

**53.** From $a^3$ subtract $2\,a^3 + a^2 + a - 1$.

**54.** From $3\,x^2 - 2\,xy + y^2$ subtract $3\,xy - 2\,y^2$.

**55.** From $5\,a^2 - 2\,ab - y^2$ subtract $2\,a^2 + 2\,ab - 2\,y^2$.

**56.** From $b$ subtract $1 + 2\,a + 3\,b + 4\,d$.

**57.** From $16 + a^3$ subtract $8 - 2\,a + a^2 + a^3$.

**58.** From $a^4 - 4\,a^3b + 6\,a^2b^2 - 4\,ab^3 + b^4$ subtract $a^4 + 4\,a^3b + 6\,a^2b^2 + 4\,ab^3 + b^4$.

**59.** From $6(a+b) + 5(b+c) - 4(c+a)$ subtract $7(a+b) - 4(c+a) + (b+c)$.

### REVIEW EXERCISES

**1.** From the sum of $a + b + c$ and $a - b + c$ subtract $a - b - 2\,c$.

**2.** From the sum of $x^2 - 4\,x + 12$ and $3\,x^2 - 3\,x - 3$ subtract $4\,x^2 + 2\,x - 7$.

**3.** From $a^3 + 2\,a^2 - 4\,a$ subtract the sum of $a^3 + a^2 - 2\,a$ and $a^2 - a + 4$.

**4.** From the difference between $x^3 + 5\,x^2 + 5\,x + 1$ and $4\,x^2 + 4\,x$ subtract $x^3 + x + 1$.

**5.** Subtract the sum of $x^2 + 2$ and $x^3 + x^2$ from $x^3 + x^2 + x + 1$.

**6.** Subtract the sum of $5\,a^2 - 6\,a + 7$ and $-2\,a^2 + 3\,a - 4$ from $2\,a^2 + 2\,a + 2$.

**7.** Subtract the sum of $6\,m^3 + 5\,m^2 + 6\,m$ and $-4\,m^3 - 5\,m^2 + 4\,m$ from $2\,m^3 + 7\,m$.

**8.** Subtract the difference of $a$ and $a + b + c$ from $a + b + c$

**9.** Subtract the sum of $x^2 + y^2$ and $x^2 - y^2$ from $2\,x^2 + 2\,y^2 + 2\,z^2$.

**10.** To the sum of $2\,a + 6\,b + 4\,c$ and $a - 4\,b - 2\,c$ add the sum of $9\,a - 6\,b + c$ and $-10\,a + 5\,b - 2\,c$.

**11.** To the sum $a^3 + a^2 + 1$ and $a^2 + a$ add the difference between $5\,a^3 + 4$ and $4\,a^3 + a^2 - a$.

**12.** What expression must be added to $7\,a^3 + 4\,a - 2$ to produce $8\,a^3 - 2\,a - 7$?

**13.** What expression must be added to $3\,a + 5\,b - c$ to produce $-2\,a - b + 2\,c$?

**14.** What expression must be subtracted from $2\,a$ to produce $-a + b$?

**15.** What must be added to $-4\,x^2 + 4\,x + 2$ to produce $0$?

If $a = x + y + z$, $b = x + y - z$, $c = x - 2\,y + z$, find:

**16.** $a + b$.            **18.** $a + b + c$.

**17.** $a - b$.            **19.** $a - b + c$.

**20.** A is $n$ years old. How old will he be 10 years hence? $2\,m - n$ years hence?

**21.** A is $2\,a$ years old. How old was he $a - b$ years ago? $a + b - c$ years ago?

## SIGNS OF AGGREGATION

**43**. By using the signs of aggregation, additions and subtractions may be written as follows:

$$a + (+ b - c + d) = a + b - c + d.$$
$$a - (+ b - c + d) = a - b + c - d.$$

Hence it is obvious that parentheses preceded by the $+$ or the $-$ sign may be removed or inserted according to the following principles:

**44.** I. *A sign of aggregation preceded by the sign $+$ may be removed or inserted without changing the sign of any term.*

II. *A sign of aggregation preceded by the sign $-$ may be removed or inserted provided the sign of every term inclosed is changed.*

$$E.g. \ a + (b - c) = a + b - c.$$
$$a + (- b + c) = a - b + c.$$
$$a - (+ b - c) = a - b + c.$$
$$a - - b - c = a + b + c.$$

**45.** If there is no sign before the first term within a parenthesis, the sign $+$ is understood.

$$a - (b - c) = a - b + c.$$

**46.** If we wish to remove several signs of aggregation, one occurring within the other, we may begin either at the innermost or outermost. The beginner will find it most convenient at every step to remove only those parentheses which contain no others.

**Ex.** Simplify $4\,a - \{(7\,a + 5\,b) - [- 6\,b + (- 2\,b - \overline{a - b})]\}$.

$$4\,a - \{(7\,a + 5\,b) - [- 6\,b + (- 2\,b - \overline{a - b})]\}$$
$$= 4\,a - \{7\,a + 5\,b - [- 6\,b + (- 2\,b - a + b)]\}$$
$$= 4\,a - \{7\,a + 5\,b - [- 6\,b - 2\,b - a + b]\}$$
$$= 4\,a - \{7\,a + 5\,b + 6\,b + 2\,b + a - b\}$$
$$= 4\,a - 7\,a - 5\,b - 6\,b - 2\,b - a + b$$
$$= - 4\,a - 12\,b. \quad Answer.$$

**EXERCISE 15 \***

Simplify the following expressions:

1. $x + (2y - z)$.

6. $a - (-a + b)$.

2. $a - (3b - 2c)$.

7. $a + (a - b) - (a + b)$.

3. $a^3 + (2a^3 - b^3 + c^3)$.

8. $a - (-a + b) + (a - 2b)$.

4. $a^2 - (a^2 + 2b^2 - c^2)$.

9. $2m - (m + n) + \overline{m - n}$.

5. $2a^2 - (4a^2 - 2b^2 + c^2)$.

10. $4x + \overline{2x - y} - (6x - y)$.

11. $2a^2 + 5a - (7 + 2a^2) + (5 - 5a)$.

12. $(7a + 5b) - (6a - 4b) + (2a - b)$.

13. $14x - (4x + y) + (x - 2y) - (x - 3y)$.

14. $(m + n + p) - (m - n - p) - \overline{m - n + p}$.

15. $m + n - (m - n) + \overline{m + n} - (-m + n)$.

16. $x^2 + [x - (x^2 - x)]$.

17. $x^2 + x - [(x + 1) - (x^2 - x)]$.

18. $a - (b - c) + [3a - \overline{2b - 6a}]$.

19. $a - [3b + \{3c - (c - a)\} - 2c]$.

20. $2x^2 - \{x^2 - (x^2 - x) - [x^2 - \overline{x^2 - x}] - x\}$.

21. $7 - (a - b) + \{a - [a - \overline{b + c}]\}$.

22. By removing parentheses, find the numerical value of:

$$1422 - [271 - \{271 + (814 - 1422)\}].$$

---

**47. Signs of aggregation may be inserted** according to § 43.

**Ex. 1.** In the following expression inclose the second and third, the fourth and fifth terms respectively in parentheses:

$$a - b + c + 2d - e$$
$$= a - (b - c) + (2d - e).$$

**Ex. 2.** Inclose in a parenthesis preceded by the sign − the last three terms of $\quad 2a + b - 5c + 2d$
$$= 2a - (-b + 5c - 2d).$$

\* See page 260.

### EXERCISE 16

In each of the following expressions inclose the last three terms in a parenthesis:

1. $a + b - c + d.$

3. $5a^2 - 7x^2 + 9x + 2.$

2. $2m - 4n + 2q - 3t.$

4. $4xy - 2x^2 - 4y^2 - 1.$

In each of the following expressions inclose the last three terms in a parenthesis preceded by the minus sign:

5. $m^2 - 2n^2 - 3p^2 + 4q^2.$

8. $5x^2 - 5x - 7 + a.$

6. $x - y + z + d.$

9. $a^3 - a^2 - a - 1.$

7. $p + q + r - s.$

$a^3 - (a^2 + a + 1)$

$p - (-q - r + s$

### EXERCISES IN ALGEBRAIC EXPRESSION

### EXERCISE 17

Write the following expressions:

1. The sum of $m$ and $n$.

2. The difference of $a$ and $b$.

NOTE. The minuend is always the first, and the subtrahend the second, of the two numbers mentioned.

3. The sum of the squares of $a$ and $b$.

4. The difference of the cubes of $m$ and $n$.

5. The difference of the cubes of $n$ and $m$.

6. The sum of the fourth powers of $a$ and $b$.

7. The product of $m$ and $n$.

8. The product of the cubes of $m$ and $n$.

9. Three times the product of the squares of $m$ and $n$.

10. The cube of the product of $m$ and $n$.

11. The square of the difference of $a$ and $b$.

12. The product of the sum and the difference of $m$ and $n$.

13. Nine times the square of the sum of $a$ and $b$ diminished by the product of $a$ and $b$.

**14.** The sum of the squares of $a$ and $b$ increased by the square root of $x$.

**15.** $x$ cube minus quantity $2\,x^2$ minus $6\,x$ plus 6.

**16.** The sum of the cubes of $a$, $b$, and $c$ divided by the dif ference of $a$ and $d$.

Write algebraically the following statements:

**17.** The sum of $a$ and $b$ multiplied by the difference of $a$ and $b$ is equal to the difference of $a^2$ and $b^2$.

**18.** The difference of the cubes of $a$ and $b$ divided by the difference of $a$ and $b$ is equal to the square of $a$ plus the product of $a$ and $b$, plus the square of $b$.

**19.** The difference of the squares of two numbers divided by the difference of the numbers is equal to the sum of the two numbers. (Let $a$ and $b$ represent the numbers.)

# CHAPTER III

## MULTIPLICATION

### MULTIPLICATION OF ALGEBRAIC NUMBERS

#### EXERCISE 18

In the annexed diagram of a balance, let us consider the forces produced at $A$ by 3 lb. weights, applied at $A$ and $B$, and let us indicate a downward pull at $A$ by a positive sign.

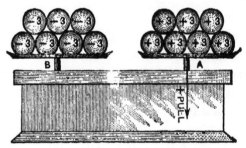

**1.** By what sign is an upward pull at $A$ represented?

**2.** What is the sign of a 3 lb. weight at $A$?

**3.** What is the sign of a 3 lb. weight at $B$?

**4.** If the two loads balance, what force is produced by the addition of five 3 lb. weights at $A$? Express this as a multiplication example.

**5.** If the two loads balance, what force is produced by taking away 5 weights from $A$? What, therefore, is $-5 \times 3$?

**6.** If the two loads balance, what force is produced by the addition of 5 weights at $B$? What, therefore, is $5 \times (-3)$?

**7.** If the two loads balance, what force is produced by taking away 5 weights from $B$? What therefore is $(-5) \times (-3)$?

**8.** If the signs obtained by the method of the preceding examples were generally true, what would be the values of

$$5 \times 4,\ 5 \times (-4),\ (-5) \times 4,\ (-5) \times (-4),$$
$$9 \times 11,\ (-9) \times 11,\ 9 \times (-11),\ (-9) \times (-11)?$$

**9.** State a rule by which the sign of the product of two factors can be obtained.

---

**48. Multiplication by a positive integer** is a repeated addition; thus, 4 multiplied by 3, or $4 \times 3 = 4 + 4 + 4 = 12$, $-4$ multiplied by 3, or

$$(-4) \times 3 = (-4) + (-4) + (-4) = -12.$$

The preceding definition, however, becomes meaningless if the multiplier is a negative number. To take a number $-7$ times is just as meaningless as to fire a gun $-7$ times.

Consequently we have to *define* the meaning of a multiplication if the multiplier is negative, and we may choose any definition that does not lead to contradictions. Practical examples, however, such as given in the preceding exercise, make it convenient to accept the following definition:

**49. Multiplication by a negative integer** is a repeated subtraction.

Thus,    $4 \times (-3) = -(4) - (4) - (4) = -12,$
$$(-4) \times (-3) = -(-4) - (-4) - (-4) = +12.$$

NOTE. This definition has the additional advantage of leading to algebraic laws for negative numbers which are identical with those for positive numbers, a result that would not be obtained by other assumptions.

In multiplying integers we have therefore four cases illustrated by the following examples:

$4 \times 3 = +12.$      $4 \times (-3) = -12.$
$(-4) \times 3 = -12.$     $(-4) \times (-3) = +12.$

**50.** We shall *assume* that the law illustrated for positive and negative integers is true for all numbers, and obtain thus the

**Law of Signs:** *The product of two numbers with like signs is positive; the product of two numbers with unlike signs is negative.*

Thus, $(-a)(+b) = -ab$; $(-a)(-b) = +ab$; etc.

### EXERCISE 19

Find the values of the following products:

1. $6 \times (-5)$.
2. $(-7) \times (-12)$.
3. $(-2) \times 6$.
4. $6 \times -7$.
5. $(-2) \times 9$.
6. $(-4) \times (-15)$.

NOTE. If no misunderstanding is possible, the parenthesis about factors is frequently omitted.

7. $-5 \times -3$.
8. $4 \cdot -7$.
9. $-3 \cdot (-4\frac{1}{3})$.
10. $-2 \cdot -\frac{1}{2}$.
11. $3 \cdot -2 \cdot -3$.
12. $6 \cdot -2 \cdot -\frac{2}{5}$.
13. $(-2)^2$.
14. $(-3)^3$.
15. $(-4)^3$.
16. $(-1)^7$.
17. $(-10)^4$.
18. $-1 \cdot -2 \cdot -3 \cdot -4 \cdot +5$.

**19.** Formulate a law of signs for a product containing an even number of negative factors.

**20.** Formulate a law of signs for a product containing an odd number of negative factors.

If $a = -2$, $b = 3$, $c = -1$, $x = 0$, and $y = 4$, find the numerical values of:

21. $4abc$.
22. $3a^2y^2$.
23. $2a^2bc$.
24. $14a^2b^3x$.
25. $-3a^2b^2c^2$.
26. $2ab^3c^4$.
27. $11a^2b^2cx$.
28. $(cy)^3$.
29. $-(abc)^2$.
30. $4a^2 + 2b^2$.
31. $2a + 3b^2 - 6c$.
32. $4a^2 - 2y^2 + x^2$.

Find the numerical value of $8\,a^3 - 12\,a^2b + 6\,ab^2 - b^3$, if

**33.** $a = 2,\quad b = -3.$      **35.** $a = -2,\quad b = 2.$

**34.** $a = 1,\quad b = -4.$      **36.** $a = 3,\quad b = 1.$

## MULTIPLICATION OF MONOMIALS

**51.** By definition, $2^3 = 2 \cdot 2 \cdot 2,$ and $2^5 = 2 \cdot 2 \cdot 2 \cdot 2 \cdot 2.$ Hence $2^3 \times 2^5 = 2 \cdot 2 \cdot 2 \times 2 \cdot 2 \cdot 2 \cdot 2 \cdot 2 = 2^8,$ *i.e.*, $2^{3+5}.$ Or in general, if $m$ and $n$ are two positive integers,

$$a^m \times a^n = (a \cdot a \cdot a \cdots \text{to } m \text{ factors}) \cdot (a \cdot a \cdot a \cdots \text{to } n \text{ factors})$$
$$= a\,a \cdots \text{to } (m + n) \text{ factors}.$$

$$a^m \times a^n = a^{m+n}.$$

This is known as

**52. The Exponent Law of Multiplication:** *The exponent of the product of several powers of the same base is equal to the sum of the exponents of the factors.*

**Ex. 1.** $5\,a^2b^3c \times -7\,a^3b^4d^4 = (5 \cdot -7) \cdot (a^2 \cdot a^3) \cdot (b^3 \cdot b^4) \cdot c \cdot d^4.$

**53.** In multiplying a product of several factors by a number, only *one* of the factors is multiplied by the number.

**Ex. 2.** $2^2 \times (2^3 \cdot 5 \cdot 7^2) = 2^5 \cdot 5 \cdot 7^2.$

**Ex. 3.** $4 \times (2 \cdot 25) = 8 \cdot 25,$ or $2 \cdot 100,$ *i.e.*, 200.

### EXERCISE 20

Express each of the following products as a power:

**1.** $m^5 \cdot m^4.$     **3.** $2^3 \cdot 2^6.$     **5.** $5^3 \cdot 5^2 \cdot 5 \cdot 5^{11}.$

**2.** $a^3 \cdot a^2.$     **4.** $3^2 \cdot 3 \cdot 3^4.$     **6.** $127^3 \cdot 127^9 \cdot 127^{11}.$

**7.** $(ab)^2 \cdot (ab)^9.$           **9.** $a^5 \cdot (-a)^7.$

**8.** $(x+y)^3 \cdot (x+y)^4 \cdot (x+y).$     **10.** $7^3 \cdot (-7)^2 \cdot 7.$

**11.** $12^{14} \cdot (-12)^{14} \cdot 12^{11}.$

Perform the operation indicated:

**12.** $3\,a \cdot 7\,abc.$     **14.** $2(7 \cdot 3 \cdot 5).$     **16.** $5(7 \cdot 11 \cdot 2).$

**13.** $4 \cdot (2 \cdot 25 \cdot 7).$     **15.** $50(11 \cdot 2 \cdot 3).$     **17.** $2(14 \cdot 50 \cdot 3).$

18. $7(5 \cdot \frac{4}{7} \cdot 2)$.

19. $11(3 \cdot 5 \cdot \frac{2}{11})$.

20. $4\,ab \cdot 5\,ab^2$.

21. $19\,m^5n^2 \cdot 2\,mn^4$.

22. $-5\,xy \cdot 3\,x^2y^2z^2$.

23. $19\,a^5b^2 \cdot (-2\,ab^4)$.

24. $11\,a^2b^3c \cdot (-4\,a^2bc^2)$.

25. $(-5\,ab^3) \cdot (-6\,a^2bc)$.

26. $6\,ef^2g \cdot (-6\,e^3f)$.

27. $-7\,p^3q^4r^4 \cdot -7\,pqt$.

28. $-4\,ab \cdot -4\,xy$.

29. $-7\,u^{10}v^{11}r^{12} \cdot (-8\,u^{20}v^{19}t^{30})$.

30. $-5\,ab \cdot 3\,bc \cdot (-2\,ac)$.

31. $9\,x^3y \cdot (-2\,a^3y^2)$.

32. $(-4\,a^2b^3)^2$.

33. $(2\,ax^3y^5)^2$.

34. $(-3\,mn^4)^2$.

35. $(-2\,a^2)^3$.

## MULTIPLICATION OF A POLYNOMIAL BY A MONOMIAL

**54.** If we had to multiply 2 yards and 3 inches by 3, the results would obviously be 6 yards and 9 inches. Similarly the quadruple of $a + 2\,b$ would be $4\,a + 8\,b$, for

$$4(a + 2\,b) = (a + 2\,b) + (a + 2\,b) + (a + 2\,b) + (a + 2\,b)$$
$$= 4\,a + 4(2\,b) = 4\,a + 8\,b.$$

**55.** This principle, called the **distributive law,** is evidently correct for any positive integral multiplier, but we shall assume it for any number.

Thus we have in general

$$a(b + c) = ab + ac.$$

**56.** To multiply a polynomial by a monomial, *multiply each term by the monomial.*

$$-3\,a^2b(6\,a^2bc + 2\,bc - 1) = -18\,a^4b^2c - 6\,a^2b^2c + 3\,a^2b.$$

### EXERCISE 21

Find the numerical values of the following expressions, by first multiplying, and then adding:

1. $2(5 + 15 + 25)$.

2. $6(10 + 20 + 30)$.

3. $3(12 + 3 + 2)$.

4. $2(6 + 5 + 10)$.

5. $12(\frac{1}{2} + \frac{1}{3} + \frac{1}{4})$.

6. $17(100 + 10 + 2)$.

7. $23(1000 + 100 + 20)$.

Express as a sum of several powers:

8. $5(5 + 5^2 + 5^7)$.
9. $6^2(6 + 6^2 + 6^3)$.
10. $7^3(7^3 + 7^2 + 7^{10})$.

11. $4^{13}(4^9 - 4^5 - 4)$.
12. $11^9(11^{10} + 11^{11} + 11^{12})$.

Perform the multiplications indicated:

13. $2\,m(m + n + p)$.
14. $-2\,mn(m^2 + n^2 - p^2)$.
15. $-2\,xy(x^2y^2 + xy + 1)$.

16. $4\,p^2q^2(-4\,pqr + 5\,pr - 7)$.
17. $-5\,x^2(-5\,x^2 - 5\,x - 5)$.
18. $-2\,xyz(7\,x^2 + 7\,y^2 - 7\,z^2)$.

19. $5\,a^2b^3c^4(-2\,ab^2c^5 + 3\,a^5b^2c - 6\,ab)$.
20. $7\,a^3b^2c(-5\,c^2 + 2\,b^2 - 3\,abc)$.
21. $-2\,mn(9\,m^3n^2 - 5\,m^2n^2 + 7\,mn)$.
22. $16\,xy^3z^5(-2\,x^5y^2z - x^5y + 4)$.
23. $-19\,p^5(p^6 - p^5 + 3\,p + 10)$.

24. By what expression must $x$ be multiplied to give
$$3\,x^3 - 4\,x^2 + 7\,x\,?$$

25. Express $3\,x^3 - 4\,x^2 + 7\,x$ as a product.
26. Find the factors of $3\,x + 3\,y + 3\,z$.
27. Find the factors of $6\,x^3 + 3\,x^2 + 3\,x^4$.
28. Find the factors of $2\,x^2y - 4\,x^3y + 8\,x^4$.
29. Find the factors of $5\,a^2b - 60\,ab^2 - 10\,ab$.
30. Find the factors of $6\,x^3y^3 - 3\,x^2y^2 + 3\,xy$.

## MULTIPLICATION OF POLYNOMIALS

**57.** Any polynomial may be written as a monomial by inclosing it within a parenthesis. Thus to multiply $a - b$ by $x + y - z$, we write $(a - b)(x + y - z)$ and apply the distributive law.

$$(a - b)(x + y - z) = x(a - b) + y(a - b) - z(a - b)$$
$$= (ax - bx) + (ay - by) - (az - bz)$$
$$= ax - bx + ay - by - az + bz.$$

**58. To multiply two polynomials,** *multiply each term of one by each term of the other and add the partial products thus formed.*

The most convenient way of adding the partial products is to place similar terms in columns, as illustrated in the following example:

Ex. 1. Multiply $2\,a - 3\,b$ by $a - 5\,b$.

$$2\,a - 3\,b$$
$$a - 5\,b$$
$$2\,a^2 - 3\,ab$$
$$- 10\,ab + 15\,b^2$$
$$2\,a^2 - 13\,ab + 15\,b^2 \quad \text{Product.}$$

**59.** If the polynomials to be multiplied contain several powers of the same letter, the work becomes simpler and more symmetrical by arranging these expressions according to either ascending or descending powers.

Ex. 2. Multiply $2 + a^3 - a - 3\,a^2$ by $2\,a - a^2 + 1$.

Arranging according to ascending powers:

*Check.* If $a = 1$

$$2 - a - 3\,a^2 + a^3 \qquad = -1$$
$$1 + 2\,a - a^2 \qquad = \ \ 2$$
$$2 - \ \ a - 3\,a^2 + a^3$$
$$+ 4.a - 2\,a^2 - 6\,a^3 + 2\,a^4$$
$$- 2\,a^2 + a^3 + 3\,a^4 - a^5$$
$$2 + 3\,a - 7\,a^2 - 4\,a^3 + 5\,a^4 - a^5 \qquad = -2$$

**60. Examples in multiplication can be checked** by numerical substitution, 1 being the most convenient value to be substituted for all letters. Since all powers of 1 are 1, this method tests only the values of the coefficients and not the values of the exponents. Since errors, however, are far more likely to occur in the coefficients than anywhere else, the student should apply this test to every example.

Perform the following multiplications and check the results:

1. $(2x - 3y)(3x + 2y)$.

2. $(4a + 7b)(2a - 3b)$.

3. $(5c - 2d)(2c - 3d)$.

4. $(2m - 3n)(7m + 8n)$.

5. $(6p + 5q)(5p + 3q)$.

6. $(2a - 5c)(2a - 6c)$.

7. $(9m - 2n)(4m + 7n)$.

8. $(x + 6y)(x - 7y)$.

9. $(6l - 7n)(11l - n)$.

10. $(13k - 2)(3k - 1)$.

11. $(6xy + 2z)(2xy - 4z)$.

12. $(8r - 7t)(5r - 3t)$.

13. $(11r + 1)(12r - 1)$.

14. $(xyz - 12)(xyz + 1)$.

15. $(abc + 7)(2abc - 3)$.

16. $(a^2 + 2a + 2)(a - 3)$.

17. $(2x^2 - x - 4)(x + 1)$.

18. $(4m^2 + m - 1)(m + 2)$.

19. $(2n^2 - 3n + 4)(n - 3)$.

20. $(3a^2 + 4a - 5)(a - 6)$.

21. $(5q^2 - 4q + 8)(1 - q)$.

22. $(9x^2 - 5 + 2x)(x - 2)$.

23. $(x^2 - x - 1)(2x + 1)$.

24. $(4a^2 - 8a + 16)(a - 2)$.

25. $(a^2 - 2ab + b^2)(a - b)$.

26. $(2a^2 - 3ab - 2b^2)(2a - 3b)$

27. $(6x^2 - 7 + 2x)(3x - 2)$.

28. $(2m^2 - 1 - 2m)(1 - m)$.

29. $(a - 3b)^2$.

30. $(4a - 1)^2$.

31. $(6a - 7)^2$.

32. $(6abc - 5)^2$.

33. $(4a^2b^2 - 3ab + 2)(2ab - 1)$.

34. $(a^2b^2 + 2abc + 4c^2)(ab - 2c)$.

35. $(m^2n^2p^2 - 2mnp + 4)(mnp + 2)$.

36. $(x + x^3 + 1 + x^2)(x - 1)$.

37. $(a^2 + 2ab + 4b^2)(a^2 - 2ab + 4b^2)$.

38. $(x^4 - x^3 + x^2 - x + 1)(x^2 + x + 1)$.

39. $(x^2 - 1)(x - 1)(x + 1)$.

40. $(m + n)(m + n)(m - n)(m - n)$.

41. $(a - 2b)^3$.

* For additional examples see page 261.

## SPECIAL CASES IN MULTIPLICATION

**61. The product of two binomials which have a common term.**

$$(x+2)(x+4) = x^2 + 2x + 4x + 8 = x^2 + 6x + 8.$$
$$(x-2)(x+4) = x^2 - 2x + 4x - 8 = x^2 + 2x - 8.$$
$$(x-2)(x-4) = x^2 - 2x - 4x + 8 = x^2 - 6x + 8.$$

**62.** *The product of two binomials which have a common term is equal to the square of the common term, plus the sum of the two unequal terms multiplied by the common term, plus the product of the two unequal terms.*

$(5a - 6b)(5a - 9b)$ is equal to the square of the common term, $25 a^2$, plus the sum of the unequal terms multiplied by the common terms, *i.e.* $(-15b)(5a) = -75 ab$, plus the product of the two unequal terms, *i.e.* $+54 b^2$. Hence the product equals $25 a^2 - 75 ab + 54 b^2$.

### EXERCISE 23

Multiply by inspection:

1. $(a+2)(a+3)$.
2. $(a-3)(a+2)$.
3. $(a-3)(a-4)$.
4. $(x+3)(x-7)$.
5. $(x-1)(x-5)$.
6. $(p-12)(p+11)$.
7. $(m+9)(m+9)$.
8. $(b-12)(b+13)$.
9. $(y-25)(y+4)$.
10. $(l+50)(l-2)$.
11. $(s-11)(s+21)$.
12. $(a-2b)(a+b)$.
13. $(a-2b)(a-3b)$.
14. $(a+2b)(a-b)$.

15. $(a-9)(a+9)$.
16. $(m-n)(m+n)$.
17. $(a^2 - 5b^2)(a^2 + 4b^2)$.
18. $(x^2y^2 + 3)(x^2y^2 - 4)$.
19. $(x^2y^2z^2 - 12)(x^2y^2z^2 + 11)$.
20. $(a^3 - 7)(a^3 - 8)$.
21. $(100 + 2)(100 + 3)$.
22. $(10 + 1)(10 + 2)$.
23. $(1000 + 5)(1000 + 4)$.
24. $(100 - 1)(100 + 2)$.
25. $(1000 - 2)(1000 + 3)$.
26. $102 \times 103$.
27. $1005 \times 1004$.
28. $99 \times 102$.

29. Find two binomials whose product equals $x^2 - 3x + 2$.

Find two binomial factors of each of the following expressions:

30. $x^2 - 5x + 6$.

31. $a^2 + 6a + 8$.

32. $a^2 - 7a + 10$.

33. $m^2 - 2m - 15$.

34. $m^2 + 2m - 15$.

35. $m^2 - 3m - 4$.

36. $n^2 - 10n + 16$.

37. $p^2 - p - 30$.

**63.** Some special cases of the preceding type of examples deserve special mention:

I.            $$(a + b)^2 = a^2 + 2\,ab + b^2.$$

II.           $$(a - b)^2 = a^2 - 2\,ab + b^2.$$

III.         $$(a + b)(a - b) = a^2 - b^2.$$

Expressed in general language:

I. *The square of the sum of two numbers is equal to the square of the first, plus twice the product of the first and the second, plus the square of the second.*

II. *The square of the difference of two numbers is equal to the square of the first, minus twice the product of the first and the second, plus the square of the second.*

III. *The product of the sum and the difference of two numbers is equal to the difference of their squares.*

The student should note that the second type (II) is only a special case of the first (I).

Ex. $(4\,x^3 + 7\,y^2)^2$ is equal to the square of the first, *i.e.* $16\,x^6$, plus twice the product of the first and the second, *i.e.* $56\,x^3y^2$, plus the square of the second, *i.e.* $49\,y^4$. Hence the required square equals $16\,x^6 + 56\,x^3y^2 + 49\,y^4$.

### EXERCISE 24

Multiply by inspection:

1. $(a + b)^2$.

2. $(a + 2)^2$.

3. $(a - x)^2$

4. $(a - 2)^2$.

5. $(p + 3)^2$.

6. $(x - 5)^2$.

7. $(p - 7)^2$.

8. $(a - 2\,b)^2$.

9. $(x + 3\,y)^2$.

10. $(2x - 3y)^2$.   13. $(p^2 + 5q)^2$.   16. $(2ab - c)^2$.

11. $(4a - 3b)^2$.   14. $(3p^2 - 9)^2$.   17. $(2xy + 3z)^2$.

12. $(a^2 - 3)^2$.   15. $(6a^2 - 7b^2)^2$.   18. $(4a^2b^2 - 5c^2)^2$.

19. $(6x^2y^2z^2 - 5)^2$.   25. $(x^2 - 11y^2)^2$.

20. $(m + n)(m - n)$.   26. $(x^2 + 11y^2)^2$.

21. $(2m + 3)(2m - 3)$.   27. $(5r^2 - 2t^2)(5r^2 + 2t^2)$.

22. $(a^2 - 7)(a^2 + 7)$.   28. $(5r^2 - 2t^2)^2$.

23. $(c^2d^2 - 5)(c^2d^2 + 5)$.   29. $(a + 5)(5 + a)$.

24. $(x^2 - 11y^2)(x^2 + 11y^2)$.   30. $(m^2 - 2n^5)(m^2 + 2n^5)$.

31. $(100 + 1)^2$.   34. $104^2$.   37. $99^2$.

32. $(100 + 2)^2$.   35. $(1000 - 1)^2$.   38. $(20 + 1)^2$.

33. $103^2$.   36. $998^2$.   39. $22^2$.

40. $(100 + 2)(100 - 2)$.   41. $99 \times 101$.   42. $998 \times 1002$.

Extract the square roots of the following expressions:

43. $x^2 + 2xy + y^2$.   45. $m^2 - 2m + 1$.   47. $n^2 - 6n + 9$.

44. $a^2 - 2ab + b^2$.   46. $n^2 + 4n + 4$.   48. $a^2 - 8ab + 16b^2$.

  49. $a^2 + 10ab + 25b^2$.

Find two binomial factors of each of the following expressions:

50. $x^2 - y^2$.   51. $a^2 - 9$.   52. $m^2 - 16$.   53. $b^2 - 25n^2$.

54. $9a^2 - 49b^2$.   56. $25a^4 - 9$.

55. $16a^2b^2c^2 - 25$.   57. $9a^2 - 30ab + 25b^2$.

---

**64. The product of two binomials whose corresponding terms are similar.**

By actual multiplication, we have

$$
\begin{array}{r}
3x + 2y \\
5x - 4y \\
\hline
15x^2 + 10xy \\
- 12xy - 8y^2 \\
\hline
15x^2 - 2xy - 8y^2
\end{array}
$$

The middle term of the result is obtained by adding the product of $5x \cdot 2y$ and $-4y \cdot 3x$. These products are frequently called the cross products, and are represented as follows:

$$3x + 2y$$
$$5x - 4y$$
$$\overline{10xy - 12xy}$$

or $\quad (3x + 2y)(5x - 4y)$.

Hence in general, *the product of two binomials whose corresponding terms are similar is equal to the product of the first two terms, plus the sum of the cross products, plus the product of the last terms.*

### EXERCISE 25

Multiply by inspection:

1. $(2a - 3)(a + 2)$.
2. $(3m + 2)(m - 1)$.
3. $(2m - 3)(3m + 2)$.
4. $(5a - 4)(4a - 1)$.
5. $(4x + y)(3x - 2y)$.
6. $(5ab - 4)(5ab - 3)$.
7. $(x^2 + 5b^2)(2x^2 - 3b^2)$.
8. $(2a^2b^2 - 7)(a^2b^2 + 5)$.
9. $(2x^2y^2 + z^2)(x^2y^2 + 2z^2)$.
10. $(6x^3 - 5)(x^3 + 3)$.
11. $(-x + 3y^2)(x - 2y^2)$.
12. $(5x^2 - x)(6x^2 + 5x)$.
13. $(10 + 2)(10 + 3)$.
14. $(100 + 3)(100 + 4)$.

**65. The square of a polynomial.**

$$(a + b + c)^2 = a^2 + b^2 + c^2 + 2ab + 2ac + 2bc.$$
$$(a - b + c - d)^2 = a^2 + b^2 + c^2 + d^2 - 2ab + 2ac - 2ad - 2bc$$
$$+ 2bd - 2cd.$$

*The square of a polynomial is equal to the sum of the squares of each term increased by twice the product of each term with each that follows it.*

The student should note that the square of each term is always positive, while the product of the terms may have plus or minus signs.

Find by inspection:

1. $(m + n + p)^2$.
2. $(x - y + z)^2$.
3. $(a + b - 5)^2$.
4. $(x - 2b - c)^2$.
5. $(a - 4b + 3c)^2$.

6. $(a + b + c - d)^2$.
7. $(x^2 - 4x + 2)^2$.
8. $(2a - 3b + 5c)^2$.
9. $(3x - 4y - z + n)^2$.
10. $(2a^2 + 3b^2 - 4c^2)^2$.

Find the square root of:

11. $x^2 + y^2 + z^2 + 2xy + 2yz + 2xz$.
12. $m^2 + n^2 + p^2 + 2mn - 2mp - 2np$.
13. $b^2 + c^2 + 1 - 2bc - 2b + 2c$.

**66.** In **simplifying a polynomial** the student should remember that a parenthesis is understood about each term. Hence, after multiplying the factors of a term, the beginner should inclose the product in a parenthesis.

Ex. Simplify $(x + 6)(x - 4) - (x - 3)(x - 5)$.

*Check.* If $x = 1$,

$(x + 6)(x - 4) - (x - 3)(x - 5) = (7 \cdot - 3) - (- 2 \cdot - 4) = - 29$
$= [x^2 + 2x - 24] - [x^2 - 8x + 15]$
$= x^2 + 2x - 24 - x^2 + 8x - 15$
$= 10x - 39.$                    $= 10 - 39.$          $= - 29.$

Simplify the following expressions, and check the answers:

1. $5(a - 2) - 5$.
2. $6 - 2(a + 7)$.
3. $4(x + 2) - 5(x - 3)$.
4. $4(x - 2) + 3(x - 7)$.
5. $2(m - n) - 3(m + n) + (m - n)$.
6. $3(b^3 + b^2) - 2(b^2 + b) - (b + b^3)$.
7. $(m + n)(m + 2n) - 3m(n + m)$.
8. $(a - 2)(a - 3) - (a - 1)(a - 4)$.

9. $(2x - y)(x - 2y) - 2(x^2 + y^2)$.

10. $(p + q)^2 - (p - q)^2$.

11. $4(m + 2) + 5(m - 3) - (m + 3)(m + 6)$.

12. $(x - 5)(x - 2) - (x - 10)(x - 1)$.

13. $(n + 5)(n - 2) + (n - 7)(n + 4) - 2(n^2 - 2)$.

14. $6(p + 2) - 7(p - 9) - 2(p + 1)(p - 1)$.

15. $(5x - 2y)(3x + 2y) - (4x - y)(x - 4y)$.

16. $3(a + b)^2 - 4(a + b)(a + 2b) + (a - b)^2$.

17. $2(x + y)(x + 2y) - 3(x + 2y)(x + 3y) - 4(x^2 - y^2)$.

18. $2(y + 3)^2 - 3(y - 2)(y - 3)$.

19. $(a + b)^2 - (b + c)^2 + (c + a)^2$.

20. $(x - 3y)(x - 2y) - [x^2 - y(x + 7y)]$.

21. $x(x + 3)^2 - [x^3 - (6x - 9)]$.

22. $(a^2 + a + 1)(a - 1) - (a^2 + 1)(a - 1)$.

# CHAPTER IV

## DIVISION

**67. Division** is the process of finding one of two factors if their product and the other factor are given.

The **dividend** is the product of the two factors, the **divisor** is the given factor, and the **quotient** is the required factor.

Thus to divide $-12$ by $+3$, we must find the number which multiplied by $+3$ gives $-12$. But this number is $-4$; hence $\dfrac{-12}{+3} = -4$.

**68.** Since

$$+a \cdot +b = +ab$$
$$+a \cdot -b = -ab$$
$$-a \cdot +b = -ab$$

and

$$-a \cdot -b = +ab,$$

it follows that

$$\frac{+ab}{+a} = +b$$

$$\frac{-ab}{+a} = -b$$

$$\frac{-ab}{-a} = +b$$

$$\frac{+ab}{-a} = -b.$$

**69.** Hence the **law of signs** is the same in division as in multiplication: *Like signs produce plus, unlike signs minus.*

**70. Law of Exponents.** It follows from the definition that $a^8 \div a^5 = a^3$, for $a^3 \times a^5 = a^8$.

Or in general, if $m$ and $n$ are positive integers, and $m$ is greater than $n$, $a^m \div a^n = a^{m-n}$, for $a^{m-n}a^n = a^m$.

**71.** *The exponent of a quotient of two powers with equal bases equals the exponent of the dividend diminished by the exponent of the divisor.*

## DIVISION OF MONOMIALS

**72.** To divide $10\,x^7y^3z$ by $-2\,x^5y^2$, we have to find the number which multiplied by $-2\,x^5y^2$ gives $10\,x^7y^3z$. This number is evidently $-5\,x^2yz$.

Therefore, $$\frac{10\,x^7y^3z}{-2\,x^5y^2} = -5\,x^2yz.$$

Hence, *the quotient of two monomials is a monomial whose coefficient is the quotient of their coefficients, preceded by the proper sign, and whose literal part is the quotient of their literal parts found in accordance with the law of exponents.*

**73.** In dividing a product of several factors by a number, only one of these factors is divided by that number. Thus $(8 \cdot 12 \cdot 20) \div 4$ equals $2 \cdot 12 \cdot 20$, or $8 \cdot 3 \cdot 20$ or $8 \cdot 12 \cdot 5$.

### EXERCISE 28

Perform the divisions indicated:

1. $-75 \div -15$.
2. $75 \div -15$.
3. $-39 \div 3$.
4. $2^{15} \div 2^{12}$.
5. $3^{19} \div 3^{18}$.

6. $\dfrac{-3^{12}}{3^{10}}$.

7. $\dfrac{-4^5}{-4^2}$.

8. $\dfrac{-7^9}{7^7}$.

9. $\dfrac{5^{11}}{-5^8}$.

10. $(3^5 \cdot 2^4) \div (3^4 \cdot 2^2)$.

11. $(2^3 \cdot 3^4 \cdot 5^7) \div (2^3 \cdot 3 \cdot 5^6)$.

12. $\dfrac{3^5 \cdot 5^6 \cdot 2^9}{3^5 \cdot 5 \cdot 2^5}$.

13. $\dfrac{2^3 \cdot 3^2 \cdot 5}{2 \cdot 3 \cdot 5}$.

14. $\dfrac{36\,a}{-12\,a}$.

15. $\dfrac{-42\,x^3}{-14\,x^2}$.

16. $\dfrac{6\,a^2b^3c}{2\,abc}$.

17. $\dfrac{-56\,a^3b^2c^5}{14\,ab^2c^3}$.

**18.** $\dfrac{80\,m^8n^9p^{10}}{-16\,m^7n^9}.$  **20.** $\dfrac{-145\,x^8y^7}{-145\,x^8y^7}.$  **22.** $\dfrac{-88\,a^{11}b^{12}}{22\,a^{11}b^{12}}.$

**19.** $\dfrac{132\,x^{10}y^3z^{14}}{-11\,x^{10}y}.$  **21.** $\dfrac{-240\,m^{40}bc^{50}}{120\,m^{39}c}.$

**23.** $(15 \cdot 25 \cdot a^2) \div 5.$  **25.** $(18 \cdot 5 \cdot 2\,a^2) \div 9\,a.$

**24.** $(7 \cdot 26\,x^2) \div 13.$  **26.** $(a + b)^7 \div (a + b)^6.$

## DIVISION OF POLYNOMIALS BY MONOMIALS

**74.** To divide $ax + bx + cx$ by $x$ we must find an expression which multiplied by $x$ gives the product $ax + bx + cx$.

But $\qquad x(a + b + c) = ax + bx + cx.$

Hence $\qquad \dfrac{ax + bx + cx}{x} = a + b + c.$

**To divide a polynomial by a monomial,** *divide each term of the dividend by the monomial and add the partial quotients thus formed.*

$E.g.\quad \dfrac{-6\,x^3y^2z^4 - 15\,xy^2z^3 + 3\,xyz^2}{-3\,xyz^2} = 2\,x^2yz^2 + 5\,yz - 1.$

### EXERCISE 29

Perform the operations indicated:

**1.** $(5^4 - 5^3 + 5^2) \div 5^2.$  **3.** $(2^4 - 2^5 - 2^7) \div 2^2.$

**2.** $(6^{99} - 6^{98} - 6^{97}) \div 6^{97}.$  **4.** $(8 \cdot 3 + 8 \cdot 5 + 8 \cdot 7) \div 8.$

**5.** $(11 \cdot 2 + 11 \cdot 3 + 11 \cdot 5) \div 11.$

**6.** $\dfrac{18\,ab - 27\,ac}{9\,a}.$  **9.** $\dfrac{5\,a^5 + 4\,a^3 - 2\,a^2}{-a}.$

**7.** $\dfrac{-14\,x^2y^2 + 21\,x^3y}{7\,x^2}.$  **10.** $\dfrac{15\,a^3b - 12\,a^2b^2 + 9\,a^2}{3\,a^2}.$

**8.** $\dfrac{27\,a^4b^4 - 9\,a^3b^3}{-3\,a^2b^2}.$  **11.** $\dfrac{-25\,m^3 + 15\,m^4 - 35\,m^5}{-5\,m^3}.$

12. $\dfrac{22\ m^4n - 33\ m^3n^2 + 55\ m^2n^3}{-11\ m^2n}$.

13. $\dfrac{-39\ x^4y^4z^4 + 26\ x^3y^3z^3 - 13\ x^2y^2z^2}{13\ x^2y^2z^2}$.

14. $\dfrac{-49\ a^5b^5c^5 + 28\ a^3b^4c^2 - 14\ a^4b^4c}{-7\ a^3b^4c}$.

15. $(115\ x^6y + 161\ x^5y^2 - 69\ x^4y^3 - 23\ x^3y^4) \div 23\ x^2y$.

16. $(52\ x^3y^3z^3 - 39\ x^2yz^4 - 65\ xyz^3 - 26\ xyz) \div 13\ xyz$.

17. $(85\ x^5 - 68\ x^4y + 51\ x^3y^2 - 34\ xy^4 + 17\ xy^5) \div -17\ x$.

## DIVISION OF A POLYNOMIAL BY A POLYNOMIAL

**75.** Let it be required to divide $25\ a - 12 + 6\ a^3 - 20\ a^2$ by $2\ a^2 + 3 - 4\ a$, or, arranging according to descending powers of $a$, divide

$$6\ a^3 - 20\ a^2 + 25\ a - 12 \text{ by } 2\ a^2 - 4\ a + 3.$$

The term containing the highest power of $a$ in the dividend (*i.e.* $6\ a^3$) is evidently the product of the terms containing respectively the highest power of $a$ in the divisor and in the quotient.

Hence the term containing the highest power of $a$ in the quotient is $\dfrac{6\ a^3}{2\ a^2}$, or $3\ a$.

If the product of $3\ a$ and $2\ a^2 - 4\ a + 3$, *i.e.* $6\ a^3 - 12\ a^2 + 9\ a$, be subtracted from the dividend, the remainder is $-8\ a^2 + 16\ a - 12$.

This remainder obviously must be the product of the divisor and the rest of the quotient. To obtain the other terms of the quotient we have therefore to divide the remainder, $-8\ a^2 + 16\ a - 12$, by $2\ a^2 - 4\ a + 3$.

We consequently repeat the process. By dividing the highest term in the new dividend $-8\ a^2$ by the highest term in the divisor $2\ a^2$, we obtain $-4$, the next highest term in the quotient.

Multiplying $-4$ by the divisor $2\ a^2 - 4\ a + 3$, we obtain the product $-8\ a^2 + 16\ a - 12$, which subtracted from the preceding dividend leaves no remainder.

Hence $3\ a - 4$ is the required quotient.

The work is usually arranged as follows :

$$\begin{array}{l|l} 6\,a^3 - 20\,i^2 + 25\,a - 12 & 2\,a^2 - 4\,a + 3 \\ \underline{6\,a^3 - 12\,a^2 + \phantom{0}9\,a} & 3\,a - 4 \\ \phantom{00} -\ 8\,a^2 + 16\,a - 12 \\ \phantom{00} -\ 8\,a^2 + 16\,a - 12 \end{array}$$

**76.** The method which was applied in the preceding example may be stated as follows :

1. *Arrange dividend and divisor according to ascending or descending powers of a common letter.*

2. *Divide the first term of the dividend by the first term of the divisor, and write the result for the first term of the quotient.*

3. *Multiply this term of the quotient by the whole divisor, and subtract the result from the dividend.*

4. *Arrange the remainder in the same order as the given expression, consider it as a new dividend, and proceed as before.*

5. *Continue the process until a remainder zero is obtained, or until the highest power of the letter according to which the dividend was arranged is less than the highest power of the same letter in the divisor.*

**77. Checks.** Numerical substitution constitutes a very convenient, but not absolutely reliable check.

An absolute check consists in multiplying quotient and divisor. The result must equal the dividend if the division was exact, or the dividend diminished by the remainder if the division was not exact.

**Ex. 1.** Divide $8\,a^3 + 8\,a - 4 + 6\,a^4 - 11\,a^2$ by $3\,a - 2$.

Arranging according to descending powers,

*Check.* If $a = b = 1$,

$$\begin{array}{l|l} 6\,a^4 + 8\,a^3 - 11\,a^2 + 8\,a - 4 & 3\,a - 2 \\ \underline{6\,a^4 - 4\,a^3} & 2\,a^3 + 4\,a^2 - a + 2 \\ \phantom{0} + 12\,a^3 - 11\,a^2 \\ \phantom{0} \underline{+ 12\,a^3 -\phantom{0} 8\,a^2} \\ \phantom{000} -\ 3\,a^2 + 8\,a \\ \phantom{000} \underline{-\ 3\,a^2 + 2\,a} \\ \phantom{00000} +\ 6\,a - 4 \\ \phantom{00000} +\ 6\,a - 4 \end{array}$$

$= 7 \div 1,$
$= 7$

**Ex. 2.** Divide $a^4 - 4 b^4 - 6 a^3b + 9 a^2b^2$ by $2 b^2 - 3 ab + a^2$.

Arranging according to descending powers of $a$, we have

$$
\begin{array}{l|l}
a^4 - 6 a^3b + 9 a^2b^2 - 4 b^4 & a^2 - 3 ab + 2 b^2 \\
a^4 - 3 a^3b + 2 a^2b^2 & a^2 - 3 ab - 2 b^2 \\
\hline
\quad - 3 a^3b + 7 a^2b^2 - 4 b^4 & \\
\quad - 3 a^3b + 9 a^2b^2 - 6 ab^3 & \\
\hline
\qquad - 2 a^2b^2 + 6 ab^3 - 4 b^4 & \\
\qquad - 2 a^2b^2 + 6 ab^3 - 4 b^4 &
\end{array}
$$

*Check.* The numerical substitution $a = 1$, $b = 1$, cannot be used in this example since it renders the divisor zero. Hence we have either to use a larger number for $a$, or multiply.

$$
(a^2 - 3 ab + 2 b^2)(a^2 - 3 ab - 2 b^2)
$$
$$
= [(a^2 - 3 ab) + 2 b^2][(a^2 - 3 ab) - 2 b^2]
$$
$$
= (a^2 - 3 ab)^2 - 4 b^4
$$
$$
= a^2 - 6 a^3b + 9 a^2b^2 - 4 b^4.
$$

### EXERCISE 30 *

Perform the operations indicated and check the answers:

1. $(x^2 - 9 x + 20) \div (x - 5)$.

2. $(y^2 - 2 y - 15) \div (y - 5)$.

3. $(15 x^2 - 46 xy + 16 y^2) \div (5 x - 2 y)$.

4. $(5 m^2 - 26 mn + 5 n^2) \div (m - 5 n)$.

5. $(12 a^2 + 10 ab - 42 b^2) \div (3 a + 7 b)$.

6. $(20 x^2 - 13 xy - 21 y^2) \div (5 x - 7 y)$.

7. $(6 k^2 - 53 k + 40) \div (6 k - 5)$.

8. $(56 x^2 + 19 x - 15) \div (8 x - 3)$.

9. $(12 c^2 - 19 cd - 18 d^2) \div (3 c + 2 d)$.

10. $(281 pq + 85 p^2 - 90 q^2) \div (17 p - 5 q)$.

11. $(6 y^2 - 37 xy + 6 x^2) \div (x - 6 y)$.

12. $(35 a^2 - 18 b^2 + 27 ab) \div (5 a + 6 b)$.

13. $(25 a^2 - 36 b^2) \div (5 a + 6 b)$.

*See page 263.

14. $(6\,a^2b^2 + 23\,ab + 20) \div (2\,ab + 5)$.

15. $(8\,x^4y^2 + 15 - 22\,x^2y) \div (2\,x^2y - 3)$.

16. $(3\,a^3 - 11\,a^2 + 9\,a - 2) \div (3\,a - 2)$.

17. $(1 + 13\,m + 47\,m^2 + 35\,m^3) \div (5\,m + 1)$.

18. $(x^3 - 8) \div (x - 2)$.

19. $(x^3 - 3\,x - 2) \div (x - 2)$.

20. $(81\,m^4 + 1 - 18\,m^2) \div (1 - 6\,m + 9\,m^2)$.

## SPECIAL CASES IN DIVISION

**78. Division of the difference of two squares.**

Since $\qquad (a + b)(a - b) = a^2 - b^2$,

$$\frac{a^2 - b^2}{a - b} = a + b, \text{ and } \frac{a^2 - b^2}{a + b} = a - b.$$

*I.e.* the difference of the squares of two numbers is divisible by the difference or by the sum of the two numbers.

**Ex. 1.** $\dfrac{9\,x^2 - 16\,y^2}{3\,x - 4\,y} = 3\,x + 4\,y.$

**Ex. 2.** $\dfrac{(a + b)^2 - (x - y)^2}{(a + b) + (x - y)} = (a + b) - (x - y) = a + b - x + y.$

### EXERCISE 31

Write by inspection the quotient of:

1. $\dfrac{m^2 - n^2}{m + n}.$

2. $\dfrac{x^2 - 1}{x - 1}.$

3. $\dfrac{c^2 - 9}{c + 3}.$

4. $\dfrac{a^2 - 16\,b^2}{a - 4\,b}.$

5. $\dfrac{4\,a^2 - 9\,b^2}{2\,a + 3\,b}.$

6. $\dfrac{36\,x^2y^2 - 49\,z^2}{6\,xy + 7\,z}.$

7. $\dfrac{169\,a^4b^2 - 81\,c^6}{13\,a^2b - 9\,c^3}.$

8. $\dfrac{64\,x^{10} - 1}{8\,x^5 - 1}.$

Find exact binomial divisors of each of the following expressions :

9. $m^4 - 1$.

10. $a^4 - b^6$.

11. $a^2 b^2 c^2 - 1$.

12. $x^{12} - y^{16}$.

13. $36 x^4 y^4 - 49$.

14. $121 a^{100} - 9 b^2$.

15. $a^{16} - 100 x^2 y^4$.

16. $1,000,000 - 1$.

# CHAPTER V

## LINEAR EQUATIONS AND PROBLEMS

**79.** The **first member** or **left side** of an equation is that part of the equation which precedes the sign of equality. The **second member** or **right side** is that part which follows the sign of equality.

Thus, in the equation $2x + 4 = x - 9$, the first member is $2x + 4$, the second member is $x - 9$.

**80.** An **identity** is an equation which is true for all values of the letters involved.

Thus, $(a + b)(a - b) = a^2 - b^2$, no matter what values we assign to $a$ and $b$. The sign of identity sometimes used is $\equiv$; thus we may write $(a + b)(a - b) \equiv a^2 - b^2$.

**81.** An **equation of condition** is an equation which is true only for certain values of the letters involved. An equation of condition is usually called an **equation**.

$x + 9 = 20$ is true only when $x = 11$; hence it is an equation of condition.

**82.** A set of numbers which when substituted for the letters in an equation produce equal values of the two members, is said to **satisfy** an equation.

Thus $x = 12$ satisfies the equation $x + 1 = 13$. $x = 20$, $y = 7$ satisfy the equation $x - y = 13$.

**83.** An **equation is employed** to discover an unknown number (frequently denoted by $x$, $y$, or $z$) from its relation to known numbers.

**84.** If an equation contains only one unknown quantity, any value of the unknown quantity which satisfies the equation is **a root** of the equation.

9 is a root of the equation $2y + 2 = 20$.

**85.** To **solve** an equation is to find its roots.

**86.** A **numerical equation** is one in which all the known quantities are expressed in arithmetical numbers; as $(7 - x)(x + 4) = x^2 - 2$.

**87.** A **literal equation** is one in which at least one of the known quantities is expressed by a letter or a combination of letters; as $x + a = bx - c$.

**88.** A **linear equation** or an equation of the **first degree** is one which when reduced to its simplest form contains only the first power of the unknown quantity; as $9x - 2 = 6x + 7$. A linear equation is also called a *simple* equation.

**89.** The process of solving equations depends upon the following principles, called **axioms**:

1. *If equals be added to equals, the sums are equal.*

2. *If equals be subtracted from equals, the remainders are equal.*

3. *If equals be multiplied by equals, the products are equal.*

4. *If equals be divided by equals, the quotients are equal.*

5. *Like powers or like roots of equals are equal.*

Note. Axiom 4 is not true if the divisor equals zero. *E.g.* $0 \times 4 = 0 \times 5$, but 4 does not equal 5.

**90.** **Transposition of terms.** A term may be transposed from one member to another by changing its sign.

Consider the equation $\qquad x + a = b$.

Subtracting $a$ from both members, $x = b - a$. (Axiom 2)

*I.e.* the term $a$ has been transposed from the left to the right member by changing its sign.

Similarly, if $\qquad\qquad x - a = b.$

Adding $a$ to both members, $\qquad x = b + a.$ $\qquad$ (Axiom 1)

The result is the same as if we had transposed $a$ from the first member to the right member and changed its sign.

**91.** *The sign of every term of an equation may be changed without destroying the equality.*

Consider the equation $\qquad -x + a = -b + c.$

Multiplying each member by $-1$, $\qquad x - a = b - c.$ $\qquad$ (Axiom 3)

## SOLUTION OF LINEAR EQUATIONS

**92. Ex. 1.** Solve the equation $6x - 5 = 4x + 1.$

Adding 5 to each term, $\qquad\qquad 6x = 4x + 1 + 5.$

Subtracting $4x$ from each term, $\quad 6x - 4x = 1 + 5.$

Uniting similar terms, $\qquad\qquad\quad 2x = 6.$

Dividing both members by 2, $\qquad\qquad x = 3.$ $\qquad$ (Axiom 4)

*Check.* When $x = 3.$

The first member, $\qquad\qquad 6x - 5 = 18 - 5 = 13.$

The second member, $\qquad\qquad 4x + 1 = 12 + 1 = 13$

Hence the answer, $x = 3$, is correct.

**93. To solve a simple equation,** *transpose the unknown terms to the first member, and the known terms to the second. Unite similar terms, and divide both members by the coefficient of the unknown quantity.*

**Ex. 2.** Solve the equation $(4 - y)(5 - y) = 2(11 - 3y) + y^2.$

Simplifying, $\qquad\qquad 20 - 9y + y^2 = 22 - 6y + y^2.$

Transposing, $\qquad\qquad\quad -9y + 6y = -20 + 22.$

Uniting, $\qquad\qquad\qquad\quad -3y = 2.$

Dividing by $-3$, $\qquad\qquad\qquad y = -\tfrac{2}{3}.$

*Check.* If $y = -\tfrac{2}{3}.$

The first member, $(4 - y)(5 - y) = (4 + \tfrac{2}{3})(5 + \tfrac{2}{3}) = \tfrac{14}{3} \cdot \tfrac{17}{3} = \tfrac{238}{9} = 26\tfrac{4}{9}.$

The second member, $2(11 - 3y) + y^2 = 2(11 + 2) + \tfrac{4}{9} = 26 + \tfrac{4}{9} = 26\tfrac{4}{9}$

Ex. 3. Solve the equation $\frac{1}{2}(x-4)=\frac{1}{3}(x+3)$.

Simplifying,　　　　　　$\frac{1}{2}x-2=\frac{1}{3}x+1$.

Transposing,　　　　　　$\frac{1}{2}x-\frac{1}{3}x=2+1$.

Uniting,　　　　　　　　$\frac{1}{6}x=3$.

Dividing by $\frac{1}{6}$,　　　　$x=18$.

*Check.* If $x=18$.

The left member　　　　$\frac{1}{2}(x-4)=\frac{1}{2}\times 14=7$.

The right member　　　$\frac{1}{3}(x+3)=\frac{1}{3}\times 21=7$.

Note. Instead of dividing by $\frac{1}{6}$ both members of the equation $\frac{1}{6}x=3$, it would be simpler to multiply both members by 6.

### EXERCISE 32 *

Solve the following equations by using the axioms only :

1. $5x=15+2x$.
2. $7x=5x+18$.
3. $3x=50-7x$.
4. $7x=16+5x$.

5. $3x-7=14$.
6. $4x+5=29$.
7. $17x+15=16x+17$.
8. $7x-3=17-3x$.

Solve the following equations by transposing, etc., and check the answers :

9. $4y-11=2y-7$.
10. $13x-9x=7-3x$.
11. $24-7y=68-11y$.

12. $9y-17+4y=35$.
13. $13y-99=7y-69$.
14. $3z-2=26-4z$.

15. $17+5x-7x=39-4x+22$.
16. $17-9x+41=12-8x-50$.
17. $14y=59-(24y+21)$.
18. $10x-(11x-50)=69-(3x+5)$.
19. $10x-(3x+3)=6-(23x-21)$.
20. $87-(28+12x)=6x-(12x-17)$.
21. $9(5x-3)=63$.
22. $5(3x-3)=9(3x-15)$.
23. $7(6x+24)=6(10x+13)$.
24. $7x+7(3x+1)=63$.

* See page 264.

25. $73 - 4\,t = 13\,t - 2\,(5\,t - 12)$.

26. $6\,(6\,x - 5) - 5\,(7\,x - 8) = 4\,(12 - 3\,x) + 1$.

27. $7\,(7\,x + 1) - 8\,(7 - 5\,x) + 24 = 12\,(4\,x - 5) + 199$.

28. $y^2 - 5 + 13\,y = 5\,(2\,y - 2) + (y^2 + 4\,y)$.

29. $x\,(x + 11) - 5\,(x - 3) = 5 - (x - x^2)$.

30. $(x - 5)\,(x + 3) = (x - 7)\,(x + 4)$.

31. $(x - 1)\,(x - 5) = (x + 7)\,(x - 3) + 14$.

32. $(x + 3)\,(x - 2) = (x - 7)\,(x + 6) + 11$.

33. $(x + 7)\,(x + 2) = (x - 11)\,(x - 2) + 5$.

34. $(x - 2)\,(x + 3) = (x - 1)\,(x + 1)$.

35. $(y + 2)\,(2\,y - 3) = (2\,y - 4)\,(y + 3)$.

36. $(4\,t - 12)\,(2\,t + 5) - (2\,t + 6)\,(4\,t - 10) = 0$.

37. $(5\,x - 7)\,(7\,x + 4) - (14\,x + 1)\,(x + 7) = 285 + 21\,x^2$.

38. $(x + 2)^2 - (x - 5)^2 = 2$.

39. $(x + 1)^2 + (x + 2)^2 = (x - 3)^2 + (x - 4)^2$.

40. $2\,(t + 1)^2 - (2\,t - 3)\,(t + 2) = 12$.

41. $(6\,u - 2)\,(u - 3) - 5\,(2\,u - 1)\,(u - 4) + 4\,u^2 - 14 = 0$.

42. $\frac{1}{2}\,x = 5 - \frac{1}{3}\,x$.　　　　44. $\frac{1}{5}\,x + 6 = \frac{1}{4}\,x + 5$.

43. $\frac{1}{4}\,x + 10 = \frac{1}{2}\,x + 5$.

## SYMBOLICAL EXPRESSIONS

**94.** Suppose one part of 70 to be $x$, and let it be required to find the other part. If the student finds it difficult to answer this question, he should first attack a similar problem stated in arithmetical numbers only, *e.g.*: One part of 70 is 25; find the other part. Evidently 45, or $70 - 25$, is the other part. Hence if one part is $x$, the other part is $70 - x$.

*Whenever the student is unable to express a statement in algebraic symbols, he should formulate a similar question stated in arithmetical numbers only, and apply the method thus found to the algebraic problem.*

**Ex. 1.** What must be added to $a$ to produce a sum $b$?

Consider the arithmetical question : What must be added to 7 to produce the sum of 12 ?

The answer is 5, or $12 - 7$.

Hence $b - a$ must be added to $a$ to give $b$.

**Ex. 2.** $x + y$ yards cost \$100; find the cost of one yard.

If 7 yards cost one hundred dollars, one yard will cost $\dfrac{\$100}{7}$.

Hence if $x + y$ yards cost \$100, one yard will cost $\dfrac{100}{x + y}$ dollars.

### EXERCISE 33

1. By how much does $a$ exceed 10 ?

2. By how much does 9 exceed $x$ ?

3. What number exceeds $a$ by 4 ?

4. What number exceeds $m$ by $n$ ?

5. What is the 5th part of $n$ ?

6. What is the $n$th part of $x$ ?

7. By how much does 10 exceed the third part of $a$ ?

8. By how much does the fourth part of $x$ exceed $b$ ?

9. By how much does the double of $b$ exceed one half of $c$ ?

10. Two numbers differ by 7, and the smaller one is $p$.
Find the greater one.

11. Divide 100 into two parts, so that one part equals $a$.

12. Divide $a$ into two parts, so that one part is 10.

13. Divide $a$ into two parts, so that one part is $b$.

14. The difference between two numbers is $d$, and the
smaller one is $s$. Find the greater one.

15. The difference between two numbers is $d$, and the
greater one is $g$. Find the smaller one.

16. What number divided by 3 will give the quotient $x^2$ ?

17. What is the dividend if the divisor is 7 and the quotient
is $x^2$ ?

**18.** What must be subtracted from $2\,b$ to give $a$?

**19.** The smallest of three consecutive numbers is $a$. Find the other two.

**20.** The greatest of three consecutive numbers is $x$. Find the other two.

**21.** A is $x$ years old, and B is $y$ years old. How many years is A older than B?

**22.** A is $y$ years old. How old was he 5 years ago? How old will he be 10 years hence?

**23.** If A's age is $x$ years, and B's age is $y$ years, find the sum of their ages 6 years hence. Find the sum of their ages 5 years ago.

**24.** A has $m$ dollars, and B has $n$ dollars. If B gave A 6 dollars, find the amount each will then have.

**25.** How many cents are in $d$ dollars? in $x$ dimes?

**26.** A has $a$ dollars, $b$ dimes, and $c$ cents. How many cents has he?

**27.** A man had $a$ dollars, and spent $b$ cents. How many cents had he left?

**28.** A room is $x$ feet long and $y$ feet wide. How many square feet are there in the area of the floor?

**29.** Find the area of the floor of a room that is 2 feet longer and 3 feet wider than the one mentioned in Ex. 28.

**30.** Find the area of the floor of a room that is 3 feet shorter and 4 feet wider than the one mentioned in Ex. 28.

**31.** A rectangular field is $x$ feet long and $y$ feet wide. Find the length of a fence surrounding the field.

**32.** What is the cost of 10 apples at $x$ cents each?

**33.** What is the cost of 1 apple if $x$ apples cost 20 cents?

**34.** What is the price of 12 apples if $x$ apples cost 20 cents?

**35.** What is the price of 3 apples if $x$ apples cost $n$ cents?

**36.** If a man walks 3 miles per hour, how many miles will he walk in $n$ hours?

**37.** If a man walks $r$ miles per hour, how many miles will he walk in $n$ hours?

**38.** If a man walks $n$ miles in 4 hours, how many miles does he walk each hour?

**39.** If a man walks $r$ miles per hour, in how many hours will he walk $n$ miles?

**40.** How many miles does a train move in $t$ hours at the rate of $x$ miles per hour?

**41.** $x$ years ago A was 20 years old. How old is he now?

**42.** A cistern is filled by a pipe in $x$ minutes. What fraction of the cistern will be filled by one pipe in one minute?

**43.** A cistern can be filled by two pipes. The first pipe alone fills it in $x$ minutes, and the second pipe alone fills it in $y$ minutes. What fraction of the cistern will be filled per second by the two pipes together?

**44.** Find 5 % of $100\,x$.      **46.** Find $a$ % of 1000.

**45.** Find 6 % of $x$.          **47.** Find $x$ % of 4.

**48.** Find $x$ % of $m$.

**49.** The numerator of a fraction exceeds the denominator by 3. If $m$ is the denominator, find the fraction.

**50.** The two digits of a number are $x$ and $y$. Find the number.

---

**95.** To express in algebraic symbols the sentence: "$a$ exceeds $b$ by as much as $b$ exceeds 9," we have to consider that in this statement "exceeds" means minus ($-$), and "by as much as" means equals ($=$). Hence we have

$a$ exceeds $b$ by as much as $c$ exceeds 9.

$$a \quad - \quad b \quad = \quad c \quad - \quad 9.$$

Similarly, the difference of the squares of $a$ and $b$ increased by 80 equals the excess of $a^3$ over 80.

$$a^2 - b^2$$

$$80 = a^3 - 80.$$

Or,    $$(a^2 - b^2) + 80 = a^3 - 80.$$

In many cases it is possible to translate a sentence word by word in algebraic symbols; in other cases the sentence has to be changed to obtain the symbols.

There are usually several different ways of expressing a symbolical statement in words, thus:

$a - b = c$ may be expressed as follows:

> The difference between $a$ and $b$ is $c$.
>
> $a$ exceeds $b$ by $c$.
>
> $a$ is greater than $b$ by $c$.
>
> $b$ is smaller than $a$ by $c$.
>
> The excess of $a$ over $b$ is $c$, etc.

### EXERCISE 34

Express the following sentences as equations:

1. The double of $a$ is 10.

2. The double of $x$ increased by 10 equals $c$.

3. The sum of $a$ and 10 equals $2x$.

4. One third of $x$ equals $c$.

5. The difference of $x$ and $y$ increased by 7 equals $a$.

6. The double of $a$ increased by one third of $b$ equals 100.

7. Four times the difference of $a$ and $b$ exceeds $c$ by as much as $d$ exceeds 9.

8. The product of the sum and the difference of $a$ and $b$ diminished by 90 is equal to the sum of the squares of $a$ and $b$ divided by 7.

9. Twenty subtracted from $2a$ gives the same result as 7 subtracted from $a$.

**10.** Nine is as much below $a$ as 17 is above $a$.

**11.** $x$ is 5 % of 450.    **13.** 100 is $x$ % of 700.

**12.** $x$ is 6 % of $m$.    **14.** 50 is $x$ % of $a$.    **15.** $m$ is $x$ % of $n$.

**16.** If A's age is $2x$, B's age is $3x - 10$, and C's age is $4x - 20$, express in algebraic symbols:

(*a*) A is twice as old as B.

(*b*) A is 4 years older than B.

(*c*) Five years ago A was $x$ years old.

(*d*) In 10 years A will be $n$ years old.

(*e*) In 3 years A will be as old as B is now.

(*f*) Three years ago the sum of A's and B's ages was 50.

(*g*) In 3 years A will be twice as old as B.

(*h*) In 10 years the sum of A's, B's, and C's ages will be 100.

**17.** If A, B, and C have respectively $2x$, $3x - 700$, and $x + 1200$ dollars, express in algebraic symbols:

(*a*) A has $5 more than B.

(*b*) If A gains $20 and B loses $40, they have equal amounts.

(*c*) If each man gains $500, the sum of A's, B's, and C's money will be $12,000.

(*d*) A and B together have $200 less than C.

(*e*) If B pays to C $100, they have equal amounts.

**18.** A sum of money consists of $x$ dollars, a second sum of $5x - 30$ dollars, a third sum of $2x + 1$ dollars. Express as equations:

(*a*) 5 % of the first sum equals $90.

(*b*) $a$ % of the second sum equals $20.

(*c*) $x$ % of the first sum equals 6 % of the third sum.

(*d*) $a$ % of the first sum exceeds $b$ % of the second sum by $900.

(*e*) 4 % of the first plus 5 % of the second plus 6 % of the third sum equals $8000.

(*f*) $x$ % of the first equals one tenth of the third sum.

## PROBLEMS LEADING TO SIMPLE EQUATIONS

**96.** The simplest kind of problems contain only one unknown number. *In order to solve them, denote the unknown number by x (or another letter) and express the given sentence as an equation. The solution of the equation gives the value of the unknown number.*

The equation can frequently be written by translating the sentence word by word into algebraic symbols; in fact, *the equation is the sentence written in algebraic shorthand.*

**Ex. 1.** Three times a certain number exceeds 40 by as much as 40 exceeds the number. Find the number.

Let $x$ = the number.
Write the sentence in algebraic symbols.
Three times a certain no. exceeds 40 by as much as 40 exceeds the no.

$$3 \quad \times \quad x \quad - \quad 40 \quad = \quad 40 \quad - \quad x$$

Or, $3x - 40 = 40 - x$.
Transposing, $3x + x = 40 + 40$.
Uniting, $4x = 80$.
$x = 20$, the required number.

*Check.* $3x$ or 60 exceeds 40 by 20; 40 exceeds 20 by 20.

**Ex. 2.** In 15 years A will be three times as old as he was 5 years ago. Find A's present age.

Let $x$ = A's present age.
The verbal statement (1) may be expressed in symbols (2).
(1) In 15 years A will be three times as old as he was 5 years ago.

$$(2) \quad x \quad + \quad 15 \quad = \quad 3 \quad \times \quad (x \quad - \quad 5)$$

Or, $x + 15 = 3(x - 5)$.
Simplifying, $x + 15 = 3x - 15$.
Transposing, $x - 3x = -15 - 15$.
Uniting, $-2x = -30$.
Dividing, $x = 15$.

*Check.* In 15 years A will be 30; 5 years ago he was 10; but $30 = 3 \times 10$.

NOTE. The student should note that $x$ stands for the *number* of years, and similarly in other examples for *number* of dollars, *number* of yards, etc.

**Ex. 3.** 56 is what per cent of 120 ?

Let $x$ = number of per cent, then the problem expressed in symbols would be

$$56 = \frac{x}{100} \cdot 120,$$

or, $\frac{6}{5} x = 56.$

Dividing, $x = 46\frac{2}{3}.$

Hence 56 is $46\frac{2}{3}\%$ of 120.

### EXERCISE 35

**1.** What number added to twice itself gives a sum of 39?

**2.** Find the number whose double increased by 14 equals 44.

**3.** Find the number whose double exceeds 40 by 10.

**4.** Find the number whose double exceeds 30 by as much as 24 exceeds the number.

**5.** A number added to 42 gives a sum equal to 7 times the original number. Find the number.

**6.** 47 diminished by three times a certain number equals twice the number plus 2. Find the number.

**7.** Forty years hence A will be three times as old as to-day. Find his present age.

**8.** Six years hence a man will be twice as old as he was 12 years ago. How old is he now?

**9.** Four times the length of the Suez Canal exceeds 180 miles by twice the length of the canal. How long is the Suez Canal?

**10.** 14 is what per cent of 500?

**11.** 50 is 4 % of what number?

**12.** What number is 7 % of 350?

**13.** Ten times the width of the Brooklyn Bridge exceeds 800 ft. by as much as 135 ft. exceeds the width of the bridge. Find the width of the Brooklyn Bridge.

**14.** A train moving at uniform rate runs in 5 hours 90 miles more than in 2 hours. How many miles per hour does it run?

**15.** A and B have equal amounts of money. If A gains $200, and B loses $100, then A will have three times as much as B. How many dollars has each?

**16.** A and B have equal amounts of money. If A gives B $200, B will have five times as much as A. How many dollars has A now?

**17.** A has $40, and B has $60. How many dollars must B give to A to make A's money equal to 4 times B's money?

**18.** A man wishes to purchase a farm containing a certain number of acres. He found one farm which contained 30 acres too many, and another which lacked 25 acres of the required number. If the first farm contained twice as many acres as the second one, how many acres did he wish to buy?

**19.** In 1800 the population of Maine equaled that of Vermont. During the following 90 years, Maine's population increased by 510,000, Vermont's population increased by 180,000, and Maine had then twice as many inhabitants as Vermont. Find the population of Maine in 1800.

---

**97.** If a problem contains **two unknown quantities,** two verbal statements must be given. In the simpler examples these two statements are given directly, while in the more complex problems they are only implied. We denote one of the unknown numbers (usually the smaller one) by $x$, and use one of the given verbal statements to express the other unknown number in terms of $x$. The other verbal statement, written in algebraic symbols, is the equation, which gives the value of $x$.

Ex. 1. One number exceeds another by 8, and their sum is 14. Find the numbers.

The problem consists of two statements:

I. One number exceeds the other one by 8.

II. The sum of the two numbers is 14.

Either statement may be used to express one unknown number in terms of the other, although in general the simpler one
should be selected.

If we select the first one, and

Let $x$ = the smaller number,

Then $x + 8$ = the greater number.

The second statement written in algebraic symbols produces
the equation

$$x + (x + 8) = 14.$$

Simplifying,　　　$x + x + 8 = 14.$

Transposing,　　　$x + x = 14 - 8.$

Uniting,　　　　　$2x = 6.$

Dividing,　　　　　$x = 3$, the smaller number.

　　　　　　　　$x + 8 = 11$, the greater number.

Another method for solving this problem is to express one unknown
quantity in terms of the other by means of statement II; viz. the sum of
the two numbers is 14.

Let　　　　　　　$x$ = the smaller number.

Then,　　　　　$14 - x$ = the larger number.

Statement I expressed in symbols is $(14 - x) - x = 8$, which leads of
course to the same answer as the first method.

Ex. 2. A has three times as many marbles as B. If A gives
25 marbles to B, B will have twice as many as A.

The two statements are:

I. A has three times as many marbles as B.

II. If A gives B 25 marbles, B will have twice as many as A.

Use the simpler statement, viz. I, to express one unknown quantity in
terms of the other.

Let　　　　　　　$x$ = B's number of marbles.

Then,　　　　　　$3x$ = A's number of marbles.

To express statement II in algebraic symbols, consider that by the
exchange A will lose, and B will gain.

Hence,　$x + 25$ = B's number of marbles after the exchange.

　　　$3x - 25$ = A's number of marbles after the exchange.

Therefore, $x + 25 = 2(3x - 25)$. (Statement II)

Simplifying, $x + 25 = 6x - 50$.

Transposing, $x - 6x = -25 - 50$.

Uniting, $-5x = -75$.

Dividing, $x = 15$, B's number of marbles.

$3x = 45$, A's number of marbles.

*Check.* $45 - 25 = 20$, $15 + 25 = 40$, but $40 = 2 \times 20$.

**98. The numbers which appear in the equation should always be expressed in the same denomination.** Never add the number of dollars to the number of cents, the number of yards to their price, etc.

Ex. 3. Eleven coins, consisting of half dollars and dimes, have a value of $3.10. How many are there of each?

The two statements are:

I. The number of coins is 11.

II. The value of the half dollars and dimes is $3.10.

Let $x =$ the number of dimes, then, from I,

$11 - x =$ the number of half dollars.

Selecting the cent as the denomination (in order to avoid fractions), we express the statement II in algebraic symbols.

$$50(11 - x) + 10x = 310.$$

Simplifying, $550 - 50x + 10x = 310$.

Transposing, $-50x + 10x = -550 + 310$.

Uniting, $-40x = -240$.

Dividing, $x = 6$, the number of dimes.

$11 - x = 5$, the number of half dollars.

*Check.* 6 dimes $= 60$ cents, 5 half dollars $= 250$ cents, their sum is $3.10.

### EXERCISE 36

1. Two numbers differ by 44, and the greater is five times the smaller. Find the numbers.

2. Two numbers differ by 60, and their sum is 70. Find the numbers.

3. The sum of two numbers is 42, and the greater is 6 times the smaller. Find the numbers.

**4.** One number is six times another number, and the greater increased by five times the smaller equals 22. Find the number.

**5.** Find two consecutive numbers whose sum equals 157.

**6.** Two numbers differ by 39, and twice the greater exceeds three times the smaller by 65. Find the numbers.

**7.** The number of volcanoes in Mexico exceeds the number of volcanoes in the United States by 2, and four times the former equals five times the latter. How many volcanoes are in the United States, and in Mexico?

**8.** A cubic foot of iron weighs three times as much as a cubic foot of aluminum. If 4 cubic feet of aluminum and 2 cubic feet of iron weigh 1600 lbs., find the weight of a cubic foot of each substance.

**9.** Divide 20 into two parts, one of which increased by 3 shall be equal to the other increased by 9.

**10.** A's age is four times B's, and in 5 years A's age will be three times B's. Find their ages.

**11.** Mount Everest is 9000 feet higher than Mt. McKinley, and twice the altitude of Mt. McKinley exceeds the altitude of Mt. Everest by 11,000 feet. What is the altitude of each mountain?

**12.** Two vessels contain together 9 pints. If the smaller one contained 11 pints more, it would contain three times as much as the larger one. How many pints does each contain?

**13.** A is 14 years older than B, and B's age is as much below 30 as A's age is above 40. What are their ages?

**14.** A line 60 inches long is divided into two parts. Twice the larger part exceeds five times the smaller part by 15 inches. How many inches are in each part?

**15.** On December 21, the night in Copenhagen lasts 10 hours longer than the day. How many hours does the day last?

**99.** If a **problem contains three unknown quantities,** three verbal statements must be given. One of the unknown numbers is denoted by $x$, and the other two are expressed in terms of $x$ by means of two of the verbal statements. The third verbal statement produces the equation.

If it should be difficult to express the selected verbal statement directly in algebraical symbols, try to obtain it by a series of successive steps.

**Ex. 1.** A, B, and C together have $\$80$, and B has three times as much as A. If A and B each gave $\$5$ to C, then three times the sum of A's and B's money would exceed C's money by as much as A had originally.

The three statements are :

I. A, B, and C together have $\$80$.

II. B has three times as much as A.

III. If A and B each gave $\$5$ to C, then three times the sum of A's and B's money would exceed C's money by as much as A had originally.

Let $\qquad$ $x =$ the number of dollars A has.

According to II, $\qquad$ $3x =$ the number of dollars B has,

and according to I, $80 - 4x =$ the number of dollars C has.

To express statement III by algebraical symbols, let us consider first the words "if A and B each gave $\$5$ to C."

$x - 5 =$ number of dollars A had after giving $\$5$.

$3x - 5 =$ number of dollars B had after giving $\$5$.

$90 - 4x =$ number of dollars C had after receiving $\$10$.

Expressing in symbols :

Three times the sum of A's and B's money exceeds C's money by A's

$$3 \quad \times \quad (x - 5 + 3x - 5) \quad - \quad (90 - 4x) = x.$$

original amount.

The solution gives $x = 8$, number of dollars A had.

$3x = 24$, number of dollars B had.

$80 - 4x = 48$, number of dollars C had.

*Check.* If A and B each gave $\$5$ to C, they would have 3, 19, and 58, respectively. $3(3 + 19)$ or 66 exceeds 58 by 8.

**Ex. 2.** A man spent $1185 in buying horses, cows, and sheep, each horse costing $90, each cow $35, and each sheep $15. The number of cows exceeded the number of horses by 4, and the number of sheep was twice as large as the number of horses and cows together. How many animals of each kind did he buy ?

The three statements are :

I. The total cost equals $1185.

II. The number of cows exceeds the number of horses by 4.

III. The number of sheep is equal to twice the number of horses and cows together.

Let $x$ = the number of horses,

then, according to II, $x + 4$ = the number of cows,

and, according to III,

$2(2x + 4)$ or $4x + 8$ = the number of sheep.

Therefore, $90x$ = the number of dollars spent for horses,

$35(x + 4)$ = the number of dollars spent for cows,

and, $15(4x + 8)$ = the number of dollars spent for sheep

Hence statement I may be written,

$$90x + 35(x + 4) + 15(4x + 8) = 1185.$$

Simplifying, $90x + 35x + 140 + 60x \times 120 = 1185.$

Transposing, $90x + 35x + 60x = -140 - 120 + 1185.$

Uniting, $185x = 925.$

Dividing, $x = 5$, number of horses.

$x + 4 = 9$, number of cows.

$4x + 8 = 28$, number of sheep.

*Check.* 5 horses, 9 cows, and 28 sheep would cost $5 \times 90 + 9 \times 35 + 28 \times 15$ or $450 + 315 + 420 = 1185$ ; $9 - 5 = 4$ ; $28 = 2(9 + 5)$.

## EXERCISE 37

**1.** Find three numbers such that the second is twice the first, the third five times the first, and the difference between the third and the second is 15

**2.** Find three numbers such that the second is twice the first, the third exceeds the second by 2, and the sum of the first and third is 20.

**3.** Find three numbers such that the second is 4 less than the first, the third is three times the second, and the sum of the first and third is 36.

**4.** Find three numbers such that the sum of the first two is 4, the third is five times the first, and the third exceeds the second by 2.

**5.** Divide 25 into three parts such that the second part is twice the first, and the third part exceeds the second by 10.

**6.** Find three consecutive numbers whose sum equals 63.

**7.** The sum of the three sides of a triangle is 28 inches, and the second one is one inch longer than the first. If twice the third side, increased by three times the second side, equals 49 inches, what is the length of each?

**8.** New York has 3,000,000 more inhabitants than Philadelphia, and Berlin has 1,000,000 more than Philadelphia (Census 1905). If the population of New York is twice that of Berlin, what is the population of each city?

**9.** The three angles of any triangle are together equal to 180°. If the second angle of a triangle is 20° larger than the first, and the third is 20° more than the sum of the second and first, what are the three angles?

**10.** In a room there were three times as many children as women, and 2 more men than women. If the number of men, women, and children together was 37, how many children were present?

**11.** A is twice as old as B, and A is 5 years younger than C. Five years ago the sum of B's and C's ages was 25 years. What are their ages?

**12.** Find three consecutive numbers such that the sum of the first and twice the last equals 22.

**13.** The gold, the copper, and the pig iron produced in one year (1906) in the United States represented together a value of $ 750,000,000. The copper had twice the value of the gold, and the value of the iron was $ 300,000,000 more than that of the copper. Find the value of each.

**14.** California has twice as many electoral votes as Colorado, and Massachusetts has one more than California and Colorado together. If the three states together have 31 electoral votes, how many has each state?

---

**100. Arrangement of Problems.** If the example contains quantities of 3 or 4 different kinds, such as length, width, and area, or time, speed, and distance, it is frequently advantageous to arrange the quantities in a systematic manner.

*E.g.* A and B start at the same hour from two towns 27 miles apart, B walks at the rate of 4 miles per hour, but stops 2 hours on the way, and A walks at the rate of 3 miles per hour without stopping. After how many hours will they meet and how many miles does A walk?

|        | Time<br>(in hours) | Rate<br>(miles per hour) | Distance<br>(miles) |
|--------|--------------------|--------------------------|---------------------|
| A . . . . | $x$             | 3                        | $3x$                |
| B . . . . | $x - 2$         | 4                        | $4(x - 2)$          |

*Explanation.* First fill in all the numbers given directly, *i.e.* 3 and 4. Let $x$ = number of hours A walks, then $x - 2$ = number of hours B walks. Since in uniform motion the distance is always the product of rate and time, we obtain $3x$ and $4(x - 2)$ for the last column. But the statement "A and B walk from two towns 27 miles apart until they meet" means the sum of the distances walked by A and B equals 27 miles.

Hence $\qquad 3x + 4(x - 2) = 27.$

Simplifying, $\quad 3x + 4x - 8 = 27.$

Uniting, $\qquad\qquad 7x = 35.$

Dividing, $\qquad\qquad x = 5,$ number of hours.

$\qquad\qquad\qquad 3x = 15,$ number of miles A walks.

Ex. 1.  The length of a rectangular field is twice its width. If the length were increased by 30 yards, and the width decreased by 10 yards, the area would be 100 square yards less. Find the dimensions of the field.

| | LENGTH (yards) | WIDTH (yards) | AREA (square yards) |
|---|---|---|---|
| First field . . . . | $2x$ | $x$ | $2x^2$ |
| Second field . . . | $2x + 30$ | $x - 10$ | $(2x + 30)(x - 10)$ |

" The area would be decreased by 100 square yards," gives

$$(2x + 30)(x - 10) = 2x^2 - 100.$$

Simplify, $\quad\quad\quad 2x^2 + 10x - 300 = 2x^2 - 100.$

Cancel $2x^2$ and transpose, $\quad\quad 10x = 200.$

$$x = 20.$$
$$2x = 40.$$

The field is 40 yards long and 20 yards wide.

*Check.* The original field has an area $40 \times 20 = 800$, the second fiel l $70 \times 10$ or 700.  But $700 = 800 - 100.$

Ex. 2.  A certain sum invested at $5\%$ brings the same interest as a sum \$200 larger at $4\%$.  What is the capital?

| PRINCIPAL (No. of dollars) | RATE % | INTEREST (No. of dollars) |
|---|---|---|
| $x$ | .05 | $.05x$ |
| $x + 200$ | .04 | $.04(x + 200)$ |

Therefore $\quad\quad\quad\quad .05x = .04(x + 200).$

Simplify, $\quad\quad\quad\quad\quad .05x = .04x + 8.$

Transposing and uniting, $\quad .01x = 8.$

Multiplying, $\quad\quad\quad\quad\quad x = 800; \quad \$800 = \text{required sum.}$

*Check.* $\quad\quad\quad \$800 \times .05 = \$40; \quad \$1000 \times .04 = \$40.$

### EXERCISE 38

1. A rectangular field is 10 yards and another 12 yards wide. The second is 5 yards longer than the first, and the sum of their areas is equal to 390 square yards. Find the length of each.

2. A rectangular field is 2 yards longer than it is wide. If its length were increased by 3 yards, and its width decreased by 2 yards, the area would remain the same. Find the dimensions of the field.

3. A certain sum invested at 5 % brings the same interest as a sum $ 50 larger invested at 4 %. Find the first sum.

4. A sum invested at 5 %, and a second sum, twice as large, invested at 4 %, together bring $ 78 interest. What are the two sums?

5. Six persons bought an automobile, but as two of them were unable to pay their share, each of the others had to pay $ 100 more. Find the share of each, and the cost of the automobile.

6. Ten yards of silk and 30 yards of cloth cost together $ 42. If the silk cost three times as much per yard as the cloth, how much did each cost per yard?

7. A man bought 6 lbs. of coffee for $ 1.55. For a part he paid 24 ¢ per pound and for the rest he paid 35 ¢ per pound. How many pounds of each kind did he buy?

8. Twenty men subscribed equal amounts to raise a certain sum of money, but four men failed to pay their shares, and in order to raise the required sum each of the remaining men had to pay one dollar more. How much did each man subscribe?

9. A sets out walking at the rate of 3 miles per hour, and two hours later B follows on horseback traveling at the rate of 5 miles per hour. After how many hours will B overtake A, and how far will each then have traveled?

10. A and B set out walking at the same time in the same direction, but A has a start of 2 miles. If A walks at the rate of $2\frac{1}{2}$ miles per hour, and B at the rate of 3 miles per hour, how far must B walk before he overtakes A?

11. A sets out walking at the rate of 3 miles per hour, and two hours later B starts from the same point, traveling by coach in the opposite direction at the rate of 6 miles per hour. After how many hours will they be 36 miles apart?

12. The distance from New York to Albany is 142 miles. If a train starts at Albany and travels toward New York at the rate of 30 miles per hour without stopping, and another train starts at the same time from New York traveling at the rate of 41 miles an hour, how many miles from New York will they meet?

# CHAPTER VI

## FACTORING

**101.** An expression is **rational with respect to a letter**, if, after simplifying, it contains no indicated root of this letter; **irrational**, if it does contain some indicated root of this letter.

$a^2 - \dfrac{1}{a} + \sqrt{b}$ is rational with respect to $a$, and irrational with respect to $b$.

**102.** An expression is **integral** with respect to a letter, if this letter does not occur in any denominator.

$\dfrac{a^2}{b} + ab + b^2$ is integral with respect to $a$, but fractional with respect to $b$.

**103.** An expression is **integral and rational**, if it is integral and rational with respect to all letters contained in it; as,

$$a^2 + 2\,ab + 4\,c^2.$$

**104.** The **factors** of an algebraic expression are the quantities which multiplied together will give the expression.

In the present chapter only integral and rational expressions are considered factors.

Although $\sqrt{a^2 - b^2} \times \sqrt{a^2 - b^2} = a^2 - b^2$, we shall not, at this stage of the work, consider $\sqrt{a^2 - b^2}$ a factor of $a^2 - b^2$.

**105.** A factor is said to be **prime**, if it contains no other factors (except itself and unity); otherwise it is **composite**.

The prime factors of $10\,a^3 b$ are 2, 5, $a$, $a$, $a$, $b$.

**106. Factoring** is the process of separating an expression into its factors. An expression is factored if written in the form of a product.

$(x^2 - 4x + 3)$ is factored if written in the form $(x - 3)(x - 1)$. It would not be factored if written $x(x - 4) + 3$, for this result is a sum, and not a product.

**107.** The factors of a monomial can be obtained by inspection

The prime factors of $12\ x^3y^2$ are 3, 2, 2, $x$, $x$, $x$, $y$, $y$.

**108.** Since factoring is the inverse of multiplication, it follows that every method of multiplication will produce a method of factoring.

*E.g.* since $(a + b)(a - b) = a^2 - b^2$, it follows that $a^2 - b^2$ can be factored, or that $a^2 - b^2 = (a + b)(a - b)$.

**109.** Factoring examples may be checked by multiplication or by numerical substitution.

### TYPE I. POLYNOMIALS ALL OF WHOSE TERMS CONTAIN A COMMON FACTOR

$$mx + my + mz = m(x + y + z). \quad (\S\ 55.)$$

**110. Ex. 1.** Factor $6\ x^3y^2 - 9\ x^2y^3 + 12\ xy^4$.

The greatest factor common to all terms is $3\ xy^2$. Divide

$$6\ x^3y^2 - 9\ x^2y^3 + 12\ xy^4 \text{ by } 3\ xy^2,$$

and the quotient is $2\ x^2 - 3\ xy + 4\ y^2$.

But, dividend = divisor × quotient.

Hence $\quad 6\ x^3y^2 - 9\ x^2y^3 + 12\ xy^4 = 3\ xy^2(2\ x^2 - 3\ xy + 4\ y^2)$.

**Ex. 2.** Factor

$$14\ a^4b^2c^2 - 21\ a^2b^4c^2 + 7\ a^2b^2c^2 = 7\ a^2b^2c^2(2\ a^2 - 3\ b^2 + 1).$$

### EXERCISE 39

Resolve into prime factors:

1. $6\,abx - 12\,cdx$.

2. $9\,x^3 - 6\,x^2$.

3. $15\,a^2b + 20\,a^2b^3$.

4. $14\,a^4b^4 - 7\,a^2b^2$.

5. $11\,m^3 + 11\,m^2 - 11\,m$.

6. $4\,x^3y + 5\,x^2y^2 - 6\,xy$.

7. $17\,x^3 - 51\,x^4 + 34\,x^5$.

8. $8\,a^2b^2 + 8\,b^2c^2 - 8\,c^2a^2$.

9. $15\,x^2y^2z^2 - 45\,x^3y^3 - 30\,x^4y$.

10. $a^4 - a^3 + a^2$.

11. $32\,a^4xy - 16\,a^3xy^2 + 48\,a^2xy^3$.

12. $9\,a^8b^7z^6 - 6\,a^7b^7z^7 + 3\,a^6b^7z^8$.

13. $12\,a^3b^3 - 18\,a^2n^2 - 24\,a^4p^3$.

14. $34\,a^3b^3c^3 - 51\,a^2b^2c^2 + 68\,abc$.

15. $22\,p^5q^2 - 55\,p^4q^3 + 77\,p^2q^5$.

16. $65\,x^3y^4z^5 - 39\,x^4y^5z^6 + 13\,x^5y^6z^7$.

17. $q^4 - q^3 - q^2 + q$.

18. $a(m+n) + b(m+n)$.

19. $3\,x^2(a+b) - 3\,y^2(a+b)$.

20. $3\,a(p+q) - 5\,b(p+q)$.

21. $13 \cdot 5 + 13 \cdot 8$.

22. $2 \cdot 3 \cdot 4 \cdot 5 + 2 \cdot 3 \cdot 4 \cdot 6$.

23. $2 \cdot 3 \cdot 5 + 2 \cdot 3 \cdot 5 \cdot 6$.

## TYPE II. QUADRATIC TRINOMIALS OF THE FORM
$$x^2 + px + q.$$

**111.** In multiplying two binomials containing a common term, *e.g.* $(x-3)$ and $(x+5)$, we had to add $-3$ and $5$ to obtain the coefficient of $x$, and to multiply $-3$ and $5$ to obtain the term which does not contain $x$ or $(x-3)(x+5) = x^2 + 2\,x - 15$.

In factoring $x^2 + 2\,x - 15$ we have, obviously, to find two numbers whose product is $-15$ and whose sum is $+2$.

Or, in general, in factoring a trinomial of the form $x^2 + px + q$, we have to find two numbers $m$ and $n$ whose sum is $p$, and whose product is $q$; and if such numbers can be found, the factored expression is $(x+m)(x+n)$.

**Ex. 1.** Factor $x^2 - 4x - 77$.

We may consider $-77$ as the product of $-1 \cdot 77$, or $-7 \cdot 11$, or $-11 \cdot 7$, or $-77 \cdot 1$, but of these only $-11$ and $7$ have a sum equal to $-4$.

Hence $\qquad x^2 - 4x - 77 = (x - 11)(x + 7)$.

Since a number can be represented in an infinite number of ways as the sum of two numbers, but only in a limited number of ways as a product of two numbers, it is advisable to consider the factors of $q$ first. If $q$ is positive, the two numbers have both the same sign as $p$. If $q$ is negative, the two numbers have opposite signs, and the greater one has the same sign as $p$.

Not every trinomial of this type, however, can be factored.

**Ex. 2.** Factor $a^2 - 11a + 30$.

The two numbers whose product is 30 and whose sum is $-11$ are $-5$ and $-6$.

Therefore $\qquad a^2 - 11a + 30 = (a - 5)(a - 6)$.

*Check.* If $a = 1$, $a^2 - 11a + 30 = 20$, and $(a - 5)(a - 6) = -4 \cdot -6 = 20$.

**Ex. 3.** Factor $x^2 + 10ax - 11a^2$.

The numbers whose product is $-11a^2$ and whose sum is $10a$ are $11a$ and $-a$.

Hence $\qquad x^2 + 10ax - 11a^2 = (x + 11a)(x - a)$.

**Ex. 4.** Factor $x^6 - 7x^3y^3 + 12y^6$.

The two numbers whose product is equal to $12y^6$ and whose sum equals $-7y^3$ are $-4y^3$ and $-3y^3$. Hence $x^6 - 7x^3y^3 + 12y^6 = (x^3 - 3y^3)(x^3 - 4y^3)$.

**112.** In solving any factoring example, the student should first determine whether all terms contain a common monomial factor.

### EXERCISE 40

Resolve into prime factors:

1. $a^2 - 7a + 12$.

2. $a^2 + 7a + 12$.

3. $m^2 - 5m + 6$.

4. $x^2 - 7x + 10$.

5. $x^2 + 8x + 15$.

6. $a^2 - 9a + 20$.

7. $x^2 - 2x - 8.$

8. $x^2 + 2x - 8.$

9. $y^2 - 6y - 16.$

10. $y^2 + 6y - 16.$

11. $y^2 - 15y + 44.$

12. $y^2 - 5y - 14.$

13. $y^2 + 4y - 21.$

14. $a^2 + 11a + 30.$

15. $x^2 - 17x + 30.$

16. $p^2 - 7p - 8.$

17. $q^2 + 5q - 24.$

18. $a^2x^2 + 7ax - 18.$

19. $a^2 - 17ab + 70b^2.$

20. $a^2 - 9ab - 22b^2.$

21. $a^4 + 8a^2 - 20.$

22. $a^2y^2 - 11ay + 24.$

23. $m^2 + 25m + 100.$

24. $3y - 4 + y^2.$

25. $a^2 - 2ab - 24b^2.$

26. $n^4 + 60 + 17n^2.$

27. $a^6 + 7a^3 - 30.$

28. $a^8 - 7a^4 - 30.$

29. $x^3 + 5x^2 + 6x.$

30. $100x^2 - 500x + 600.$

31. $6a^4 - 18a^3 + 12a^2.$

32. $x^2y - 4xy - 21y.$

33. $m^2a^2 - 4ma^2 - 21a^2.$

34. $10x^2y^2 - 70x^2y - 180x^2.$

35. $200x^2 + 400x + 200.$

36. $4a^2 - 48ab + 44b^2.$

## TYPE III. QUADRATIC TRINOMIALS OF THE FORM
$$px^2 + qx + r.$$

**113.** According to § 66,
$$(4x + 3)(5x - 2) = 20x^2 + 7x - 6.$$
$20x^2$ is the product of $4x$ and $5x$.

$-6$ is the product of $+3$ and $-2$.

$+7x$ is the sum of the cross products.

Hence in factoring $6x^2 - 13x + 5$, we have to find two binomials whose corresponding terms are similar, such that

The first two terms are factors of $6x^2$.

The last two terms are factors of $5$,

and the sum of the cross products equals $-13x$.

By actual trial we find which of the factors of $6x^2$ and $5$ give the correct sum of cross products.

If we consider that the factors of $+5$ must have like signs, and that they must be negative, as $-13\,x$ is negative, all possible combinations are contained in the following:

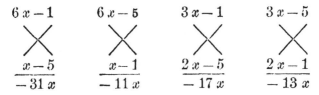

$$
\begin{array}{cccc}
6\,x-1 & 6\,x-5 & 3\,x-1 & 3\,x-5 \\
x-5 & x-1 & 2\,x-5 & 2\,x-1 \\
\hline
-31\,x & -11\,x & -17\,x & -13\,x
\end{array}
$$

Evidently the last combination is the correct one, or

$$6\,x^2 - 13\,x + 5 = (3\,x - 5)(2\,x - 1).$$

**114.** In actual work it is not always necessary to write down all possible combinations, and after a little practice the student should be able to find the proper factors of simple trinomials at the first trial. The work may be shortened by the following considerations:

1. *If $p$ and $r$ are positive, the second terms of the factors have the same sign as $q$.*

2. *If $p$ is positive, and $r$ is negative, then the second terms of the factors have opposite signs.*

If a combination should give a sum of cross products, which has the same absolute value as the term $qx$, but the opposite sign, exchange the signs of the second terms of the factors.

3. *If $px^2 + qx + r$ does not contain any monomial factor, none of the binomial factors can contain a monomial factor.*

Ex. Factor $3\,x^2 - 83\,x + 54$.

The factors of the first term consist of one pair only, viz. $3\,x$ and $x$, and the signs of the second terms are minus. 54 may be considered the product of the following combinations of numbers: $1 \times 54$, $2 \times 27$, $3 \times 18$, $6 \times 9$, $9 \times 6$, $18 \times 3$, $27 \times 2$, $54 \times 1$. Since the first term of the first factor ($3\,x$) contains a 3, we have to reject every combination of factors of 54 whose first factor contains a 3. Hence only $1 \times 54$ and $2 \times 27$ need be considered.

$$3\,x - 1$$
$$x - 54$$
$$\overline{-163\,x}$$

$$3\,x - 2$$
$$x - 27$$
$$\overline{-83\,x}$$

Therefore $3\,x^2 - 83\,x + 54 = (3\,x - 2)(x - 27)$.

**115.** The type $px^2 + qx + r$ is the most important of the trinomial types, since all others (II, IV) are special cases of it. In all examples of this type, the expressions should be arranged according to the ascending or the descending powers of some letter, and the monomial factors should be removed.

### EXERCISE 41

Resolve into prime factors:

1. $2\,x^2 + 9\,x - 5$.
2. $4\,a^2 - 9\,a + 2$.
3. $3\,x^2 - 8\,x + 4$.
4. $5\,m^2 - 26\,m + 5$.
5. $6\,n^2 + 5\,n - 4$.
6. $3\,x^2 + 13\,x + 4$.
7. $3\,x^2 + 3\,x - 6$.
8. $12\,y^2 - y - 6$.
9. $2\,y^2 + y - 3$.
10. $2\,t^2 - 17\,t - 9$.
11. $10\,a^2 - 19\,a + 6$.
12. $9\,y^2 + 32\,y - 16$.
13. $2\,m^2 + 7\,m + 3$.
14. $10\,x^2 - 9\,x - 7$.
15. $12\,x^2 - 17\,x + 6$.
16. $6\,n^2 + 13\,n + 2$.
17. $2\,y^2 + 17\,y - 9$.
18. $14\,a^2 + a - 4$.

19. $3\,x^2 + 11\,xy - 4\,y^2$.
20. $15\,x^2 - 77\,xy + 10\,y^2$.
21. $10\,a^2 - 23\,ab + 12\,b^2$.
22. $6\,x^2 - 13\,xy + 6\,y^2$.
23. $12\,x^2 - 7\,xy - 10\,y^2$.
24. $4\,x^2 + 14\,x + 12$.
25. $2\,x^3 + 11\,x^2 + 12\,x$.
26. $5\,x^3 + 14\,x^2 + 8\,x$.
27. $20\,x^2 - 50\,xy + 20\,y^2$.
28. $100\,x^2 - 200\,xy + 100\,y^2$.
29. $5\,a^6 - 9\,a^5 - 2\,a^4$.
30. $90\,x^2y^2 - 260\,xy^2 - 30\,y^2$.
31. $90\,a^2 - 300\,ab + 250\,b^2$.
32. $8\,y^3 - 3\,y^2 - 4\,y^4$.
33. $40\,a^2 - 90\,a^2n^2 + 20\,a^2n^4$.
34. $144\,x^2 - 290\,xy + 144\,y^2$.
35. $4\,x^4 + 8\,x^2y^2 + 3\,y^4$.

## TYPE IV. THE SQUARE OF A BINOMIAL

$$x^2 \pm 2\,xy + y^2.$$

**116.** Expressions of this form are special cases of the preceding type, and may be factored according to the method used for that type. In most cases, however, it is more convenient to factor them according to § 65.

$$x^2 + 2\,xy + y^2 = (x+y)^2.$$
$$x^2 - 2\,xy + y^2 = (x-y)^2.$$

A trinomial belongs to this type, *i.e.* it is a perfect square, when two of its terms are perfect squares, and the remaining term is equal to twice the product of the square roots of these terms.

The student should note that a term, in order to be a perfect square, must have a positive sign.

$16\,x^2 - 24\,xy + 9\,y^2$ is a perfect square, for $2\sqrt{16\,x^2} \times \sqrt{9\,y^2} = 24\,xy$.
Evidently $16\,x^2 - 24\,xy + 9\,y^2 = (4\,x - 3\,y)^2$.

*To factor a trinomial which is a perfect square, connect the square roots of the terms which are squares by the sign of the remaining term, and indicate the square of the resulting binomial.*

### EXERCISE 42

Determine whether or not the following expressions are perfect squares, and factor whenever possible:

1. $m^2 + 2\,mn + n^2.$
2. $c^2 - 2\,cd - d^2.$
3. $q^2 - 10\,q + 25.$
4. $x^2 - 10\,x + 16.$
5. $a^2 - 4\,ab + 4\,b^2.$
6. $a^2 - 10\,ab - 25.$
7. $m^2 - 14\,mn + 49\,n^2.$
8. $9 + 6\,a^2b^2 + a^4.$
9. $a^2 + 10\,ab + b^4.$
10. $m^4 + 6\,m^2n + 9\,n^2.$
11. $y^2 - 16\,y + 64.$
12. $9\,a^2 - 12\,ab + 4\,b^2.$
13. $25\,x^2 - 20\,xy + 4\,y^2.$
14. $16\,a^2 - 26\,ab + 9\,b^2.$

**15.** $16\,a^2 - 24\,ab + 9\,b^2$.

**16.** $36\,x^4 + 60\,x^2 + 25$.

**17.** $225\,x^2y^2 - 60\,xy + 4$.

**18.** $10\,a^2 - 20\,ab + 10\,b^2$.

**19.** $x^4 - 2\,x^3y + x^2y^2$.

**20.** $m^3 - 6\,m^2 + 9\,m$.

Make the following expressions perfect squares by supplying the missing terms:

**21.** $a^2 - 6\,ab + (\quad)$.

**22.** $x^2 - 8\,x + (\quad)$.

**23.** $a^2 + 6\,a + (\quad)$.

**24.** $9\,a^2 - (\quad) + 16\,b^2$.

**25.** $144\,a^2 + (\quad) + 81\,b^2$.

**26.** $100\,m^4 + (\quad) + 49$.

**27.** $64\,a^4 - 48\,a^2 + (\quad)$.

**28.** $4\,m^2 - 20\,m + (\quad)$.

**29.** $4\,a^2 + 12\,a + (\quad)$.

**30.** $16\,x^2 - (\quad) + 25$.

## TYPE V. THE DIFFERENCE OF TWO SQUARES

$$x^2 - y^2.$$

**117.** According to § 65,

$$a^2 - b^2 = (a + b)(a - b),$$

i.e. *the difference of the squares of two numbers is equal to the product of the sum and the difference of the two numbers.*

Ex. 1. $\quad 4\,x^6y^8 - 9\,z^6 = (2\,x^3y^4 + 3\,z^3)(2\,x^3y^4 - 3\,z^3).$

Ex. 2. $\quad 16\,a^8 - 64\,b^{10} = 16(a^8 - 4\,b^{10})$
$$= 16(a^4 + 2\,b^5)(a^4 - 2\,b^5).$$

Ex. 3. $\quad a^4 - b^4 = (a^2 + b^2)(a^2 - b^2)$
$$= (a^2 + b^2)(a + b)(a - b).$$

Note. $a^2 + b^2$ is prime.

### EXERCISE 43

Resolve into prime factors:

**1.** $x^2 - y^2$.

**2.** $a^2 - 9$.

**3.** $36 - b^2$.

**4.** $4\,a^2 - 1$.

**5.** $1 - 49\,a^2$.

**6.** $81 - t^2$.

**7.** $100\,a^2 - b^2$.

**8.** $a^2b^2 - 121$.

**9.** $9\,x^2 - 16\,y^2$.

10. $25\,x^2y^2 - 81\,z^2$.
11. $49\,a^2y^2 - 64$.
12. $25\,a^4 - 1$.
13. $100\,a^2b^2 - c^4$.
14. $169\,a^4 - 100$.
15. $225\,a^2 - 16\,b^4$.

16. $a^4 - b^4$.
17. $a^4b^4 - 81$.
18. $1 - x^8$.
19. $10\,a^2 - 10\,b^2$.
20. $13\,a^2x - 13\,b^2x$.
21. $x^3 - xy^2$.

22. $x^2 - y^8$.
23. $75\,x^3y^8 - 48\,xy^8$.
24. $242\,x^4y - 2\,y$.
25. $144\,x^2 - x^2y^4$.
26. $100^2 - 3^2$.
27. $9991$.

**118. One or both terms are squares of polynomials.**

Ex. 1. Factor $a^2 - (c + d)^2$.
$$a^2 - (c + d)^2 = (a + c + d)(a - c - d).$$

Ex. 2. Resolve into prime factors and simplify
$$(4\,a + 3\,b)^2 - (2\,a - 5\,b)^2.$$

$(4\,a + 3\,b)^2 - (2\,a - 5\,b)^2 = [(4\,a + 3\,b) + (2\,a - 5\,b)][(4\,a + 3\,b) - (2\,a - 5\,b)]$
$$= [4\,a + 3\,b + 2\,a - 5\,b][4\,a + 3\,b - 2\,a + 5\,b]$$
$$= (6\,a - 2\,b)(2\,a + 8\,b)$$
$$= 2(3\,a - b) \cdot 2(a + 4\,b)$$
$$= 4(3\,a - b)(a + 4\,b).$$

**EXERCISE 44**

Resolve into prime factors:

1. $(m + n)^2 - p^2$.
2. $(m - n)^2 - p^2$.
3. $(m + n)^2 - 16\,p^2$.
4. $x^2 - (y + z)^2$.
5. $16\,x^2 - (y + z)^2$.
6. $25\,a^2 - (b - c)^2$.
7. $(m + 2\,n)^2 - 36\,p^2$.

8. $(m - 3\,n)^2 - (a + b)^2$.
9. $(2\,a - 5\,b)^2 - (5\,c - 9\,d)^2$.
10. $(a + b)^2 - b^2$.
11. $x^2 - (x - y)^2$.
12. $(x + 3\,y)^2 - 9\,y^2$.
13. $(2\,a + 5)^2 - (3\,a - 4)^2$.
14. $(2\,x + 3\,y)^2 - (3\,x - 2\,y)^2$.

## TYPE VI. GROUPING TERMS

**119.** By the introduction of parentheses, polynomials can frequently be transformed into bi- and trinomials, which may be factored according to types I–VI.

*A. After grouping the terms, we find that the new terms contain a common factor.*

**Ex. 1.** Factor $ax + bx + ay + by$.

$$ax + bx + ay + by = x(a + b) + y(a + b)$$
$$= (a + b)(x + y).$$

**Ex. 2.** Factor $x^3 - 5x^2 - x + 5$.

$$x^3 - 5x^2 - x + 5 = x^2(x - 5) - (x - 5)$$
$$= (x - 5)(x^2 - 1)$$
$$= (x - 5)(x + 1)(x - 1).$$

### EXERCISE 45

Resolve into prime factors:

1. $ax + bx + ay + by$.

2. $ma - na + mb - nb$.

3. $2an - 3bn + 2aq - 3bq$.

4. $4cx + 4cy - 5dx - 5dy$.

5. $10ax - 5ay - 6bx + 3by$.

6. $x^3 + x^2 + 2x + 2$.

7. $x^3 + x^2 - x - 1$.

8. $m^2x^2 + n^2x^2 - m^2y^2 - n^2y^2$.

9. $3a^2x - 4b^2x + 3a^2y - 4b^2y$.

10. $a^3 - a^2b + ab^2 - b^3$.

11. $c^3 - 7c^2 + 2c - 14$.

12. $a^5 - a^4x - ab + bx$.

*B. By grouping, the expression becomes the difference of two squares.*

**Ex. 1.** Factor $9x^2 - y^2 - 4z^2 + 4yz$.

$$9x^2 - y^2 - 4z^2 + 4yz = 9x^2 - (y^2 - 4yz + 4z^2)$$
$$= (3x)^2 - (y - 2z)^2$$
$$= (3x + y - 2z)(3x - y + 2z).$$

Ex. 2. Factor $4\,a^2 - b^2 + 9\,x^2 - 4\,y^2 - 12\,ax + 4\,by$.

Arranging the terms,

$4\,a^2 - b^2 + 9\,x^2 - 4\,y^2 - 12\,ax + 4\,by$
$= 4\,a^2 - 12\,ax + 9\,x^2 - b^2 + 4\,by - 4\,y^2$
$= (4\,a^2 - 12\,ax + 9\,x^2) - (b^2 - 4\,by + 4\,y^2)$
$= (2\,a - 3\,x)^2 - (b - 2\,y)^2$
$= (2\,a - 3\,x + b - 2\,y)(2\,a - 3\,x - b + 2\,y).$

### EXERCISE 46

Resolve into prime factors:

1. $x^2 + 2\,xy + y^2 - q^2.$
2. $1 - a^2 - 2\,ab - b^2.$
3. $a^2 - 4\,ab + 4\,b^2 - 25.$
4. $36\,x^2 - y^2 + 6\,yz - 9\,z^2.$
5. $9\,m^2 - 6\,mn + n^2 - a^2b^2.$
6. $a^2 + 2\,ab + b^2 - x^2 - 2\,xy - y^2.$
7. $a^2 - 10\,ab + 25\,b^2 - x^2 + 4\,xy - 4\,y^2.$
8. $x^4 - x^2 - 2\,x - 1.$

## SUMMARY OF FACTORING

**I.** First find monomial factors common to all terms.

**II.** Binomials are factored by means of the formula

$$a^2 - b^2 = (a + b)(a - b).$$

**III.** Trinomials are factored by the method of cross products, although frequently the particular cases II and IV are more convenient.

**IV.** Polynomials are reduced to the preceding cases by grouping terms.

### EXERCISE 47

#### MISCELLANEOUS EXAMPLES *

Resolve into prime factors:

1. $m^2 - 16.$
2. $8\,m^2 + 16.$
3. $m^2 - m - 2.$
4. $6\,a^4 + 37\,a^2 + 6.$
5. $6\,a^4 - 12\,a^2 + 6.$
6. $2\,xy + x^2 + y^2.$
7. $a^3 - 9\,a^2.$
8. $4\,x^2 - 10\,xy + 4\,y^2.$
9. $m^2 - 6\,m + 9 - n^2.$

* See page 266.

10. $x^3 - xy^2$.

11. $10\,a^2 - 32\,ab + 6\,b^2$.

12. $4\,a^4 - 4\,b^4$.

13. $49\,a^4 - 42\,a^3 + 9\,a^2$.

14. $20\,a^4b^4 - 90\,a^2b^2 - 50$.

15. $9 - a^2 - 12\,ab - 36\,b^2$.

16. $2\,x^3 - 7\,x^2 - 4\,x + 14$.

17. $1 - n^8$.

18. $n^8 - n^6y^2$.

19. $5\,x^2 - 50\,x + 45$.

20. $a^6 - a^3 - 156$.

21. $18\,x^3 - 12\,xy - 3\,x^4 + 2\,x^2y$.

22. $3\,a^2 + 6\,ab + 3\,b^2 - 48$.

23. $80\,x^4 - 310\,x^2 - 40$.

24. $a^5 - a$.

25. $a^3 + a^2 + a + 1$.

26. $a^4 - 2\,a^3 + a^2 - 1$.

27. $42\,x^2 - 85\,xy + 42\,y^2$.

28. $10\,m^6 - 43\,m^3 - 9$.

29. $25\,a^2b^2 + 25\,ab - 24$.

30. $13\,c^2 - 13\,c - 156$.

31. $a^3 - 225\,ab^2$.

32. $20\,n^2 + 2\,ns - 42\,s^2$.

33. $(a + b)^2 - 64$.

34. $256 - (x - 2\,y)^2$.

35. $2\,a^4 - 128$.

36. $x^2y^2z^2 - 51\,xyz + 50$.

37. $17\,x^2 + 17\,xy - 102\,y^2$.

38. $a^5 - 2\,a^3 + a^2b - 2\,b$.

39. $a^2 + 2\,ab$.

40. $3\,p^8 - 39\,p^5 + 108\,p^2$.

41. $3\,x^4 - 3\,x^2 - 36$.

# CHAPTER VII

## HIGHEST COMMON FACTOR AND LOWEST COMMON MULTIPLE

### HIGHEST COMMON FACTOR

**120.** The **highest common factor** (H. C. F.) of two or more expressions is the algebraic factor of highest degree common to these expressions; thus $a^6$ is the H. C. F. of $a^7$ and $a^6b^7$.

Two expressions which have no common factor except unity are *prime to one another*.

**121.** The H. C. F. of two or more monomials whose factors are prime can be found by inspection.

The H. C. F. of $a^4$ and $a^2b$ is $a^2$.

The H. C. F. of $a^2b^3c^4$, $a^3b^2c^4$, and $a^4b^3$ is $a^2b^2$.

The H. C. F. of $(a + b)^3$ and $(a + b)^2(a - b)^4$ is $(a + b)^2$.

**122.** If the expressions have numerical coefficients, find by arithmetic the greatest common factor of the coefficients, and prefix it as a coefficient to H. C. F. of the algebraic expressions. Thus the H. C. F. of $6\,x^4yz$, $12\,x^3y^3z$, and $60\,x^3y^3$ is $6\,x^3y$.

The student should note that the power of each factor in the H. C. F. is the lowest power in which that factor occurs in any of the given expressions.

Find the H. C. F. of:

1. $4\,a^3b^2$, $6\,a^2b^3$.

2. $15\,a^2b^3c^4$, $25\,a^3b^4c^2$.

3. $13\,x^3yz^2$, $39\,x^4y^2u^3$.

4. $3^3 \cdot 2^2$, $3^2 \cdot 2^3$, $3 \cdot 2^4$.

5. $2 \cdot 5^2 \cdot 3^3$, $2^2 \cdot 5^3 \cdot 3^4$, $2^3 \cdot 5^4 \cdot 3^2$.

6. $7^3 \cdot 5^3$, $7^2 \cdot 5$, $7 \cdot 11^2$.

7. $6 mn^2u^4$, $12 m^2nu^3$, $30 mu^5$.

8. $39 x^4y^3z^5$, $52 x^3y^5z^4$, $65 x^5y^3z^4$.

9. $38 x^4y^2$, $95 y^4z^2$, $57 x^2z^4$.

10. $225 a^4b^3c^2d$, $75 ab^2cd$, $-15 bcd^2$.

11. $4 a^9$, $8 a^{10}$, $16 a^{11}$, $24 a^8b$.

12. $4(m + n)^3$, $5(m + n)^2$, $7(m + n)^2(m - n)$.

13. $6(m+1)^3(m+2)$, $8(m+1)^2(m+3)$, $4(m+1)^2$.

14. $6 x^3(x - y)^5$, $9 x^2(x - y)^4$, $12 x(x - y)^3$.

**123.** **To find the H. C. F. of polynomials**, resolve each poly-nomial into prime factors, and apply the method of the preceding article.

Ex. 1. Find the H. C. F. of $x^2 - 4 xy + 4 y^2$, $x^2 - 3 xy + 2 y^2$, and $x^2 - 7 xy + 10 y^2$.

$$x^2 - 4 xy + 4 y^2 = (x - 2 y)^2.$$
$$x^2 - 3 xy + 2 y^2 = (x - 2 y)(x - y).$$
$$x^2 - 7 xy + 10 y^2 = (x - 2 y)(x - 5 y).$$

Hence the H. C. F. $= x - 2 y.$

### EXERCISE 49

Find the H. C. F. of:

1. $4 a^3b^4$, $8 a^6b^3 - 12 a^3b^6$.

2. $15 x^4y^3z^2$, $10 x^3z^2 - 5 x^3y^2$.

3. $25 m^2n$, $15 m^3n^2 - 10 m^2n^3$.

4. $3 ax^4 - 3 bx^4$, $6 mx^4 - 6 nx^4$.

5. $6 a^2 - 6 ab$, $5 ab - 5 b^2$.

6. $x^2 - y^2$, $4 x^2y + 4 xy^2$.

7. $a^2 - b^2$, $a^2 + 2 ab + b^2$.

8. $x^2 - 5 x + 6$, $x^2 - 6 x + 8$.

9. $x^2 + 6 x + 9$, $x^2 + 3 x$, $xy + 3 y$.

10. $x^2 - 7 x + 12$, $x^2 - 8 x + 16$, $x^2 - 5 x + 4$.

11. $x^2 + 5 x + 6$, $x^2 - 9$, $x^2 + x - 6$.

12. $a^2 - 8 a + 16$, $a^3 - 16 a$, $a^2 - 3 a - 4$.

13. $a^2 + 2 a - 3$, $a^2 + 7 a + 12$, $a^3 - 9 a$.

14. $y^2 + 3 y - 54$, $y^2 + y - 42$, $y^2 + 2 y - 48$.

15. $2 a^2 + 5 a + 2$, $4 a^2 + 4 a + 1$, $2 a^2 + a$.

16. $a^2 + 4 ab + 3 b^2$, $a^2 + 2 ab - 3 b^2$, $a^2 + 9 ab + 18 b^2$.

## LOWEST COMMON MULTIPLE

**124.** A **common multiple** of two or more expressions is an expression which can be divided by each of them without a remainder.

Common multiples of $3 x^2$ and $5 y$ are $30 x^2 y$, $60 x^2 y^2$, $300 x^2 y$, etc.

**125.** The **lowest common multiple** (L. C. M.) of two or more expressions is the common multiple of lowest degree; thus, $x^3 y^3$ is the L. C. M. of $x^3 y$ and $x y^3$.

**126.** If the expressions have a numerical coefficient, find by arithmetic their least common multiple and prefix it as a coefficient to the L. C. M of the algebraic expressions.

The L. C. M. of $3 a^3 b^3$, $2 a^2 b^2 c^3$, $6 c^6$ is $6 a^3 b^3 c^6$.
The L. C. M. of $12(a + b)^3$ and $(a + b)^2(a - b)^2$ is $12(a + b)^3(a - b)^2$.

**127.** Obviously the power of each factor in the L. C. M. is equal to the highest power in which it occurs in any of the given expressions.

**128.** To find the L. C. M. of several expressions which are not completely factored, resolve each expression into prime factors and apply the method for monomials.

**Ex. 1.** Find the L. C. M. of $4 a^2 b^2$ and $4 a^4 - 4 a^2 b^3$.

$$4 a^2 b^2 = 4 a^2 b^2.$$
$$4 a^4 - 4 a^2 b^3 = 4 a^2(a^2 - b^3).$$

Hence, L. C. M. $\qquad = 4 a^2 b^2(a^2 - b^3).$

**Ex. 2.** Find the L. C. M. of $a^3 - b^2$, $a^2 + 2 ab + b^2$, and $b - a$.

$$a^2 - b^2 = (a + b)(a - b).$$
$$a^2 + 2 ab + b^2 = (a + b)^2.$$
$$b - a = -(a - b).$$

Hence the L. C. M. $\qquad = (a + b)^2(a - b).$

**Note.** The L. C. M. of the last example is also $-(a + b)^2(a - b)$. In general, each set of expressions has two lowest common multiples, which have the same absolute value, but opposite signs.

Find the L. C. M. of :

1. $a, a^2, a^3$.

2. $x^2y, xy^2, y^3$.

3. $2x^3, 4x^2, 8x$.

4. $3xy^3, 8x^3y, 24x$.

5. $6a^3b, 5a^2b^2, 15b^3, 30a$.

6. $7a^3, 3a^2, 6a, x$.

7. $4a^5bcd, 20ab^5cd^2, 40abc^5, 8d^5$.

8. $9a, 3ab, 3(a+b)$.

9. $2(m+n)^2, 3(m+n)4m^2$

10. $(a-2)(a-3)^2, (a-3)(a-4)^2, (a-4)(a-2)^2$.

11. $3a, 2a^2, 2a^2b - 2ab^2$.

12. $6x+6y, 5x-5y, x^2-y^2$.

13. $a+b, a-b, a^2-b^2$.   14. $a^2-4, a^2+4a+4$.

15. $a^2-b^2, a^2+2ab+b^2, 2a-2b$.

16. $x^2-3x+2, x^2-5x+6$.

17. $a+2, a+3, a+1$.

18. $2a-1, 4a^2-1, 4a+2$.

19. $x^2+3x+2, x^2+5x+6, x^2+4x+4$.

20. $5x^2+5x, 3x^2-3, 15x$.

21. $a^2+ab, ab+b^2, a^2-ab-6b^2$.

22. $x^4-1, x^2-1, x-1$.

23. $x^2-7x+10, x^2-10x+16, x^2-5x+6$.

24. $ax+ay-bx-by, 3a-3b, 2x+2y$.

(For additional examples see page 268.)

# CHAPTER VIII

## FRACTIONS

### REDUCTION OF FRACTIONS

**129.** A **fraction** is an indicated quotient; thus $\dfrac{a}{b}$ is identical with $a \div b$. The dividend $a$ is called the *numerator* and the divisor $b$ the *denominator*. The numerator and the denominator are the *terms* of the fraction.

**130.** All operations with fractions in algebra are identical with the corresponding operations in arithmetic. Thus, the value of a fraction is not altered by multiplying or dividing both its numerator and its denominator by the same number; the product of two fractions is the product of their numerators divided by the product of their denominators, etc.

In arithmetic, however, only positive integral numerators and denominators are considered, but we shall *assume* that the arithmetic principles are generally true for all algebraic numbers.

**131.** *If both terms of a fraction are multiplied or divided by the same number, the value of the fraction is not altered.*

Thus
$$\frac{a}{b} = \frac{ma}{mb}, \text{ and } \frac{mx}{my} = \frac{x}{y}.$$

**132.** A fraction is in its **lowest terms** when its numerator and its denominator have no common factors.

**Ex. 1.** Reduce $\dfrac{6\,xy^2z^4}{9\,x^2y^2z^3}$ to its lowest terms.

Remove successively all common divisors of numerator and denominator, as $3$, $x$, $y^2$, and $z^3$ (or divide the terms by their H. C. F. $3\,xy^2z^3$).

Hence
$$\frac{6\,xy^2z^4}{9\,x^2y^2z^3} = \frac{2\,z}{3\,x}.$$

**133.** To reduce a fraction to its lowest terms, *resolve numerator and denominator into their factors, and cancel all factors that are common to both.* **Never cancel terms of the numerator or the denominator;** *cancel factors only.*

**Ex. 2.** Reduce $\dfrac{a^3 - 6\,a^2 + 8\,a}{6\,a^4 - 24\,a^2}$ to its lowest terms.

$$\frac{a^3 - 6\,a^2 + 8\,a}{6\,a^4 - 24\,a^2} = \frac{a(a^2 - 6\,a + 8)}{6\,a^2(a^2 - 4)}$$

$$= \frac{a(a-2)(a-4)}{6\,a^2(a+2)(a-2)}$$

$$= \frac{a-4}{6\,a(a+2)}.$$

**Ex. 3.** Reduce $\dfrac{a^2 + ab - 2\,b^2}{b^2 - a^2}$ to its lowest terms.

$$\frac{a^2 + ab - 2\,b^2}{b^2 - a^2} = \frac{a^2 + ab - 2\,b^2}{-(a^2 - b^2)}$$

$$= -\frac{a^2 + ab - 2\,b^2}{a^2 - b^2}$$

$$= -\frac{(a-b)(a+2\,b)}{(a+b)(a-b)}$$

$$= -\frac{a+2\,b}{a+b}.$$

**EXERCISE 51 \***

Reduce to lowest terms:

1. $\dfrac{9 \cdot 5^3}{3 \cdot 5^2}$.

2. $\dfrac{2 \cdot 3^2 \cdot 7^3}{2 \cdot 3^3 \cdot 7^2}$.

3. $\dfrac{12\,a^4}{4\,a^3}$.

4. $\dfrac{36\,x^3y^3}{18\,x^2y^2z}$.

5. $\dfrac{225\,a^3b^3c^5}{25\,a^3b^3c^7}$.

6. $\dfrac{39\,a^2b^3c^4}{13\,a^4b^3c^2}$.

\* See page 268.

7. $\dfrac{11\,a}{22\,a^2bc}$.

8. $\dfrac{14\,a^2b^2}{42\,a^2x^2}$.

9. $\dfrac{3\,a+3\,b}{3\,x+3\,y}$.

10. $\dfrac{7\,x+14\,y}{9\,x+18\,y}$.

11. $\dfrac{5\,a^2b-10\,ab^2}{15\,a^2b+20\,ab^2}$.

12. $\dfrac{m^2n-mn^2}{3\,m-3\,n}$.

13. $\dfrac{a^2+2\,ab+b^2}{a^2-b^2}$.

14. $\dfrac{x^2-3\,xy+2\,y^2}{x^2-4\,xy+4\,y^2}$.

15. $\dfrac{m^2-7\,m+12}{m^2-5\,m+6}$.

16. $\dfrac{b^2+4\,b-5}{b^2+5\,b-6}$.

17. $\dfrac{4\,x^2-25\,y^2}{4\,x-10\,y}$.

18. $\dfrac{x^2+7\,x+12}{x^2+8\,x+15}$.

19. $\dfrac{a^2-8\,a-9}{a^2-9\,a-10}$.

20. $\dfrac{m^2-7\,m+12}{m^2-2\,m-3}$.

21. $\dfrac{n^2-13\,n+22}{n^2-9\,n-22}$.

22. $\dfrac{a^2-10\,ab+16\,b^2}{a^2+6\,ab-16\,b^2}$.

23. $\dfrac{a^3-3\,a^2-4\,a}{a^3-8\,a^2+16\,a}$.

24. $\dfrac{3\,x^2-10\,xy+3\,y^2}{9\,x^2-6\,xy+y^2}$.

25. $\dfrac{a^4-b^4}{6\,a^2b-6\,b^3}$.

26. $\dfrac{a^4-9\,a^2}{3\,a^2-10\,a+3}$.

27. $\dfrac{12\,m^2-7\,mn+n^2}{15\,m^2-8\,mn+n^2}$.

28. $\dfrac{3\,a^3-6\,a^2n-6\,an^2}{3\,a^3-7\,a^2n+2\,an^2}$.

29. $\dfrac{ax+ay-mx-my}{bx+by-nx-ny}$.

30. $\dfrac{a^6-b^6}{2\,a^3-2\,b^3}$.

31. $\dfrac{a^4-b^4}{a^2-2\,ab+b^2}$.

32. $\dfrac{x^3-5\,x^2+4\,x}{-x^3-3\,x^2+4\,x}$.

**134.** **Reduction of fractions to equal fractions of lowest common denominator.** Since the terms of a fraction may be multiplied by any quantity without altering the value of the fraction, we may use the same process as in arithmetic for reducing fractions to the lowest common denominator.

Ex. 1. Reduce $\dfrac{7\,a}{6\,b^2x^2}$, $\dfrac{2\,b}{3\,a^3}$, and $\dfrac{5\,xy}{4\,ab^2}$ to their lowest common denominator.

The L. C. M. of $6\,b^2x^2$, $3\,a^3$, and $4\,ab^2$ is $12\,a^3b^2x^2$.

To reduce $\dfrac{7\,a}{6\,b^2x^2}$ to a fraction with the denominator $12\,a^3b^2x^2$, numerator and denominator must be multiplied by $\dfrac{12\,a^3b^2x^2}{6\,b^2x^2}$ or $2\,a^3$.

Similarly, multiplying the terms of $\dfrac{2\,b}{3\,a^3}$ by $4\,b^2x^2$, and the terms of $\dfrac{5\,xy}{4\,ab^2}$ by $3\,a^2x^2$, we have $\dfrac{14\,a^4}{12\,a^3b^2x^2}$, $\dfrac{8\,b^3x^2}{12\,a^3b^2x^2}$, and $\dfrac{15\,a^2x^3y}{12\,a^3b^2x^2}$.

**135.** *To reduce fractions to their lowest common denominator, take the L. C. M. of the denominators for the common denominator. Divide the L. C. M. by the denominator of each fraction, and multiply each quotient by the corresponding numerator.*

Ex. Reduce $\dfrac{3\,x}{(x+3)(x-3)}$, $\dfrac{3\,x-2}{(x-3)(x-1)}$, and $\dfrac{5}{(x-1)(x+3)}$ to their lowest common denominator.

The L. C. D. $=(x+3)(x-3)(x-1)$.

Dividing this by each denominator, we have the quotients $(x-1)$, $(x+3)$, and $(x-3)$.

Multiplying these quotients by the corresponding numerators and writing the results over the common denominator, we have

$$\dfrac{3\,x^2-3\,x}{(x+3)(x-3)(x-1)}, \quad \dfrac{3\,x^2+7\,x-6}{(x+3)(x-3)(x-1)}, \quad \text{and} \quad \dfrac{5\,x-15}{(x+3)(x-3)(x-1)}.$$

NOTE. Since $a = \dfrac{a}{1}$, we may extend this method to integral expressions.

### EXERCISE 52

Reduce the following to their lowest common denominator:

1. $\dfrac{3\,m^2}{4\,n^3}$, $\dfrac{5}{3\,n^2}$.

3. $2$, $\dfrac{a^2}{b^2c^2}$.

5. $\dfrac{1}{x}$, $\dfrac{1}{y}$, $\dfrac{1}{z}$.

2. $\dfrac{22\,a^2}{3\,x^2y^2}$, $\dfrac{5\,a}{2\,y^2z^2}$.

4. $\dfrac{x^2}{m^2}$, $\dfrac{1}{m}$, $\dfrac{5}{2\,m^3}$.

6. $\dfrac{5\,c}{2\,ab^2}$, $\dfrac{2\,b}{3\,ac^2}$, $\dfrac{5}{6\,a^2}$.

7. $3\,a$, $\dfrac{2\,x}{a^3}$, $\dfrac{5\,a}{6\,x}$.

11. $\dfrac{2\,a-1}{9\,a^2-1}$, $\dfrac{3}{3\,a-1}$.

8. $\dfrac{5\,a^5b^2}{2\,x^4y^2z^4}$, $\dfrac{3\,zb^3}{7\,a^2x^3y^3}$, $\dfrac{2\,a^3}{3\,xyz}$.

12. $\dfrac{6}{a+b}$, $\dfrac{5}{a-b}$, $\dfrac{4\,a}{a^2-b^2}$.

9. $\dfrac{1}{x-y}$, $\dfrac{4}{3\,x-3\,y}$.

13. $\dfrac{3}{x^2-2\,x+1}$, $\dfrac{5}{x^2-4\,x+3}$.

10. $\dfrac{3\,b}{ax-ay}$, $\dfrac{2\,a}{bx-by}$.

14. $\dfrac{6\,a+1}{a-5}$, $\dfrac{6\,a-1}{a+5}$, $\dfrac{2\,a^2}{a^2-25}$.

15. $\dfrac{3}{2\,x-2}$, $\dfrac{5}{x^2-1}$, $\dfrac{1}{3\,x+3}$.

16. $\dfrac{a+2}{a^2-4\,a+3}$, $\dfrac{a+1}{a^2-5\,a+6}$, $\dfrac{a+3}{a^2-3\,a+2}$.

## ADDITION AND SUBTRACTION OF FRACTIONS

**136.** Since $\dfrac{a}{c}+\dfrac{b}{c}=\dfrac{a+b}{c}$ (Art. 74), **fractions having a common denominator are added or subtracted** *by dividing the sum or the difference of the numerators by the common denominator.*

**137.** If the given fractions have different denominators, they must be reduced to equal fractions which have the lowest common denominator before they can be added (or subtracted).

H

**Ex. 1.** Simplify $\dfrac{2\,a+3\,b}{2(2\,a-3\,b)}+\dfrac{6\,a-b}{4(2\,a+3\,b)}$.

The L. C. D. is $4(2\,a-3\,b)(2\,a+3\,b)$.

Multiplying the terms of the first fraction by $2(2\,a+3\,b)$, the terms of the second by $(2\,a-3\,b)$, and adding, we obtain

$$\frac{2\,a+3\,b}{2(2\,a-3\,b)}+\frac{6\,a-b}{4(2\,a+3\,b)}=\frac{2(2\,a+3\,b)(2\,a+3\,b)+(2\,a-3\,b)(6\,a-b)}{4(2\,a-3\,b)(2\,a+3\,b)}$$

$$=\frac{8\,a^2+24\,ab+18\,b^2+12\,a^2-20\,ab+3\,b^2}{4(2\,a-3\,b)(2\,a+3\,b)}$$

$$=\frac{20\,a^2+4\,ab+21\,b^2}{4(2\,a-3\,b)(2\,a+3\,b)}.$$

**138**. The results of addition and subtraction should be reduced to their lowest terms.

**Ex. 2.** Simplify $\dfrac{a+2\,b}{a^2-ab}+\dfrac{2\,a+b}{a^2-3\,ab+2\,b^2}-\dfrac{a-3\,b}{a^2-2\,ab}$.

$$a^2-ab=a(a-b).$$
$$a^2-3\,ab+2\,b^2=(a-b)(a-2\,b).$$
$$a^2-2\,ab=a(a-2\,b).$$

Hence the L. C. D. $= a(a-b)(a-2\,b).$

$$\frac{a+2\,b}{a^2-ab}+\frac{2\,a+b}{a^2-3\,ab+2\,b^2}-\frac{a-3\,b}{a^2-2\,ab}$$

$$=\frac{(a+2\,b)(a-2\,b)+a(2\,a+b)-(a-3\,b)(a-b)}{a(a-b)(a-2\,b)}$$

$$=\frac{a^2-4\,b^2+(2\,a^2+ab)-(a^2-4\,ab+3\,b^2)}{a(a-b)(a-2\,b)}$$

$$=\frac{a^2-4\,b^2+2\,a^2+ab-a^2+4\,ab-3\,b^2}{a(a-b)(a-2\,b)}$$

$$=\frac{2\,a^2+5\,ab-7\,b^2}{a(a-b)(a-2\,b)}=\frac{(2\,a+7\,b)(a-b)}{a(a-b)(a-2\,b)}$$

$$=\frac{2\,a+7\,b}{a(a-2\,b)}.$$

NOTE. In simplifying a term preceded by the minus sign, *e.g.* $-(a-3\,b)(a-b)$, the student should remember that parentheses are understood about terms (§ 66); hence he should, in the beginning, write the product in a parenthesis, as $-(a^2-4\,ab+3\,b^2)$.

Simplify :

1. $\dfrac{3a+5}{4} + \dfrac{2a-4}{5}.$

2. $\dfrac{9m+7n}{3} - \dfrac{6m-5n}{2}.$

3. $\dfrac{2x+3y}{9} - \dfrac{3x-y}{15} - \dfrac{x+2y}{45}.$

4. $\dfrac{a}{b} + \dfrac{x}{y}.$

8. $\dfrac{2a+3b}{4a} - \dfrac{5a-7b}{10b}.$

5. $\dfrac{1}{m} - \dfrac{1}{n}.$

9. $\dfrac{6a-11b}{13a} - \dfrac{15a-2b}{11b}.$

6. $\dfrac{5}{3a^2} + \dfrac{3}{2ab} - \dfrac{2}{b^2}.$

10. $\dfrac{1}{a} + \dfrac{1}{b} + \dfrac{1}{c}.$

7. $\dfrac{3a-4b}{4a} - \dfrac{2a-3b}{3b}.$

11. $\dfrac{3x+4y}{5x^2} - \dfrac{y-2z}{7yz} + \dfrac{5y}{3xy}.$

12. $\dfrac{3u-2v}{12uv} - \dfrac{5u+3v}{18v} - \dfrac{5uv-8v}{30u}.$

13. $\dfrac{a}{2} + \dfrac{1}{b}.$

19. $\dfrac{a}{a+b} + \dfrac{b}{a-b}.$

14. $\dfrac{a}{3} + \dfrac{1}{a} + \dfrac{1}{a^2}.$

20. $2 + \dfrac{a-2b}{a+b}.$

15. $\dfrac{3}{pq} + \dfrac{2}{p} - \dfrac{3}{q}.$

21. $\dfrac{1+m}{1-m} - \dfrac{1-m}{1+m}.$

16. $\dfrac{1}{m+2} + \dfrac{1}{m+3}.$

22. $\dfrac{3x}{x+1} + \dfrac{2x}{x+2}.$

17. $\dfrac{1}{m-3} - \dfrac{1}{m-2}.$

23. $\dfrac{2a-1}{a-2} - \dfrac{a-2}{a+3}.$

18. $\dfrac{3}{x+2} + \dfrac{2}{x+3}.$

24. $\dfrac{6a+5}{3a+3} - \dfrac{2a-7}{a+1}.$

25. $\dfrac{1}{(x+1)(x+2)} - \dfrac{1}{(x+2)(x+3)}.$

**\*** See page 270.

**26.** $\dfrac{1}{x^2 - 3x + 2} - \dfrac{x - 3}{x - 2}.$     **27.** $\dfrac{x - 2y}{x + 3y} - \dfrac{x - 4y}{x - 3y}.$

**28.** $\dfrac{5x}{x - 9} - \dfrac{6x}{x - 3} + \dfrac{x^2 - 2x}{x^2 - 12x + 27}.$

**29.** $\dfrac{2}{a - 4} - \dfrac{4}{3a - 9} + \dfrac{1}{a}.$     **30.** $\dfrac{a - 1}{2} + \dfrac{1}{a + 1}.$

**31.** $\dfrac{a + 2b}{a^2 - 4ab + 4b^2} + \dfrac{1}{a - 2b}.$

**32.** $\dfrac{4}{x - 1} - \dfrac{3}{x - 3} + \dfrac{x - 15}{x^2 - 4x + 3}.$

**33.** $\dfrac{2m}{1 - m^2} + \dfrac{1}{1 + m} + \dfrac{1}{1 - m}.$

**34.** $\dfrac{b}{a - b} - \dfrac{b}{a + 2b} - \dfrac{3b}{a^2 + ab - 2b^2}.$

**35.** $\dfrac{4}{2x^2 - 5x + 2} + \dfrac{x - 2}{x^2 - 5x + 6}.$

**36.** $\dfrac{3x - 1}{(x - 1)^3} + \dfrac{1}{(x - 1)^2} + \dfrac{2}{x - 1}.$

**37.** $\dfrac{a}{a^2 - 2ab + b^2} - \dfrac{b}{a^2 - b^2} + \dfrac{1}{a + b}.$

**38.** $a + 1 + \dfrac{a^2 - 2a - 1}{a + 1}.$    $\left(\text{HINT: Let } a + 1 = \dfrac{a + 1}{1}.\right)$

**39.** $\dfrac{x + 1}{x - 1} + 1.$     **42.** $a - b - \dfrac{a^3 - 2b^3}{a^2 + ab + b^2}.$

**40.** $\dfrac{y^3}{x - y} + x + y.$     **43.** $a + \dfrac{1}{a} + \dfrac{1}{a^2}.$

**41.** $m - 2 - \dfrac{m^2 + 6}{m - 3}.$     **44.** $a + b - \dfrac{a + b}{a - b} + \dfrac{a - b}{2}.$

**45.** $\dfrac{x^2}{x^2 - 7x + 12} - \dfrac{4x^2 - 2}{4x^2 - 17x + 4} + 1.$

**139.** To reduce a fraction to an integral or mixed expression.

$$\frac{a+b}{c} = \frac{a}{c} + \frac{b}{c}.$$ (§ 74)

Hence $\dfrac{5\,a^2 - 15\,a - 7}{5\,a} = \dfrac{5\,a^2}{5\,a} - \dfrac{15\,a}{5\,a} - \dfrac{7}{5\,a} = a - 3 - \dfrac{7}{5\,a}$

**Ex. 1.** Reduce $\dfrac{4\,x^3 - 2\,x^2 + 4\,x - 17}{2\,x - 3}$ to a mixed expression.

$$
\begin{array}{l}
4\,x^3 - 2\,x^2 + 4\,x - 17\ \big|\ \underline{2\,x - 3} \\
\underline{4\,x^3 - 6\,x^2}\qquad\quad\ \ \big|\ 2\,x^2 + 2\,x + 5 \\
\qquad\ 4\,x^2 + 4\,x \\
\qquad\ \underline{4\,x^2 - 6\,x} \\
\qquad\qquad + 10\,x - 17 \\
\qquad\qquad \underline{+ 10\,x - 15} \\
\qquad\qquad\qquad\ - 2
\end{array}
$$

Therefore $\dfrac{4\,x^3 - 2\,x^2 + 4\,x - 17}{2\,x - 3} = \dfrac{(2\,x^2 + 2\,x + 5)(2\,x - 3) - 2}{(2\,x - 3)}$

$$= 2\,x^2 + 2\,x + 5 - \frac{2}{2\,x - 3}.$$

### EXERCISE 54

Reduce each of the following fractions to a mixed or integral expression:

**1.** $\dfrac{6\,a^2 - 5\,a + 7}{a}.$

**2.** $\dfrac{9\,a^2 - 6\,a + 2}{3\,a}.$

**3.** $\dfrac{m^2 - 5\,m + 6}{m - 4}.$

**4.** $\dfrac{n^2 + 7\,n + 14}{n + 3}.$

**5.** $\dfrac{a^2 + 2\,a + 6}{a + 1}.$

**6.** $\dfrac{2\,x^2 - 5\,x + 15}{2\,x - 1}.$

**7.** $\dfrac{a^2 + 1}{a + 1}.$

**8.** $\dfrac{x^3 + 2\,x^2 - 3\,x + 5}{x - 3}.$

## MULTIPLICATION OF FRACTIONS

**140.** **Fractions are multiplied** *by taking the product of the numerators for the numerator, and the product of the denominators for the denominator;* or, expressed in symbols:

$$\frac{a}{b} \cdot \frac{c}{d} = \frac{ac}{bd}.$$

**141.** Since $\frac{a}{1} = a$, we may extend any principle proved for fractions to integral numbers, *e.g.* $\frac{a}{b} \times c = \frac{ac}{b}.$

*To multiply a fraction by an integer, multiply the numerator by that integer.*

**142.** *Common factors in the numerators and the denominators should be canceled before performing the multiplication.* (In order to cancel common factors, each numerator and denominator has to be factored.)

**Ex. 1.** Simplify $\dfrac{4\,a^3b}{3\,cd^5} \times \dfrac{6\,c^2d^2}{5\,ax} \times -\dfrac{7\,x^4y}{2\,ab^2}.$

$$\text{The expression} = -\frac{4 \cdot 6 \cdot 7\, a^3bc^2d^2x^4y}{3 \cdot 5 \cdot 2 \cdot a \cdot ab^2cd^5x}$$

$$= -\frac{28\,acx^3y}{5\,bd^3}.$$

**Ex. 2.** Simplify $\dfrac{x^2 - y^2}{x^2 - 3\,xy + 2\,y^2} \times \dfrac{xy - 2\,y^3}{x^2 + xy} \times \dfrac{x^2 - xy}{(x - y)^2}.$

$$= \frac{(x - y)(x + y)}{(x - y)(x - 2\,y)} \times \frac{y(x - 2\,y)}{x(x + y)} \times \frac{x(x - y)}{(x - y)(x - y)}$$

$$= \frac{y}{x - y}.$$

Find the following products:

1. $\dfrac{2^3}{5^2} \times \dfrac{5^3}{2^4}$.

3. $\dfrac{a^2b^3}{48} \cdot \dfrac{36}{a^3b^2}$.

5. $\dfrac{21\,m^3}{34\,ab^2} \cdot \dfrac{17\,ab}{14\,m^4}$.

2. $\dfrac{3^2}{5^3} \cdot \dfrac{5^2}{3^3}$.

4. $\dfrac{xy^2z^3}{7\,d} \cdot \dfrac{14\,d^2}{x^3y^2z}$.

6. $\dfrac{7\,a^2z}{4\,x^2z} \cdot \dfrac{x^2y^2}{7\,az^3}$.

7. $\dfrac{3\,ac^2}{7\,b} \cdot \dfrac{14\,b^2}{5\,c} \cdot \dfrac{10\,a^3}{3\,bc^2}$.

11. $\dfrac{4\,a+8\,b}{7\,a-21\,b} \cdot \dfrac{a-3\,b}{4}$.

8. $\dfrac{3\,a+3\,b}{5\,a-5\,b} \cdot \dfrac{10\,(a-b)}{a+b}$.

12. $\dfrac{a^2-2\,ab+b^2}{a^2-12\,ab+20\,b^2} \cdot \dfrac{a-10\,b}{a-b}$.

9. $\dfrac{3\,a^2b}{4\,ac^2} \cdot \left(-\dfrac{6\,a^2}{ab^2}\right) \cdot \dfrac{5\,b^2c}{6\,ab}$.

13. $\dfrac{x+1}{x+2} \cdot \dfrac{x+2}{x+3}$.

10. $\dfrac{9\,m^2-25\,n^2}{3\,m+5\,n} \cdot \dfrac{1}{3\,m-5\,n}$.

14. $\dfrac{a^2+3\,a-4}{a^2-3\,a-4} \cdot \dfrac{a^2-5\,a+4}{a^2+5\,a+4}$.

15. $\dfrac{x^2-1}{3\,x+1} \cdot \dfrac{3\,x^2-4\,x+1}{x^2+x} \cdot \dfrac{3\,x^2+x}{(x-1)^2}$.

16. $\dfrac{x^2y-xy^2}{x^2y+xy^2} \cdot \dfrac{x^3-x^2y}{x\,(x+y)}$.

17. $\dfrac{a^2-5\,a-6}{a^2+5\,a-6} \cdot \dfrac{a^2+8\,a+12}{a^2-5\,a-6} \cdot \dfrac{a-1}{a+2}$.

18. $\dfrac{x^2+6\,x-7}{x^2+15\,x+56} \cdot \dfrac{x-5}{x-1} \cdot \dfrac{x^2+18\,x+80}{x^2+5\,x-50}$.

## DIVISION OF FRACTIONS

**143.** **To divide an expression by a fraction,** *invert the divisor and multiply it by the dividend.* Integral or mixed divisors should be expressed in fractional form before dividing.

**144.** The **reciprocal of a number** is the quotient obtained by dividing 1 by that number.

The reciprocal of $a$ is $\dfrac{1}{a}$.

The reciprocal of $\frac{3}{4}$ is $1 \div \frac{3}{4} = \frac{4}{3}$.

The reciprocal of $\dfrac{a+b}{x}$ is $1 \div \dfrac{a+b}{x} = \dfrac{x}{a+b}$.

Hence the reciprocal of a fraction is obtained by inverting the fraction, and the principle of division may be expressed as follows:

**145.** *To divide an expression by a fraction, multiply the expression by the reciprocal of the fraction.*

Ex. 1. Divide $\dfrac{x^3 - y^3}{x^3 + xy^2}$ by $\dfrac{x^2 - y^2}{x^2y + y^3}$.

$$\frac{x^3 - y^3}{x^3 + xy^2} \div \frac{x^2 - y^2}{x^2y + y^3} = \frac{x^3 - y^3}{x^3 + xy^2} \times \frac{x^2y + y^3}{x^2 - y^2}$$

$$= \frac{(x - y)(x^2 + xy + y^2)}{x(x^2 + y^2)} \times \frac{y(x^2 + y^2)}{(x + y)(x - y)}$$

$$= \frac{y(x^2 + xy + y^2)}{x(x + y)}.$$

**EXERCISE 56 \***

Simplify the following expressions:

1. $\dfrac{7\,m^3n}{5\,x^2y} \div \dfrac{7\,m^3n}{5\,xy^2}$.

2. $\dfrac{14\,m^4x}{13\,ab^2} \div \dfrac{5\,m^5}{2\,a^2b^2}$.

3. $\dfrac{17\,x^2y^2}{u^2} \div 17\,x^3$.

4. $\dfrac{9\,a}{b} \div 18$.

5. $\dfrac{a^2 + a}{a^2 - a} \div \dfrac{a + 1}{a - 1}$.

6. $\dfrac{a^2 + 2\,ab + b^2}{a^2 - 2\,ab + b^2} \div \dfrac{a + b}{a - b}$.

**7.** $\dfrac{x^2-5x+4}{x^2+5x+4} \div \dfrac{x-4}{x+4}.$

**9.** $\dfrac{x^2-5x+4}{x^2+5x+4} \div \dfrac{x^2-3x-4}{x^2+3x-4}.$

**8.** $\dfrac{a^2+ab}{a^2-ab} \div \dfrac{a^2+2ab+b^2}{a}.$

**10.** $\dfrac{x^2-7x+10}{x^2-4x+4} \div \dfrac{x^2-5x}{x+3}.$

**11.** $\dfrac{m^2+6m-7}{m^2+15m+56} \times \dfrac{m^2+18m+80}{m^2+5m-50} \div \dfrac{m-1}{m-5}$

**12.** $\dfrac{a^2+a-2}{a^2-2a-8} \cdot \dfrac{a^2+a-20}{a^2-25} \div \dfrac{a-1}{a^2-5a}.$

**13.** $\dfrac{6x^2y-4xy^2}{45x^2-20y^2} \cdot \dfrac{15x+10y}{x^2y^2} \div \dfrac{x+y}{xy}.$

**14.** $\dfrac{4x^2-5x-6}{2x^2+3x-2} \cdot \dfrac{12x^2+5x-2}{8x^2-6x-9} \div \dfrac{3x^2-4x-4}{2x^2+x-6}.$

**15.** $\dfrac{a^2+ab}{a^2-ab} \cdot \dfrac{a-b}{a^3+a^2b+ab^2} \div \dfrac{a+b}{a^2+ab+b^2}.$

## COMPLEX FRACTIONS

**146.** A **complex fraction** is a fraction whose numerator or denominator, or both, are fractional.

**Ex. 1.** Simplify $\dfrac{\dfrac{a}{b}+\dfrac{b}{c}+\dfrac{c}{a}}{\dfrac{a}{b}-\dfrac{b}{c}+\dfrac{c}{a}}.$

$$\dfrac{\dfrac{a}{b}+\dfrac{b}{c}+\dfrac{c}{a}}{\dfrac{a}{b}-\dfrac{b}{c}+\dfrac{c}{a}} = \dfrac{\dfrac{a^2c+ab^2+bc^2}{abc}}{\dfrac{a^2c-ab^2+bc^2}{abc}}$$

$$= \dfrac{a^2c+ab^2+bc^2}{abc} \times \dfrac{abc}{a^2c-ab^2+bc^2}$$

$$= \dfrac{a^2c+ab^2+bc^2}{a^2c-ab^2+bc^2}.$$

**147.** In many examples the easiest mode of simplification is multiply both the numerator and the denominator of the mplex fraction by the L. C. M. of their denominators.

If the numerator and denominator of the preceding examples e multiplied by $abc$, the answer is directly obtained.

Ex. 2. Simplify $\dfrac{\dfrac{x+y}{x-y}+\dfrac{x-y}{x+y}}{\dfrac{1}{x-y}-\dfrac{1}{x+y}}$.

Multiplying the terms of the complex fraction by $+y)(x-y)$, the expression becomes

$$\frac{(x+y)^2+(x-y)^2}{(x+y)-(x-y)}=\frac{2\,x^2+2\,y^2}{2\,y}$$

$$=\frac{x^2+y^2}{y}.$$

## EXERCISE 57

Simplify:

1. $\dfrac{\dfrac{x^2}{y}}{x}$.

2. $\dfrac{\dfrac{x^2}{y^2}}{x}$.

3. $\dfrac{\dfrac{x}{y}}{\dfrac{x^2}{y^2}}$.

4. $\dfrac{\dfrac{a}{b}}{\dfrac{c}{b}}$.

5. $\dfrac{a+\dfrac{b}{c}}{b+\dfrac{a}{c}}$.

7. $\dfrac{a}{c-\dfrac{b}{a}}$.

9. $\dfrac{\dfrac{x}{2}+\dfrac{y}{3}}{\dfrac{x}{3}+\dfrac{y}{2}}$.

6. $\dfrac{a}{\dfrac{x}{a}+\dfrac{y}{a}}$.

8. $\dfrac{m+\dfrac{n}{p}}{m-\dfrac{n}{p}}$.

10. $\dfrac{4+\dfrac{2}{x}}{2+\dfrac{3}{x}}$.

11. $\dfrac{\dfrac{m}{3}+2\,n}{\dfrac{m}{3}-5\,n}.$

15. $\dfrac{\dfrac{1}{x+1}+\dfrac{1}{x-1}}{\dfrac{1}{x+1}-\dfrac{1}{x-1}}.$

19. $\dfrac{\dfrac{1}{x^2}+\dfrac{2}{x}+1}{\dfrac{4}{x^2}+\dfrac{5}{x}+1}.$

12. $\dfrac{a-\dfrac{b}{2}}{\dfrac{a}{2}+b}.$

16. $\dfrac{\dfrac{a}{b}-\dfrac{b}{a}}{\dfrac{a}{b}+\dfrac{b}{a}-2}.$

20. $\dfrac{1}{1+\dfrac{1}{2+\dfrac{1}{2}}}.$

13. $\dfrac{\dfrac{a+b}{n}}{\dfrac{a}{m}+\dfrac{b}{n}}.$

17. $\dfrac{\dfrac{1}{a+2}}{1-\dfrac{1}{a+1}}.$

21. $\dfrac{a}{1+\dfrac{1}{a+\dfrac{1}{a}}}.$

14. $\dfrac{\dfrac{x^2-y^2}{y^2}}{\dfrac{x-y}{y^3}}.$

18. $1+\dfrac{1}{1+\dfrac{1}{a}}.$

22. $\dfrac{\dfrac{x+1}{x-1}+\dfrac{x-1}{x+1}}{\dfrac{x+1}{x-1}-\dfrac{x-1}{x+1}}.$

(For additional examples see page 273.)

# CHAPTER IX

## FRACTIONAL AND LITERAL EQUATIONS

### FRACTIONAL EQUATIONS

**148. Clearing of fractions.** If an equation contains fractions, these may be removed by multiplying each term by the L. C. M. of the denominator.

**Ex. 1.** Solve $\dfrac{2x}{6} - \dfrac{x-3}{3} = 12 - \dfrac{x+4}{2} - x.$

Multiplying each term by 6 (Axiom 3, § 89),

$$2x - 2(x-3) = 72 - 3(x+4) - 6x.$$

Removing parentheses, $2x - 2x + 6 = 72 - 3x - 12 - 6x.$

Transposing, $\quad 2x - 2x + 3x + 6x = -6 + 72 - 12.$

Uniting, $\qquad\qquad\qquad\qquad 9x = 54.$

$$x = 6.$$

*Check.* If $x = 6$, each member is reduced to **1.**

**Ex. 2.** Solve $\dfrac{5}{x-1} - \dfrac{14}{x+1} = -\dfrac{9}{x+3}.$

Multiplying by $(x-1)(x+1)(x+3)$,

$$5(x+1)(x+3) - 14(x-1)(x+3) = -9(x+1)(x-1).$$

Simplifying, $5x^2 + 20x + 15 - 14x^2 - 28x + 42 = -9x^2 + 9.$

Transposing, $\quad 5x^2 - 14x^2 + 9x^2 + 20x - 28x = -15 - 42 + 9.$

Uniting, $\qquad\qquad\qquad\qquad\qquad -8x = -48,$

$$x = 6.$$

*Check.* If $x = 6$, each member is reduced to $-1$.

Solve the following equations:

**1.** $\dfrac{x-3}{4} = \dfrac{x+7}{6}$.

**3.** $\dfrac{x-7}{2} = \dfrac{5-x}{5}$.

**2.** $\dfrac{x+4}{3} = \dfrac{x-1}{2}$.

**4.** $\dfrac{24-2x}{2} + x = \dfrac{41-x}{3}$.

**5.** $\dfrac{3x-4}{2} + \dfrac{2x+1}{6} = \dfrac{7-7x}{12}$.

**6.** $\dfrac{x-1}{20} + \dfrac{x+9}{10} = \dfrac{x+3}{6}$.

**8.** $\dfrac{x}{2} + \dfrac{x}{3} + \dfrac{x}{6} = 5$.

**7.** $\dfrac{4x+4}{3} = x + 4$.

**9.** $x + \dfrac{x}{3} + \dfrac{x}{4} = 19$.

**10.** $\dfrac{x+4}{x-4} = 9$.

**12.** $\dfrac{3}{x} = 12$.

**14.** $\dfrac{7}{x} + 5 = 19$.

**11.** $\dfrac{1}{x} = 2$.

**13.** $\dfrac{12}{7x} = 2$.

**15.** $\dfrac{20}{x+1} = 5$.

**16.** $\dfrac{15}{x} = \dfrac{10}{x} + 1$.

**18.** $x + \dfrac{x}{2} + \dfrac{x}{3} = 11$.

**17.** $\dfrac{1}{x} + \dfrac{2}{x} + \dfrac{3}{x} = 1$.

**19.** $\dfrac{3x}{4} + \dfrac{x}{2} + \dfrac{7x}{16} - x = \dfrac{1}{8}$.

**20.** $\dfrac{y+2}{3} + \dfrac{y-2}{3} + \dfrac{y-3}{4} = \dfrac{4y+6}{6}$.

**21.** $\dfrac{y-1}{5} + \dfrac{y-2}{4} + \dfrac{y-3}{3} = 3$.

**22.** $\dfrac{u+2}{7} + \dfrac{u-2}{3} - \dfrac{u+1}{6} = 1$.

**23.** $\dfrac{t+1}{3} + \dfrac{t-1}{7} - \dfrac{t-4}{4} = 0$.

**24.** $\dfrac{y-3}{y+3} = \dfrac{1}{2}$.

**29.** $\dfrac{2x-4}{3x+5} = \dfrac{5}{2}$.

**25.** $\dfrac{x+4}{x} = \dfrac{x+6}{x+1}$.

**30.** $\dfrac{6}{x-1} + \dfrac{10}{x+1} = \dfrac{60}{x^2-1}$.

**26.** $4x+1 = \dfrac{6x+1}{x+1}$.

**31.** $\dfrac{19-12x}{2-9x} = \dfrac{8x-7}{6x+7}$.

**27.** $\dfrac{2x-12}{3x} = 2$.

**32.** $\dfrac{6}{x+3} + \dfrac{2}{x-3} = \dfrac{20}{x^2-9}$.

**28.** $\dfrac{x+1}{x-4} = \dfrac{x+7}{x-1}$.

**33.** $\dfrac{3}{x-2} - \dfrac{2}{x+2} = \dfrac{13}{x^2-4}$.

**34.** $\dfrac{7}{2x-5} = \dfrac{13}{6x^2-15x} - \dfrac{4}{3x}$.

**35.** $\dfrac{3}{3x-2} - \dfrac{4}{3x^2-2x} = \dfrac{2}{3x-2}$.

**36.** $\dfrac{51}{2x+3} - \dfrac{23x+26}{4x^2-9} = \dfrac{22}{2x-3}$.

**37.** $\dfrac{1}{x+2} - \dfrac{1}{x+8} = \dfrac{6x}{x^2+10x+16}$.

**38.** $\dfrac{x+29}{x+9} = \dfrac{2x+5}{x-2} - 1$.

**39.** $\dfrac{7x-11}{x-11} - \dfrac{4x-2}{x+1} = 3$.

**40.** $\dfrac{25x-26}{x^2-x-20} + \dfrac{3}{x-5} = \dfrac{7}{x+4}$.

**149. If two or more denominators are monomials, and the remaining one a polynomial, it is advisable first to remove the monomial denominators only, and after simplifying the resulting equation to clear of all denominators.

**Ex. 1.** Solve $\dfrac{16\,x+3}{10}-\dfrac{2\,x-5}{5\,x-1}=\dfrac{8\,x-1}{5}$.

Multiplying each term by 10, the L. C. M. of the monomial denominators,

$$16\,x+3-\frac{20\,x-50}{5\,x-1}=16\,x-2.$$

Transposing and uniting, $\qquad 5=\dfrac{20\,x-50}{5\,x-1}$.

Multiplying by $5\,x-1$, $\qquad 25\,x-5=20\,x-50.$

Transposing and uniting, $\qquad 5\,x=-45.$

Dividing, $\qquad x=-9.$

*Check.* If $x=-9$, each member is reduced to $-\frac{71}{5}$.

Solve the following equations:

**41.** $\dfrac{4\,x+3}{3}-\dfrac{2\,x+11}{5\,x-2}=\dfrac{8\,x+5}{6}$.

**42.** $\dfrac{21\,x+4}{9}+\dfrac{7\,x-13}{6\,x+3}=\dfrac{7\,x+3}{3}$.

**43.** $\dfrac{24\,x+13}{15}-\dfrac{8\,x+2}{5}=\dfrac{2\,x-7}{7\,x-16}$.

**44.** $\dfrac{10\,x+6}{5}-\dfrac{4\,x+7}{6\,x+11}=\dfrac{6\,x+1}{3}$.

**45.** $\dfrac{6\,x+10}{5}+\dfrac{3\,x+5}{2\,x-25}=\dfrac{18\,x+43}{15}$.

**46.** $\dfrac{17\,x-9}{14}-\dfrac{34\,x-67}{28}=\dfrac{13\,x+7}{6\,x+14}$.

**47.** $\dfrac{7\,x+4}{9}=\dfrac{14\,x+9}{18}+\dfrac{2}{3}-\dfrac{7\,x-29}{5\,x-12}$.

## LITERAL EQUATIONS

**150.** **Literal equations** (§ 88) are solved by the same method as numerical equations.

When the terms containing the unknown quantity cannot be actually added, they are united by factoring.

Thus,
$$ax + bx = (a + b)x.$$

$$x + m^2x - mnx = (1 + m^2 - mn)\,x.$$

**Ex. 1.** Solve $\dfrac{x-a}{b} + \dfrac{x-b}{a} = -\dfrac{3\,b+a}{a}$.

Clearing of fractions, $\qquad ax - a^2 + bx - b^2 = -3\,b^2 - ab.$

Transposing, $\qquad\qquad ax + bx = a^2 + b^2 - 3\,b^2 - ab.$

Uniting, $\qquad\qquad (a + b)x = a^2 - ab - 2\,b^2.$

Dividing, $\qquad\qquad x = \dfrac{a^2 - ab - 2\,b^2}{a + b}$

Reducing to lowest terms, $\qquad x = a - 2\,b.$

**151.** It frequently occurs that the unknown letter is not expressed by $x$, $y$, or $z$.

**Ex. 2.** If $\dfrac{2(a-b)}{3\,a-c} = \dfrac{2\,a+b}{3(a-c)}$, find $a$ in terms of $b$ and $c$.

Multiplying by $3(a-c)(3\,a-c)$

$$6(a - b)(a - c) = (2\,a + b)(3\,a - c).$$

$$6\,a^2 - 6\,ab - 6\,ac + 6\,bc = 6\,a^2 - 2\,ac + 3\,ab - bc.$$

Transposing all terms containing $a$ to one member,

$$-6\,ab - 6\,ac + 2\,ac - 3\,ab = -6\,bc - bc.$$

Simplifying, $\qquad -9\,ab - 4\,ac = -7\,bc.$

Uniting the $a$, and multiplying by $-1$, $a(9\,b + 4\,c) = 7\,bc.$

Dividing, $\qquad\qquad a = \dfrac{7\,bc}{9\,b + 4\,c}.$

Solve the following equations:

1. $m + x = n - x.$

2. $m + 3x = 8 - 4x.$

3. $a + 2b + 3x = 2a + b + 2x.$

4. $mx = n.$  5. $mx - p = n.$

6. $3(x - a) = 2b.$

7. $a(b - x) = c.$

8. $4(a - x) = 3(b - x).$

9. $3(2a + x) = 2(3a - x).$

10. $ax + bx = c.$

11. $mx + m = ax + n.$

12. $ax + bx = 5.$

13. $ax + bx = 6 - cx.$

14. $(m + n)x = 2a + (m - n)x.$

15. $(m - n)x = px + q.$

16. $\dfrac{x}{m} = n.$  17. $\dfrac{m}{x} = n.$

18. $\dfrac{a}{x} + \dfrac{b}{x} = 1.$

19. $\dfrac{a}{x} - 1 = \dfrac{b}{x} - 3.$

20. $\dfrac{m + nx}{a} + b = \dfrac{ab - mx}{a}.$

21. $\dfrac{x}{a} + \dfrac{x}{b} = c.$

22. $\dfrac{x}{a} + \dfrac{x + 1}{b} = \dfrac{x - 1}{a} + 5.$

23. $\dfrac{a}{1 - x} = b.$  24. $\dfrac{b}{1 + x} = b.$

25. $\dfrac{x + 1}{x - 1} = \dfrac{m}{n}.$

26. $\dfrac{m + x}{m - x} = \dfrac{m + n}{m - n}.$

27. $\dfrac{x - a}{x - 2} + \dfrac{x - b}{x - 1} = 2.$

28. If $s = vt$, solve for $v$.

29. If $s = vt$, solve for $t$.

30. If $s = \dfrac{g}{2}t^2$, solve for $g$.

31. If $\dfrac{1 + a}{1 - a} = \dfrac{b}{c}$, solve for $a$.

32. If $\dfrac{1}{f} = \dfrac{1}{p} + \dfrac{1}{q}$, solve for $f$.

33. Solve the same equation for $p$.

34. The formula for simple interest (§ 30, Ex. 5) is $i = \dfrac{prn}{100}$, $i$ denoting the interest, $p$ the principal, $r$ the number of %, and $n$ the number of years. Find the formula for:

(a) The principal.

(b) The rate.

(c) The time, in terms of other quantities.

I

**35.** (*a*) Find a formula expressing degrees of Fahrenheit (*F*) in terms of degrees of centigrade (*C*) by solving the equation $C = \dfrac{5}{9}(F - 32)$.

(*b*) Express in degrees Fahrenheit 40° C., 100° C., − 20° C.

**36.** If *C* is the circumference of a circle whose radius is *R*, then $C = 2\pi R$. Find *R* in terms of *C* and $\pi$.

## PROBLEMS LEADING TO FRACTIONAL AND LITERAL EQUATIONS

**152.** Ex. 1. When between 3 and 4 o'clock are the hands of a clock together?

At 3 o'clock the hour hand is 15 minute spaces ahead of the minute hand, hence the question would be formulated : After how many minutes has the minute hand moved 15 spaces more than the hour hand?

Let $\qquad x =$ the required number of minutes after 3 o'clock,

then $\qquad x =$ the number of minute spaces the minute hand moves over,

and $\qquad \dfrac{x}{12} =$ the number of minute spaces the hour hand moves over.

Therefore $x - \dfrac{x}{12} =$ the number of minute spaces the minute hand moves more than the hour hand.

Or $\qquad\qquad\qquad\qquad x - \dfrac{x}{12} = 15.$

Uniting, $\qquad\qquad\qquad \dfrac{11\,x}{12} = 15.$

Multiplying by 12, $\qquad\quad 11\,x = 180.$

Dividing, $\qquad\qquad x = \tfrac{180}{11} = 16\tfrac{4}{11}$ minutes after 3 o'clock.

Ex. 2. A can do a piece of work in 3 days and B in 2 days. In how many days can both do it working together?

If we denote the required number of days by *x* and the piece of work by 1, then A would do each day $\tfrac{1}{3}$ and B $\tfrac{1}{2}$, while in *x* days they would do respectively $\dfrac{x}{3}$ and $\dfrac{x}{2}$, and hence the sentence written in algebraic symbols is $\dfrac{x}{3} + \dfrac{x}{2} = 1.$

A more symmetrical but very similar equation is obtained by writing in symbols the following sentence : " The work done by A in one day plus the work done by B in one day equals the work done by both in one day."

Let $\qquad x =$ the required number of days.

Then $\qquad \dfrac{1}{x} =$ the part of the work both do in one day.

Therefore, $\qquad \dfrac{1}{3} + \dfrac{1}{2} = \dfrac{1}{x}.$

Solving, $\qquad x = \frac{6}{5}$, or $1\frac{1}{5}$, the required number of days.

**Ex. 3.** The speed of an express train is $\frac{9}{5}$ of the speed of an accommodation train. If the accommodation train needs 4 hours more than the express train to travel 180 miles, what is the rate of the express train?

| | Time (hours) | Rate (miles per hour) | Distance (miles) |
|---|---|---|---|
| Express train | $180 \div \dfrac{9x}{5} = \dfrac{100}{x}$ | $\dfrac{9x}{5}$ | 180 |
| Accommodation train | $180 \div x = \dfrac{180}{x}$ | $x$ | 180 |

Therefore, $\qquad \dfrac{180}{x} = \dfrac{100}{x} + 4.$ $\qquad\qquad$ (1)

Clearing, $\qquad 180 = 100 + 4x.$

Transposing, $\qquad 4x = 80.$

Hence $\qquad x = 20.$

$\qquad\qquad \frac{9}{5}x = 36 =$ rate of express train.

*Explanation :* If $x$ is the rate of the accommodation train, then $\dfrac{9x}{5}$ is the rate of the express train. But in uniform motion Time $= \dfrac{\text{Distance}}{\text{Rate}}$ Hence the rates can be expressed, and the statement, "The accommodation train needs 4 hours more than the express train," gives the equation (1).

## EXERCISE 60

**1.** Find a number whose third and fourth parts added together make 21.

**2.** Find the number whose fourth part exceeds its fifth part by 3.

**3.** Two numbers differ by 6, and one half the greater exceeds the smaller by 2. Find the numbers.

**4.** The sum of two numbers is 30, and one is $\frac{3}{7}$ of the other. What are the numbers?

**5.** Find two consecutive numbers such that $\frac{1}{4}$ of the greater increased by $\frac{1}{3}$ of the smaller equals 9.

**6.** Two numbers differ by 3, and $\frac{1}{6}$ of the greater is equal to $\frac{1}{5}$ of the smaller. Find the numbers.

**7.** Twenty years ago A's age was $\frac{1}{3}$ of his present age. Find A's age.

**8.** The sum of the ages of a father and his son is 50, and 10 years hence the son's age will be $\frac{2}{5}$ of the father's age. Find their present ages.

**9** A post is a fifth of its length in the ground, one half of its length in water, and 9 feet above water. What is the length of the post?

**10** A man left $\frac{2}{3}$ of his property to his wife, $\frac{1}{5}$ to his daughter, and the remainder, which was $4000, to his son. How much money did the man leave?

**11.** A man lost $\frac{2}{3}$ of his fortune and $500, and found that he had $\frac{1}{4}$ of his original fortune left. How much money had he at first?

**12.** After spending $\frac{1}{4}$ of his money and $10, a man had left $\frac{1}{3}$ of his money and $15. How much money had he at first?

**13.** The speed of an accommodation train is ¾ of the speed of an express train. If the accommodation train needs 1 hour more than the express train to travel 120 miles, what is the rate of the express train? (§ 152, Ex. 3.)

**14.** An express train starts from a certain station two hours after an accommodation train, and after traveling 150 miles overtakes the accommodation train. If the rate of the express train is ⅔ of the rate of the accommodation train, what is the rate of the latter?

**15.** At what time between 4 and 5 o'clock are the hands of a clock together? (§ 152, Ex. 1.)

**16.** At what time between 7 and 8 o'clock are the hands of a clock together?

**17.** At what time between 7 and 8 o'clock are the hands of a clock in a straight line and opposite?

**18.** A man has invested ½ of his money at 4%, ⅓ at 5%, and the remainder at 6%. How much money has he invested if his annual interest therefrom is $560?

**19.** A has invested capital at 4½% and B has invested $5000 more at 4%. They both derive the same income from their investments. How much money has each invested?

**20.** An ounce of gold when weighed in water loses $\frac{1}{19}$ of an ounce, and an ounce of silver $\frac{2}{21}$ of an ounce. How many ounces of gold and silver are there in a mixed mass weighing 20 ounces in air, and losing 1½ ounces when weighed in water?

**21.** A can do a piece of work in 3 days, and B in 4 days. In how many days can both do it working together? (§ 152, Ex. 2.)

**22.** A can do a piece of work in 2 days, and B in 6 days. In how many days can both do it working together?

**23.** A can do a piece of work in 4 days, and B in 12 days. In how many days can both do it working together?

**153.** The last three questions and their solutions differ only in the numerical values of the two given numbers. Hence, by taking for these numerical values two general algebraic numbers, *e.g. m* and *n*, it is possible to solve all examples of this type by one example. Answers to numerical questions of this kind may then be found by numerical substitution. The problem to be solved, therefore, is :

A can do a piece of work in *m* days and B in *n* days. In how many days can both do it working together ?

If we let $x =$ the required number of days, and apply the method of § 170, Ex. 2, we obtain the equation $\dfrac{1}{m} + \dfrac{1}{n} = \dfrac{1}{x}$.

Solving, $x = \dfrac{mn}{m+n}$.

Therefore both working together can do it in $\dfrac{mn}{m+n}$ days.

To find the numerical answer, if A can do this work in 6 days and B in 3 days, make $m = 6$ and $n = 3$. Then $\dfrac{6 \cdot 3}{6+3} = 2$, *i.e.* they can both do it in 2 days.

---

Solve the following problems :

**24.** In how many days can A and B working together do a piece of work if each alone can do it in the following number of days :

     (*a*) A in 5, B in 5.
     (*b*) A in 6, B in 30.
     (*c*) A in 4, B in 16.
     (*d*) A in 6, B in 12.

**25.** Find three consecutive numbers whose sum is 42.

**26.** Find three consecutive numbers whose sum is 57.

The last two examples are special cases of the following problem :

**27.** Find three consecutive numbers whose sum equals *m*. Find the numbers if $m = 24$; 30,009; 918,414.

**28.** Find two consecutive numbers the difference of whose squares is 11.

**29.** Find two consecutive numbers the difference of whose squares is 21.

**30.** If each side of a square were increased by 1 foot, the area would be increased by 19 square feet. Find the side of the square.

The last three examples are special cases of the following one:

**31.** The difference of the squares of two consecutive numbers is $m$; find the smaller number. By using the result of this problem, solve the following ones:

Find two consecutive numbers the difference of whose squares is (a) 51, (b) 149, (c) 16,721, (d) 1,000,001.

**32.** Two men start at the same hour from two towns, 88 miles apart, the first one traveling 3 miles per hour, and the second 5 miles per hour. After how many hours do they meet, and how many miles does each travel?

**33.** Two men start at the same time from two towns, $d$ miles apart, the first traveling at the rate of $m$, the second at the rate of $n$ miles per hour. After how many hours do they meet, and how many miles does each travel?

Solve the problem if the distance, the rate of the first, and the rate of the second are, respectively:

(a) 60 miles, 3 miles per hour, 2 miles per hour.

(b) 35 miles, 2 miles per hour, 5 miles per hour.

(c) 64 miles, $3\frac{1}{2}$ miles per hour, $4\frac{1}{2}$ miles per hour.

**34.** A cistern can be filled by two pipes in $m$ and $n$ minutes respectively. In how many minutes can it be filled by the two pipes together? Find the numerical answer, if $m$ and $n$ are, respectively, (a) 20 and 5 minutes, (b) 8 and 56 minutes, (c) 6 and 3 hours.

# CHAPTER X

## RATIO AND PROPORTION

### RATIO

**154.** The **ratio** of two numbers is the quotient obtained by dividing the first number by the second.

Thus the ratio of $a$ and $b$ is $\dfrac{a}{b}$ or $a \div b$. The ratio is also frequently written $a:b$, the symbol : being a sign of division. (In most European countries this symbol is employed as the usual sign of division.) The ratio of $12:3$ equals 4, $6:12 = .5$, etc.

**155.** A ratio is used to compare the magnitude of two numbers.

Thus, instead of writing "$a$ is 5 times as large as $b$," we may write $a:b = 5$.

**156.** The first term of a ratio is the **antecedent**, the second term the **consequent**.

In the ratio $a:b$, $a$ is the antecedent, $b$ is the consequent. The numerator of any fraction is the antecedent, the denominator the consequent.

**157.** The ratio $\dfrac{b}{a}$ is the **inverse** of the ratio $\dfrac{a}{b}$.

**158.** *Since a ratio is a fraction, all principles relating to fractions may be applied to ratios.* *E.g.* a ratio is not changed if its terms are multiplied or divided by the same number, etc.

Ex. 1. Simplify the ratio $2\frac{1}{2} : 3\frac{1}{3}$.

$$2\frac{1}{2} : 3\frac{1}{3} = \frac{5}{2} : \frac{10}{3} = \frac{5}{2} \times \frac{3}{10} = \frac{3}{4} = 3:4.$$

A somewhat shorter way would be to multiply each term by 6.

Ex. 2. Transform the ratio $5 : 3\frac{3}{4}$ so that the first term will equal 1.

$$5 : 3\tfrac{3}{4} = \frac{5}{5} : \frac{3\frac{3}{4}}{5} = 1 : \frac{3}{4}.$$

## EXERCISE 61

Find the value of the following ratios:

1. $72 : 18$.      3. $6\frac{2}{3} : 16\frac{2}{3}$.      5. $\$24 : \$8$.

2. $\frac{1}{3} : \frac{1}{9}$.      4. $4\frac{4}{9} : 5\frac{5}{7}$.      6. $5\frac{2}{3}$ hours : $8\frac{1}{2}$ hours.

Simplify the following ratios:

7. $3\frac{1}{2} : 4\frac{1}{2}$.      9. $7\frac{1}{7} : 4\frac{4}{11}$.      11. $16\,x^2y : 24\,xy^2$.

8. $3\frac{1}{3} : 1\frac{2}{3}$.      10. $27\,ab : 18\,ab$.      12. $64\,x^3y : 48\,xy^3$.

13. $5(m+n)z : 15(m+n)z^2$.      14. $x^2 - y^2 : (x+y)^2$.

Transform the following ratios so that the antecedents equal unity:

15. $16 : 64$.      16. $7\frac{2}{5} : 6\frac{1}{6}$.      17. $\frac{1}{2} : 1$.      18. $16\,a^2 : 24\,ab$.

---

**159.** A **proportion** is a statement expressing the equality of two ratios.

$\frac{3}{4} = \frac{6}{8}$ or $a : b = c : d$ are proportions.

**160.** The first and fourth terms of a proportion are the **extremes**, the second and third terms are the **means**. The last term is the **fourth proportional** to the first three.

In the proportion $a : b = c : d$, $a$ and $d$ are the extremes, $b$ and $c$ the means. The last term $d$ is the fourth proportional to $a$, $b$, and $c$.

**161.** If the means of a proportion are equal, either mean is the **mean proportional** between the first and the last terms, and the last term the **third proportional** to the first and second terms.

In the proportion $a : b = b : c$, $b$ is the mean proportional between $a$ and $c$, and $c$ is the third proportional to $a$ and $b$.

**162.** Quantities of one kind are said to be *directly proportional* to quantities of another kind, if the ratio of any two of the first kind is equal to the ratio of the corresponding two of the other kind.

If 4 ccm. of iron weigh 30 grams, then 6 ccm. of iron weigh 45 grams, or 4 ccm. : 6 ccm. = 30 grams : 45 grams. Hence the weight of a mass of iron is proportional to its volume.

Note. Instead of "directly proportional" we may say, briefly, "proportional."

Quantities of one kind are said to be *inversely proportional* to quantities of another kind, if the ratio of any two of the first kind is equal to the inverse ratio of the corresponding two of the other kind.

If 6 men can do a piece of work in 4 days, then 8 men can do it in 3 days, or 6 : 8 equals the inverse ratio of 4 : 3, *i.e.* 3 : 4. Hence the number of men required to do some work, and the time necessary to do it, are inversely proportional.

**163.** *In any proportion the product of the means is equal to the product of the extremes.*

Let $\qquad a : b = c : d,$

or $\qquad \dfrac{a}{b} = \dfrac{c}{d}.$

Clearing of fractions, $\quad ad = bc.$

**164.** *The mean proportional between two numbers is equal to the square root of their product.*

Let the proportion be $\quad a : b = b : c.$

Then $\qquad b^2 = ac.$  (§ 163.)

Hence $\qquad b = \sqrt{ac}.$

**165.** *If the product of two numbers is equal to the product of two other numbers, either pair may be made the means, and the other pair the extremes, of a proportion.* (Converse of § 163.)

If $mn = pq$, and we divide both members by $nq$, we have

$$\frac{m}{q} = \frac{p}{n}.$$

**Ex. 1.** Find $x$, if $6 : x = 12 : 7$.

$$12x = 42. \quad (\S \ 163.)$$

Hence $\qquad x = \frac{42}{12} = 3\frac{1}{2}$.

**Ex. 2.** Determine whether the following proportion is true or not:

$$8 : 5 = 7 : 4\frac{3}{8}.$$

$8 \times 4\frac{3}{8} = 35$, and $5 \times 7 = 35$; hence the proportion is true.

**166.** If $a : b = c : d$, then

I. $b : a = d : c$. (Frequently called **Inversion**.)

II. $a : c = b : d$. (Called **Alternation**.)

III. $a + b : b = c + d : d$. (**Composition**.)

Or $\quad a + b : a = c + d : c$.

IV. $a - b : b = c - d : d$. (**Division**.)

V. $a + b : a - b = c + d : c - d$. (**Composition and Division**.)

Any of these propositions may be proved by a method which is illustrated by the following example:

To prove $\qquad \dfrac{a - b}{b} = \dfrac{c - d}{d}.$

This is true if $\qquad ad - bd = bc - bd.$

Or if $\qquad ad = bc.$

But $\qquad ad = bc. \quad (\S \ 163.)$

Hence $\qquad \dfrac{a - b}{b} = \dfrac{c - d}{d}.$

**167.** These transformations are used to simplify proportions.

I. Change the proportion $4 : 5 = x : 6$ so that $x$ becomes the last term.

By inversion $5 : 4 = 6 : x$.

II. Alternation shows that a proportion is not altered when its antecedents or its consequents are multiplied or divided by the same number.

*E.g.* to simplify $48 : 21 = 32 : 7\,x$, divide the antecedents by 16, the consequents by 7, $\qquad 3 : 3 = 2 : x.$

Or $\qquad\qquad\qquad 1 : 1 = 2 : x,$ *i.e.* $x = 2.$

III. To simplify the proportion $5 : 6 = 4 - x : x.$
Apply composition, $11 : 6 = 4 : x.$

IV. To simplify the proportion $8 : 3 = 5 + x : x.$
Apply division, $\qquad\qquad 5 : 3 = 5 : x.$
Divide the antecedents by 5, $1 : 3 = 1 : x.$

V. To simplify $\dfrac{m + 3\,n}{m - 3\,n} = \dfrac{m + x}{m - x}.$

Apply composition and division, $\dfrac{2\,m}{6\,n} = \dfrac{2\,m}{2\,x}.$

Or $\qquad\qquad\qquad\qquad\qquad \dfrac{m}{3\,n} = \dfrac{m}{x}.$

Dividing the antecedents by $m$, $\quad \dfrac{1}{3\,n} = \dfrac{1}{x}.$

Note. A parenthesis is understood about each term of a proportion.

### EXERCISE 62

Determine whether the following proportions are true:

1. $5\frac{1}{3} : 8 = 2 : 3.$
2. $3\frac{1}{2} : \frac{5}{4} = 7 : 2\frac{1}{2}.$
3. $15 : 22 = 10\frac{1}{2} : 15.$
4. $11 : 5 = 12 : 5\frac{5}{11}.$
5. $8\,xy : 17 = \frac{1}{2}\,xy : 1\frac{1}{16}.$

Simplify the following proportions, and determine whether they are true or not:

6. $120 : 42 = 20 : 7.$
7. $72 : 50 = 180 : 125.$
8. $18 : 19 = 24 : 25.$
9. $6\frac{3}{7} : 13\frac{1}{3} = 5\frac{2}{5} : 11\frac{1}{5}.$
10. $m^2 - n^2 : (m - n)^2 = (m + n)^2 : m^2 - n^2.$

Determine the value of $x$:

11. $40 : 28 = 15 : x$.

12. $112 : 42 = 10 : x$.

13. $63 : x = 135 : 20$.

14. $x : 15 = 1\frac{5}{7} : 18$.

15. $21 : 4x = 72 : 96$.

16. $2.8 : 1.6 = 35 : x$.

17. $4 a^2 : 15 ab = 2 a : x$.

18. $16 n^2 : x = 28 n : 70 m$.

Find the fourth proportional to:

19. $1, 3, 5$.

20. $2, 4, 6$.

21. $3, 3\frac{1}{3}, \frac{3}{5}$.

22. $m, n, p$.

23. $m^2, mp, mq$.

Find the third proportional to:

24. $9$ and $12$.

25. $16$ and $28$.

26. $14$ and $21$.

27. $a^2$ and $ab$.

28. $1$ and $a$.

29. $a$ and $1$.

Find the mean proportional to:

30. $4$ and $16$.

31. $\frac{4}{3}$ and $\frac{25}{3}$.

32. $2 a$ and $18 a$.

33. $8 a^2$ and $2 b^2$.

34. $m+1$ and $m-1$.

35. Form two proportions commencing with $5$ from the equation $6 \times 10 = 5 \times 12$.

36. If $ab = xy$, form two proportions commencing with $b$.

Find the ratio of $x : y$, if:

37. $6 x = 7 y$.

38. $9 x = 2 y$.

39. $6 x = y$.

40. $mx = ny$.

41. $(a + b)x = cy$.

42. $x : 5 = y : 2$.

43. $x : m = y : n$.

44. $2 : 3 = y : x$.

45. $7 : y = 2 : x$.

46. $y : b = x : a$.

47. $y : 1 = x : a^2$.

Transform the following proportions so that only one term contains $x$:

48. $2 : 3 = 4 - x : x$.

49. $6 : 5 = 15 - x : x$.

50. $a : 2 = 5 - x : x$.

51. $22 : 3 = 2 + x : x$.

52. $19 : 18 = 3 + x : x$.

53. $2 : 5 = x : 3 + x$.

**54.** State the following propositions as proportions :

(*a*) Triangles (*T* and *T'*) of equal altitudes are to each other as their basis (*b* and *b'*).

(*b*) The circumferences (*C* and *C'*) of two circles are to each other as their radii (*R* and *R'*).

(*c*) The volume of a body of gas (*V*) is inversely proportional to the pressure (*P*).

(*d*) The areas (*A* and *A'*) of two circles are to each other as the squares of their radii (*R* and *R'*).

(*e*) The number of men (*m*) is inversely proportional to the number of days (*d*) required to do a certain piece of work.

**55.** State whether the quantities mentioned below are directly or inversely proportional :

(*a*) The number of yards of a certain kind of silk, and the total cost.

(*b*) The time a train needs to travel 10 miles, and the speed of the train.

(*c*) The length of a rectangle of constant width, and the area of the rectangle.

(*d*) The sum of money producing $60 interest at 5%, and the time necessary for it.

(*e*) The distance traveled by a train moving at a uniform rate, and the time.

**56.** A line 11 inches long on a certain map corresponds to 22 miles. A line $7\frac{1}{2}$ inches long represents how many miles ?

**57.** The areas of circles are proportional to the squares of their radii. If the radii of two circles are to each other as 4 : 7, and the area of the smaller circle is 8 square inches, what is the area of the larger?

**58.** The temperature remaining the same, the volume of a gas is inversely proportional to the pressure. A body of gas under a pressure of 15 pounds per square inch has a volume of 16 cubic feet. What will be the volume if the pressure is 12 pounds per square inch ?

**59.** The number of miles one can see from an elevation of *h* miles is very nearly the mean proportional between *h* and the diameter of the earth (8000 miles). What is the greatest distance a person can see from an elevation of 5 miles? From the Metropolitan Tower (700 feet high)? From Mount McKinley (20,000 feet high)?

---

**168.** *When a problem requires the finding of* **two numbers** **which are to each other as** m : n, *it is advisable to represent these unknown numbers by* **mx** *and* **nx**.

**Ex. 1.** Divide 108 into two parts which are to each other as 11 : 7.

Let                        $11 x =$ the first number,

then                      $7 x =$ the second number.

Hence              $11 x + 7 x = 108,$

or                        $18 x = 108.$

Therefore              $x = 6.$

Hence              $11 x = 66$ is the first number,

and                        $7 x = 42$ is the second number.

**Ex. 2.** A line *AB*, 4 inches long, is produced to a point *C*, so that $(AC):(BC)=7:5$. Find *AC* and *BC*.

Let                $AC = 7 x.$

Then              $BC = 5 x.$

Hence              $AB = 2 x.$

Or                $2 x = 4.$

$x = 2.$

Therefore          $7 x = 14 = AC.$

$5 x = 10 = BC.$

### EXERCISE 63

**1.** Divide 44 in the ratio 2 : 9.

**2.** Divide 45 in the ratio 3 : 7.

**3.** Divide 39 in the ratio 1 : 5.

**4.** A line 24 inches long is divided in the ratio 3 : 5. What are the parts?

**5.** Brass is an alloy consisting of two parts of copper and one part of zinc. How many ounces of copper and zinc are in 10 ounces of brass?

**6.** Gunmetal consists of 9 parts of copper and one part of tin. How many ounces of each are there in 22 ounces of gunmetal?

**7.** Air is a mixture composed mainly of oxygen and nitrogen, whose volumes are to each other as 21 : 79. How many cubic feet of oxygen are there in a room whose volume is 4500 cubic feet?

**8.** The total area of land is to the total area of water as 7 : 18. If the total surface of the earth is 197,000,000 square miles, find the number of square miles of land and of water.

**9.** Water consists of one part of hydrogen and 8 parts of oxygen. How many grams of hydrogen are contained in 100 grams of water?

**10.** Divide 10 in the ratio $a : b$.

**11.** Divide 20 in the ratio $1 : m$.

**12.** Divide $a$ in the ratio $3 : 7$.

**13.** Divide $m$ in the ratio $x : y$.

**14.** The three sides of a triangle are 11, 12, and 15 inches, and the longest is divided in the ratio of the other two. How long are the parts?

**15.** The three sides of a triangle are respectively $a$, $b$, and $c$ inches. If $c$ is divided in the ratio of the other two, what are its parts?

(For additional examples see page 279.)

# CHAPTER XI

## SIMULTANEOUS LINEAR EQUATIONS

**169.** An equation of the first degree containing two or more unknown numbers can be satisfied by any number of values of the unknown quantities.

If $$2x - 3y = 5,$$ (1)

then $$y = \frac{2x - 5}{3},$$ (2)

*I.e.* if $x = 0$, $y = -\frac{5}{3}$.

If $x = \frac{1}{2}$, $y = -\frac{4}{3}$.

If $x = 1$, $y = -1$, etc.

Hence, the equation is satisfied by an infinite number of sets of values. Such an equation is called *indeterminate*.

However, if there is given another equation, expressing a different relation between $x$ and $y$, such as

$$x + y = 10,$$ (3)

these unknown numbers can be found.

From (3) it follows $y = 10 - x$, and since the equations have to be satisfied by the same values of $x$ and $y$, the two values of $y$ must be equal.

Hence $$\frac{2x - 5}{3} = 10 - x.$$ (4)

The root of (4) is $x = 7$, which substituted in (2) gives $y = 3$.

Therefore, if both equations are to be satisfied by the same values of $x$ and $y$, there is only one solution.

**170.** A **system of simultaneous equations** is a group of equations that can be satisfied by the same values of the unknown numbers.

$x + 2y = 5$ and $7x - 3y = 1$ are simultaneous equations, for they are satisfied by the values $x = 1$, $y = 2$. But $2x - y = 5$ and $4x - 2y = 6$ are not simultaneous, for they cannot be satisfied by any value of $x$ and $y$. The first set of equations is also called *consistent*, the last set *inconsistent*.

**171. Independent equations** are equations representing different relations between the unknown quantities; such equations cannot be reduced to the same form.

$5x + 5y = 50$, and $3x + 3y = 30$ can be reduced to the same form; viz. $x + y = 10$. Hence they are not independent, for they express the same relation. Any set of values satisfying $5x + 5y = 50$ will also satisfy the equation $3x + 3y = 30$.

**172.** *A system of two simultaneous equations containing two unknown quantities is solved by combining them so as to obtain one equation containing only one unknown quantity.*

**173.** The process of combining several equations so as to make one unknown quantity disappear is called **elimination**.

**174.** The two **methods of elimination** most frequently used are:

I. *By Addition or Subtraction.*

II. *By Substitution.*

## ELIMINATION BY ADDITION OR SUBTRACTION

**175. Ex. 1.** Solve $\begin{cases} 3x + 2y = 13, & (1) \\ 2x - 7y = -8. & (2) \end{cases}$

Multiply (1) by 2, $\qquad 6x + 4y = 26.$ $\qquad\qquad$ (3)

Multiply (2) by 3, $\qquad 6x - 21y = -24.$ $\qquad$ (4)

Subtract (4) from (3), $\qquad 25y = 50.$

Therefore, $\qquad\qquad\qquad y = 2.$

Substitute this value of $y$ in either of the given equations, preferably the simpler one (1),    $3x + 4 = 13.$

Therefore    $x = 3.$

$y = 2.$

In general, eliminate the letter whose coefficients have the lowest common multiple.

*Check.*    $3 \cdot 3 + 2 \cdot 2 = 9 + 4 = 13,$

$2 \cdot 3 - 7 \cdot 2 = 6 - 14 = -8.$

Ex. 2.    Solve $\begin{cases} 5x - 3y = 47, & (1) \\ 13x + 5y = 135. & (2) \end{cases}$

| | | |
|---|---|---|
| Multiply (1) by 5, | $25x - 15y = 235.$ | (3) |
| Multiply (2) by 3, | $39x + 15y = 405.$ | (4) |
| Add (3) and (4), | $64x = 640.$ | |
| Therefore | $x = 10.$ | (5) |
| Substitute (5) in (1), | $50 - 3y = 47.$ | |
| Transposing, | $-3y = -3.$ | |
| Therefore | $y = 1.$ | |
| | $x = 10.$ | |
| *Check.* | $5 \cdot 10 - 3 \cdot 1 = 47,$ | |
| | $13 \cdot 10 + 5 \cdot 1 = 135.$ | |

**176.** Hence to eliminate by addition or subtraction:

*Multiply, if necessary, the equations by such numbers as will make the coefficients of one unknown quantity equal.*

*If the signs of these coefficients are like, subtract the equations; if unlike, add the equations.*

**EXERCISE 64**

Solve the following systems of equations and check the answers:

1. $\begin{cases} x + y = 47, \\ x - y = 13. \end{cases}$     3. $\begin{cases} 5x + 7y = 31, \\ 5x - 3y = 1. \end{cases}$

2. $\begin{cases} 3x + y = 33, \\ 2x - y = 2. \end{cases}$     4. $\begin{cases} 7x + 5y = 22, \\ 10x - 2y = 4. \end{cases}$

**5.** $\begin{cases} x + 2y = 13, \\ 3x - y = 11. \end{cases}$

**6.** $\begin{cases} 5x - 4y = 3, \\ 7x + y = 24. \end{cases}$

**7.** $\begin{cases} 3x + 7y = 7, \\ 5x + 10y = 5. \end{cases}$

**8.** $\begin{cases} 5x - y = 13, \\ x - 5y = 17. \end{cases}$

**9.** $\begin{cases} 2x - 11y = 30, \\ 3x + 3y = 6. \end{cases}$

**10.** $\begin{cases} 2x + 3y = 41, \\ 3x + 2y = 39. \end{cases}$

**11.** $\begin{cases} 2x - 3y = -2, \\ 3x + y = 19. \end{cases}$

**12.** $\begin{cases} 7x + y = 42, \\ 5x - 2y = 11. \end{cases}$

**13.** $\begin{cases} 5x - 4y = 29, \\ 3x - 3y = 15. \end{cases}$

**14.** $\begin{cases} 3x - y = 10, \\ 7x + 3y = 50. \end{cases}$

**15.** $\begin{cases} 2x + 3y = 30, \\ 8x - 15y = -60. \end{cases}$

**16.** $\begin{cases} 7x + 4y = 69, \\ 4x + 3y = 43. \end{cases}$

**17.** $\begin{cases} 6x - 4y = 0, \\ 5x + 2y = 8. \end{cases}$

**18.** $\begin{cases} 10x + 2y = 40, \\ 30x - 3y = -15. \end{cases}$

**19.** $\begin{cases} x - y = 6, \\ x + y = 50. \end{cases}$

**20.** $\begin{cases} 3x + 5y = 16, \\ x - 5y = 2. \end{cases}$

**21.** $\begin{cases} 3x - y = 0, \\ 11x - 2y = 95. \end{cases}$

**22.** $\begin{cases} 13x - 12y = 111, \\ 19x - 25y = 13. \end{cases}$

**23.** $\begin{cases} 4x + 3y = 5, \\ 5x - 6y = 55. \end{cases}$

**24.** $\begin{cases} \frac{1}{3}x + \frac{1}{13}y = 3\frac{1}{2}, \\ 3x - 4y = 1. \end{cases}$

**25.** $\begin{cases} .9x + 1.4y = 8.3, \\ 3.9x - 3.5y = -2.3. \end{cases}$

## ELIMINATION BY SUBSTITUTION

**177.** Solve
$$\begin{cases} 2x - 7y = -8, & (1) \\ 3x + 2y = 13. & (2) \end{cases}$$

Transposing $-7y$ in (1) and dividing by 2, $\quad x = \dfrac{7y-8}{2}.$

Substituting this value in (2), $\quad 3\left(\dfrac{7y-8}{2}\right) + 2y = 13.$

Clearing of fractions, $\qquad 21y - 24 + 4y = 26.$

$$25y = 50.$$

Therefore $\qquad\qquad\qquad\qquad y = 2.$

This value substituted in either (1) or (2) gives $x = 3.$

**178.** Hence to eliminate by substitution :

*Find in one equation the value of an unknown quantity in terms of the other. Substitute this value for one unknown quantity in the other equation, and solve the resulting equation.*

**EXERCISE 65**

Solve by substitution :

**1.** $\begin{cases} x + 7y = 21, \\ 2x - 3y = 25. \end{cases}$

**7.** $\begin{cases} 2x - 3y = 11, \\ 6x + 5y = 19. \end{cases}$

**2.** $\begin{cases} 25x - 12y = -1, \\ 7x - 4y = 1. \end{cases}$

**8.** $\begin{cases} x - y = 5, \\ 3x + 5y = -9. \end{cases}$

**3.** $\begin{cases} 3x + 2y = 5, \\ 5x - 4y = 12. \end{cases}$

**9.** $\begin{cases} 5x = 2y + 10, \\ 3x = 4y - 8. \end{cases}$

**4.** $\begin{cases} 9x - 2y = 8, \\ x + 5y = 27. \end{cases}$

**10.** $\begin{cases} 9y = 15 - 3x, \\ 3y = x + 1. \end{cases}$

**5.** $\begin{cases} 2x + 17y = 57, \\ 7x - 3y = 12. \end{cases}$

**11.** $\begin{cases} 11x = 5y + 12, \\ 3x - 2y = 2. \end{cases}$

**6.** $\begin{cases} 6x + 7y = 40, \\ 2x + 5y = 22. \end{cases}$

**12.** $\begin{cases} 19x + y = 0, \\ 14x - 2y = 52. \end{cases}$

**179.** Whenever one unknown quantity can be removed with-out clearing of fractions, it is advantageous to do so; in most cases, however, the equation must be cleared of fractions and simplified before elimination is possible.

Ex.   Solve
$$\begin{cases} \dfrac{x+2}{3} + \dfrac{y+3}{4} = 3, & (1) \\[2ex] \dfrac{x+3}{2} - \dfrac{y+2}{7} = 1. & (2) \end{cases}$$

Multiplying (1) by 12 and (2) by 14,

$$4x + 8 + 3y + 9 = 36. \tag{3}$$
$$7x + 21 - 2y - 4 = 14. \tag{4}$$

From (3),           $4x + 3y = 19.$       (5)

From (4),           $7x - 2y = -3.$     (6)

Multiplying (5) by 2 and (6) by 3,

$$8x + 6y = 38. \tag{7}$$
$$21x - 6y = -9. \tag{8}$$

Adding (7) and (8),       $29x = 29.$

$$x = 1.$$

Substituting in (6),      $7 - 2y = -3.$

$$y = 5.$$

*Check.*   $\dfrac{1+2}{3} + \dfrac{5+3}{4} = 1 + 2 = 3,$

$$\dfrac{1+3}{2} - \dfrac{5+2}{7} = 2 - 1 = 1.$$

**EXERCISE 66**

Solve by any method, and check the answers:

**1.** $\begin{cases} 2(x+2) + 3(y+3) = 26, \\ 4(x+4) + 5(y+5) = 64. \end{cases}$

**2.** $\begin{cases} 8(x-8) - 9(y-9) = 26, \\ 6(x-6) - 7(y-7) = 18. \end{cases}$

**3.** $\begin{cases} x + \dfrac{2y}{3} = 4, \\[2ex] \dfrac{x}{2} + 3y = 10. \end{cases}$

**4.** $\begin{cases} \dfrac{3x}{4} + \dfrac{4y}{5} = 7, \\[2mm] \dfrac{3x}{2} - \dfrac{y}{5} = 5. \end{cases}$

**5.** $\begin{cases} 2(y-3) = 2(3-x), \\ 3(y-1) = 4(3-x). \end{cases}$

**6.** $\begin{cases} 9(y-1) - 5(x-2) = 2, \\ 3(y-1) + 4(x-2) = 29. \end{cases}$

**7.** $\begin{cases} 4(5x-1) + 5(6y-1) = 121, \\ 2(3x-1) + 3(4y-1) = 43. \end{cases}$

**8.** $\begin{cases} 5(x-2) + 3(y-3) = 27, \\ 3x + 5(y+1) = 55. \end{cases}$

**9.** $\begin{cases} \dfrac{x+y}{3} + \dfrac{x-y}{2} = 5, \\[2mm] \dfrac{x+y}{2} + \dfrac{x-y}{4} = \dfrac{13}{2}. \end{cases}$

**10.** $\begin{cases} \dfrac{5x}{4} + \dfrac{2y}{3} = 28, \\[2mm] \dfrac{x}{2} + \dfrac{5y}{4} = 23. \end{cases}$

**11.** $\begin{cases} \dfrac{2x-5}{4} + 2y = 5, \\[2mm] \dfrac{2x-5}{4} + 3y = 7. \end{cases}$

**12.** $\begin{cases} \dfrac{x}{6} + 3y = 15, \\[2mm] \dfrac{x}{4} + \dfrac{2y}{3} = 11. \end{cases}$

**13.** $\begin{cases} \dfrac{3x-2}{4-2y} = \dfrac{4}{3}, \\[2mm] x+y = 2. \end{cases}$

**14.** $\begin{cases} \dfrac{15}{2x+y} = \dfrac{8}{y-2x}, \\[2mm] 18x - 100 = \dfrac{3y+44}{7}. \end{cases}$

**15.** $\begin{cases} \dfrac{6x-6}{5y-5} = 2, \\[2mm] \dfrac{x+3y}{2x-18} = 8. \end{cases}$

**16.** $\begin{cases} \dfrac{14x+18}{45-y} = 8, \\[2mm] \dfrac{13y+6}{x-10} = 25. \end{cases}$

**17.** $\begin{cases} \dfrac{x}{y} = \dfrac{3}{4}, \\[2mm] \dfrac{x-1}{y+2} = \dfrac{1}{2}. \end{cases}$

**18.** $\begin{cases} \dfrac{x+4}{y+1} = 2, \\[2mm] \dfrac{x+2}{y-1} = 3. \end{cases}$

19. $\begin{cases} \dfrac{3x+y-4}{2x+y+1} = \dfrac{1}{2}, \\[2mm] \dfrac{2x+y-9}{x+2y+7} = -\dfrac{4}{11}. \end{cases}$

20. $\begin{cases} \dfrac{x+1}{y+1} = \dfrac{3}{4}, \\[2mm] \dfrac{y+1}{x+y} = \dfrac{4}{5}. \end{cases}$

21. $\begin{cases} \dfrac{x+2}{3} - \dfrac{y+3}{4} = \dfrac{2(x-y)}{5}, \\[2mm] \dfrac{x-2}{4} - \dfrac{y-2}{3} = 2y-9. \end{cases}$

22. $\begin{cases} \dfrac{3x+2y}{5} + \dfrac{7x-4y}{3} = x+1, \\[2mm] \dfrac{4x-y}{3} + \dfrac{3x+y}{4} = y+1. \end{cases}$

23. $\begin{cases} x:y = 2:3, \\ x+1 = y-1. \end{cases}$

24. $\begin{cases} (x+1):(y+1) = 3:4, \\ 2x+3y = 13. \end{cases}$

25. $\begin{cases} (x+1)(y+1) = (x+3)(y-2), \\ (x+3)(y+1) = (x+1)(y+3). \end{cases}$

26. $\begin{cases} x:y = 2:3, \\ (x+1):(y+2) = 3:5. \end{cases}$

---

**180.** In many equations it is advantageous at first not to consider $x$ and $y$ as unknown quantities, but some expressions involving $x$, and $y$, *e.g.* $\dfrac{1}{x}$ and $\dfrac{1}{y}$.

Ex. 1. Solve
$$\begin{cases} \dfrac{3}{x} + \dfrac{8}{y} = 3, & (1) \\[2mm] \dfrac{15}{x} - \dfrac{4}{y} = 4. & (2) \end{cases}$$

$2 \times (2)$,
$$\dfrac{30}{x} - \dfrac{8}{y} = 8. \qquad (3)$$

$(1) + (3)$,
$$\dfrac{33}{x} = 11.$$

Clearing of fractions,
$$33 = 11\,x.$$
$$x = 3.$$

Substituting $x = 3$ in (1),
$$1 + \dfrac{8}{y} = 3.$$
$$\dfrac{8}{y} = 2.$$

Therefore
$$y = 4.$$

*Check.* $\frac{3}{3} + \frac{8}{4} = 1 + 2 = 3$; $\frac{15}{3} - \frac{4}{4} = 5 - 1 = 4$.

Examples of this type, however, can also be solved by the regular method.

Clearing (1) and (2) of fractions,
$$3\,y + 8\,x = 3\,xy. \qquad (4)$$
$$15\,y - 4\,x = 4\,xy. \qquad (5)$$
$2 \times (5)$,
$$30\,y - 8\,x = 8\,xy. \qquad (6)$$
$(4) + (6)$,
$$33\,y = 11\,xy. \qquad (7)$$
Dividing by $11\,y$,
$$3 = x, \text{ etc.}$$

**EXERCISE 67**

Solve:

1. $$\begin{cases} \dfrac{1}{x} + \dfrac{1}{y} = \dfrac{3}{2}, \\[2mm] \dfrac{1}{x} - \dfrac{1}{y} = \dfrac{1}{2}. \end{cases}$$

3. $$\begin{cases} \dfrac{5}{x} + \dfrac{10}{y} = 4\tfrac{1}{2}, \\[2mm] \dfrac{10}{x} - \dfrac{5}{y} = 4. \end{cases}$$

2. $$\begin{cases} \dfrac{4}{x} + \dfrac{3}{y} = 3, \\[2mm] \dfrac{1}{x} - \dfrac{1}{y} = \dfrac{1}{6}. \end{cases}$$

4. $$\begin{cases} \dfrac{4}{x} - \dfrac{6}{y} = -10, \\[2mm] \dfrac{6}{x} + \dfrac{5}{y} = 27. \end{cases}$$

5. $\begin{cases} \dfrac{4}{x} - \dfrac{6}{y} = 5, \\ \dfrac{5}{x} + \dfrac{3}{y} = 1. \end{cases}$

10. $\begin{cases} \dfrac{4}{x} - \dfrac{7}{y} = 2, \\ \dfrac{12}{x} - \dfrac{25}{y} = -2. \end{cases}$

6. $\begin{cases} \dfrac{6}{x} + \dfrac{8}{y} = 4, \\ \dfrac{9}{x} - \dfrac{4}{y} = 2. \end{cases}$

11. $\begin{cases} \dfrac{9}{y} - \dfrac{5}{x} = 2, \\ \dfrac{3}{y} + \dfrac{4}{x} = 29. \end{cases}$

7. $\begin{cases} \dfrac{2}{x} - \dfrac{5}{y} = \dfrac{3}{2}, \\ \dfrac{12}{x} + \dfrac{10}{y} = 1. \end{cases}$

12. $\begin{cases} \dfrac{4}{y} = \dfrac{3}{x} + 4, \\ \dfrac{5}{y} = \dfrac{4}{x} + 3. \end{cases}$

8. $\begin{cases} \dfrac{5}{x} - \dfrac{2}{y} = 7, \\ \dfrac{2}{x} + \dfrac{2}{y} = 0. \end{cases}$

13. $\begin{cases} \dfrac{21}{x} - \dfrac{9}{y} = -3, \\ \dfrac{3}{x} + \dfrac{5}{y} = 31. \end{cases}$

9. $\begin{cases} \dfrac{3}{x} - \dfrac{3}{y} = \dfrac{1}{2}, \\ \dfrac{1}{x} - \dfrac{3}{y} = \dfrac{7}{6}. \end{cases}$

14. $\begin{cases} \dfrac{1}{x} + \dfrac{1}{4\,y} = 4, \\ \dfrac{1}{3\,x} - \dfrac{1}{y} = -3. \end{cases}$

## LITERAL SIMULTANEOUS EQUATIONS

**181. Ex. 1.** Solve $\begin{cases} ax + by = c, \\ mx + ny = p. \end{cases}$     (1) (2)

| | | |
|---|---|---|
| (1) × $n$, | $anx + bny = cn.$ | (3) |
| (2) × $b$, | $bmx + bny = bp.$ | (4) |
| (3) − (4), | $anx - bmx = cn - bp.$ | |

Uniting, $\qquad (an - bm)x = cn - bp.$

Dividing, $\qquad x = \dfrac{cn - bp}{an - bm}.$ $\qquad\qquad$ (5)

(1) × $m$, $\qquad amx + bmy = cm.$ $\qquad\qquad$ (6)

(2) × $a$, $\qquad amx + any = ap.$ $\qquad\qquad$ (7)

(7) − (6), $\qquad any - bmy = ap - cm.$

Uniting, $\qquad (an - bm)y = ap - cm.$

$$y = \dfrac{ap - cm}{an - bm}.$$

### EXERCISE 68

1. $\begin{cases} x + y = m, \\ x - y = n. \end{cases}$

2. $\begin{cases} ax + y = m, \\ bx + y = n. \end{cases}$

3. $\begin{cases} mx + ny = n, \\ nx + my = m. \end{cases}$

4. $\begin{cases} mx + y = a, \\ x + y = b. \end{cases}$

5. $\begin{cases} x + my = 1, \\ x - ny = -1. \end{cases}$

6. $\begin{cases} ax + by = 1, \\ cx + dy = 1. \end{cases}$

7. $\begin{cases} ax + by = c, \\ dx + ey = f. \end{cases}$

8. $\begin{cases} ax + by = 2\,ab, \\ x + y = a + b. \end{cases}$

9. $\begin{cases} ax + by = 5\,ab, \\ 2\,x + 3\,y = 9\,a + 4\,b. \end{cases}$

10. $\begin{cases} px = qy, \\ x + y = c. \end{cases}$

11. $\begin{cases} \dfrac{x + a}{y} = 3\,a, \\ \dfrac{y + 1}{x} = \dfrac{1}{a}. \end{cases}$

12. Find $a$ and $s$ in terms of $n$, $d$, and $l$ if

$$\begin{cases} s = \dfrac{n}{2}(a + l), \\ l = a + (n - 1)\,d. \end{cases}$$

13. From the same simultaneous equations find $d$ in terms of $a$, $n$, and $l$.

14. From the same equations find $s$ in terms of $a$, $d$, and $l$.

## SIMULTANEOUS EQUATIONS INVOLVING MORE THAN TWO UNKNOWN QUANTITIES

**182.** To solve equations containing three unknown quantities three simultaneous independent equations must be given.

*By eliminating one unknown quantity from any pair of equations, and the same unknown quantity from another pair, the problem is reduced to the solution of two simultaneous equations containing two unknown quantities.*

Similarly, four equations containing four unknown quantities are reduced to three equations containing three unknown quantities, etc.

**Ex. 1.** Solve the following system of equations:

$$2x - 3y + 4z = 8, \tag{1}$$
$$3x + 4y - 5z = -4, \tag{2}$$
$$4x - 6y + 3z = 1. \tag{3}$$

Eliminate $y$.

Multiplying (1) by 4, $\quad 8x - 12y + 16z = \phantom{-}32$
Multiplying (2) by 3, $\quad \underline{9x + 12y - 15z = -12}$

Adding, $\qquad 17x \qquad + \quad z = \phantom{-}20 \tag{4}$

Multiplying (2) by 3, $\quad 9x + 12y - 15z = -12$
Multiplying (3) by 2, $\quad \underline{8x - 12y + \phantom{0}6z = \phantom{-}2}$

Adding, $\qquad 17x \qquad - \quad 9z = -10 \tag{5}$

Eliminating $x$ from (4) and (5).

$(4) - (5), \qquad\qquad\qquad 10z = 30.$
Therefore $\qquad\qquad\qquad\qquad z = 3. \tag{6}$
Substitute this value in (4), $\quad 17x + 3 = 20.$
Therefore $\qquad\qquad\qquad\qquad x = 1. \tag{7}$

Substituting the values of $x$ and $z$ in (1),

$$2 - 3y + 12 = 8.$$
$$3y = 6.$$
Hence $\qquad\qquad\qquad\qquad y = 2. \tag{8}$

*Check.* $\quad 2 \cdot 1 - 3 \cdot 2 + 4 \cdot 3 = 8; \; 3 \cdot 1 + 4 \cdot 2 - 5 \cdot 3 = -4;$
$\qquad\quad 4 \cdot 1 - 6 \cdot 2 + 3 \cdot 3 = 1.$

## EXERCISE 69

1.
$$\begin{cases} 2\,x + 4\,y + 6\,z = 28, \\ 10\,x + y + z = 15, \\ 5\,x + 4\,y + 3\,z = 22. \end{cases}$$

8.
$$\begin{cases} 2\,x - y + z = 5, \\ 3\,x + 2\,y - 4\,z \doteq -4, \\ x - 2\,y + 2\,z = 4. \end{cases}$$

2.
$$\begin{cases} 2\,x + 3\,y + 5\,z = 47, \\ x + y + z = 12, \\ 5\,x - 2\,y + 2\,z = 7. \end{cases}$$

9.
$$\begin{cases} x + y + z = 4, \\ x - y + z = -4, \\ x + y - z = -4. \end{cases}$$

3.
$$\begin{cases} 3\,x + 2\,y - z = 16, \\ x + 2\,y + 3\,z = 32, \\ 2\,x + y + 2\,z = 25. \end{cases}$$

10.
$$\begin{cases} 6\,x - 5\,y + 18\,z = 18, \\ 2\,x + y + 4\,z = -1, \\ 2\,x - 5\,y + 10\,z = 14. \end{cases}$$

4.
$$\begin{cases} x + 5\,y - 3\,z = 10, \\ 4\,x - 3\,y + 4\,z = 10, \\ 5\,x + 4\,y - 2\,z = 21. \end{cases}$$

11.
$$\begin{cases} 2\,x + 2\,y + z = 35, \\ x + 3\,y + 4\,z = 42, \\ 3\,x + y + 2\,z = 40. \end{cases}$$

5.
$$\begin{cases} 14\,x - 6\,y - z = 6, \\ 7\,x - 4\,y + 2\,z = 10, \\ 2\,x - 3\,y - 3\,z = -17. \end{cases}$$

12.
$$\begin{cases} x + 7\,y - 9\,z = 26, \\ 3\,x - 4\,y + 4\,z = 52, \\ 2\,x - 3\,y - 4\,z = 6. \end{cases}$$

6.
$$\begin{cases} 6\,x - 7\,y - 6\,z = 3, \\ 9\,x - 14\,y + 9\,z = 12, \\ 18\,x - 20\,y + 15\,z = 45. \end{cases}$$

13.
$$\begin{cases} x + 2\,y + z = 14, \\ x - y + 2\,z = 6, \\ 2\,x + y - z = -8. \end{cases}$$

7.
$$\begin{cases} 2\,x + 4\,y - 5\,z = -5, \\ x - 3\,y + 4\,z = 7, \\ 3\,x - 6\,y + 3\,z = 0. \end{cases}$$

14.
$$\begin{cases} 4\,x + 2\,y + z = 87, \\ 2\,x + y - 4\,z = -33, \\ 2\,x - 3\,y + z = 0. \end{cases}$$

15. $\begin{cases} 2x + 3y + 4z = 3, \\ 4x - 9y + 12z = 2, \\ 6x - 3y + 8z = 4. \end{cases}$

23. $\begin{cases} x + y + z = 99, \\ y - 3x = 0, \\ 5y - 3z = 0. \end{cases}$

16. $\begin{cases} x + y + z = 9, \\ x + 2y + 4z = -5, \\ x + 3y + 9z = -33. \end{cases}$

24. $\begin{cases} \dfrac{x}{4} + \dfrac{y}{3} - \dfrac{z}{5} = 8, \\ \dfrac{x}{4} + \dfrac{y}{5} + \dfrac{z}{10} = 16, \\ \dfrac{x}{6} - \dfrac{y}{15} + \dfrac{z}{10} = 6. \end{cases}$

17. $\begin{cases} 7x + 6y + 7z = 100, \\ x - 2y + z = 0, \\ 3x + y - 2z = 0. \end{cases}$

18. $\begin{cases} 3x + 2y = 12, \\ 4x - 8y - z = -20, \\ 6x - 6y + 2z = 2. \end{cases}$

25. $\begin{cases} \dfrac{x + y}{x + z} = 2, \\ \dfrac{y + z}{x + y} = \dfrac{2}{3}, \\ x + y + z = 13. \end{cases}$

19. $\begin{cases} x + y + z = 9, \\ 2x - 5y = -11, \\ y + 2z = 11. \end{cases}$

26. $\begin{cases} .2x + .3y + .4z = 2, \\ .3x + .4y + .5z = 2.6, \\ .4x + .3y - z = -2. \end{cases}$

20. $\begin{cases} x + y = 5, \\ y + z = -1, \\ z + x = -2. \end{cases}$

27. $\begin{cases} y + \dfrac{x}{2} = 41, \\ x + \dfrac{z}{4} = 20\frac{1}{2}, \\ y + \dfrac{z}{5} = 34. \end{cases}$

21. $\begin{cases} x + y + z = 26, \\ 7x = 11z, \\ 7y = 8z. \end{cases}$

22. $\begin{cases} 3x = 7y - 18, \\ 2y = 7z - 2, \\ 4z = 5x - 32. \end{cases}$

28. $\begin{cases} \dfrac{x}{3} + \dfrac{y}{6} + \dfrac{z}{9} = -1, \\ \dfrac{x}{6} + \dfrac{y}{9} + \dfrac{z}{12} = -1, \\ \dfrac{x}{9} + \dfrac{y}{12} + \dfrac{z}{15} = -1. \end{cases}$

**29.**
$$\begin{cases} \dfrac{1}{x}+\dfrac{1}{y}+\dfrac{1}{z}=\dfrac{3}{2}, \\[2mm] \dfrac{1}{y}+\dfrac{1}{z}=1, \\[2mm] \dfrac{1}{z}+\dfrac{1}{x}=1. \end{cases}$$

**30.**
$$\begin{cases} x+y+z=9, \\ x:y=2:3, \\ x:z=1:2. \end{cases}$$

**31.**
$$\begin{cases} x+y=2\,m, \\ y+z=2\,p, \\ z+x=2\,n. \end{cases}$$

## PROBLEMS LEADING TO SIMULTANEOUS EQUATIONS

**183.** Problems involving several unknown quantities must contain, either directly or implied, as many verbal statements as there are unknown quantities. Simple examples of this kind can usually be solved by equations involving only one unknown quantity. (§ 99.)

In complex examples, however, it is advisable to represent every unknown quantity by a different letter, and to express every verbal statement as an equation.

**Ex. 1.** The sum of three digits of a number is 8. The digit in the tens' place is $\frac{1}{3}$ of the sum of the other two digits, and if 396 be added to the number, the first and the last digits will be interchanged. Find the number.

Obviously it is difficult to express two of the required digits in terms of the other ; hence we employ 3 letters for the three unknown quantities.

Let            $x =$ the digit in the hundreds' place,

                      $y =$ the digit in the tens' place,

and          $z =$ the digit in the units' place.

Then    $100x + 10y + z =$ the number.

The three statements of the problem can now be readily expressed in symbols :

$$x + y + z = 8. \tag{1}$$

$$y = \tfrac{1}{3}(x + z). \tag{2}$$

$$100x + 10y + z + 396 = 100z + 10y + x. \tag{3}$$

The solution of these equations gives $x = 1$, $y = 2$, $z = 5$.

Hence the required number is 125.

*Check.*    $1 + 2 + 5 = 8$; $2 = \tfrac{1}{3}(1 + 5)$; $125 + 396 = 521$.

**Ex. 2.** If both numerator and denominator of a fraction be increased by one, the fraction is reduced to $\frac{2}{3}$; and if both numerator and denominator of the reciprocal of the fraction be diminished by one, the fraction is reduced to 2. Find the fraction.

Let $\qquad x =$ the numerator,

and $\qquad y =$ the denominator;

then $\dfrac{x}{y} =$ the fraction. By expressing the two statements in symbols,

we obtain,

$$\frac{x+1}{y+1} = \frac{2}{3}, \tag{1}$$

and

$$\frac{y-1}{x-1} = 2. \tag{2}$$

These equations give $x = 3$ and $y = 5$. Hence the fraction is $\frac{3}{5}$.

*Check.* $\dfrac{3+1}{5+1} = \dfrac{4}{6} = \dfrac{2}{3}; \dfrac{5-1}{3-1} = \dfrac{4}{2} = 2.$

**Ex. 3.** A, B, and C travel from the same place in the same direction. B starts 2 hours after A and travels one mile per hour faster than A. C, who travels 2 miles an hour faster than B, starts 2 hours after B and overtakes A at the same instant as B. How many miles has A then traveled?

|  | TIME (Hours) | RATE (Miles per hour) | DISTANCE (Miles) |
|---|---|---|---|
| A . . . . . | $x$ | $y$ | $xy$ |
| B . . . . . | $x-2$ | $y+1$ | $xy + x - 2y - 2$ |
| C . . . . . | $x-4$ | $y+3$ | $xy + 3x - 4y - 12$ |

Since the three men traveled the same distance,

$$xy = xy + x - 2y - 2. \tag{1}$$
$$xy = xy + 3x - 4y - 12. \tag{2}$$
Or $\qquad x - 2y = 2. \tag{3}$
$$3x - 4y = 12. \tag{4}$$
$(4) - 2 \times (3)$, $\qquad x = 8.$
From (3) $\qquad y = 3.$
Hence $xy = 24$ miles, the distance traveled by A.

*Check.* $8 \times 3 = 24, 6 \times 4 = 24, 4 \times 6 = 24.$

## EXERCISE 70

**1.** Four times a certain number increased by three times another number equals 33, and the second increased by 2 equals three times the first. Find the numbers.

**2.** Five times a certain number exceeds three times another number by 11, and the second one increased by 5 equals twice the first number. Find the numbers.

**3.** Half the sum of two numbers equals 4, and the fourth part of their difference equals 1. Find the numbers.

**4.** If 4 be added to the numerator of a fraction, its value is $\frac{3}{4}$. If 3 be added to the denominator, the fraction is reduced to $\frac{1}{3}$. Find the fraction.

**5.** If the numerator and the denominator of a fraction be increased by 3, the fraction equals $\frac{1}{2}$. If 1 be subtracted from both terms, the value of the fraction is $\frac{3}{10}$. Find the fraction.

**6.** If the numerator of a fraction be trebled, and its denominator diminished by one, it is reduced to $\frac{3}{4}$. If the denominator be doubled, and the numerator increased by 4, the fraction is reduced to $\frac{1}{2}$. Find the fraction.

**7.** A fraction is reduced to $\frac{1}{4}$, if its numerator and its denominator are increased by 1, and twice the numerator increased by the denominator equals 15. What is the fraction?

**8.** The sum of the digits of a number of two figures is 6, and if 18 is added to the number the digits will be interchanged. What is the number? (See Ex. 1, § 183.)

**9.** If 27 is added to a number of two digits, the digits will be interchanged, and four times the first digit exceeds the second digit by 3. Find the number.

**10.** The sum of the three digits of a number is 9, and the sum of the first two digits exceeds the third digit by 3. If 9 be added to the number, the last two digits are interchanged. Find the number.

L

11. Twice A's age exceeds the sum of B's and C's ages by 30, and B's age is $\frac{1}{2}$ the sum of A's and C's ages. Ten years ago the sum of their ages was 90. Find their present ages.

12. Ten years ago A was as old as B will be 5 years hence; and 5 years ago B was as old as C is now. If the sum of their ages is 55, how old is each now?

13. A man invested $5000, a part at 6 % and the remainder at 5 %, bringing a total yearly interest of $260. What was the amount of each investment?

14. A man invested $750, partly at 5 % and partly at 4 %, and the 5 % investment brings $15 more interest than the 4 % investment. What was the amount of each investment?

15. A sum of $10,000 is partly invested at 6 %, partly at 5 %, and partly at 4 %, bringing a total yearly interest of $530. The 6 % investment brings $70 more interest than the 5 % and 4 % investments together. How much money is invested at 6 %, 5 %, and 4 %, respectively?

16. A sum of money at simple interest amounted in 6 years to $8000, and in 8 years to $8500. What was the sum of money and the rate of interest?

17. A sum of money at simple interest amounted in 2 years to $990, and in 5 years to $1125. What was the sum and the rate of interest?

18. The sums of $1500 and $2000 are invested at different rates and their annual interest is $190. If the rates of interest were exchanged, the annual interest would be $195. Find the rates of interest.

19. Three cubic centimeters of gold and two cubic centimeters of silver weigh together 78 grams. Two cubic centimeters of gold and three cubic centimeters of silver weigh together 69$\frac{1}{2}$ grams. Find the weight of one cubic centimeter of gold and one cubic centimeter of silver.

**20.** A farmer sold a number of horses, cows, and sheep, for $740, receiving $100 for each horse, $50 for each cow, and $15 for each sheep. The number of sheep was twice the number of horses and cows together. How many did he sell of each if the total number of animals was 24?

**21.** The sum of the 3 angles of a triangle is 180°. If one angle exceeds the sum of the other two by 20°, and their difference by 60°, what are the angles of the triangle?

**22.** On the three sides of a triangle $ABC$, respectively, three points, $D$, $E$, and $F$, are taken so that $AD = AF$, $BD = BE$, and $CE = CF$. If $AB = 6$ inches, $BC = 7$ inches, and $AC = 5$ inches, what is the length of $AD$, $BE$, and $CF$?

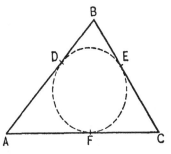

NOTE. If a circle is inscribed in the triangle $ABC$ touching the sides in $D$, $E$, and $F$ (see diagram), then $AD = AF$, $BD = BE$, and $CE = CF$.

**23.** A circle is inscribed in triangle $ABC$ touching the three sides in $D$, $E$, and $F$. Find the parts of the sides if $AB = 9$, $BC = 7$, and $CA = 8$.

**24.** In the annexed diagram angle $a$ = angle $b$, angle $c$ = angle $d$, and angle $e$ = angle $f$. If angle $ABC = 60°$, angle $BAC = 50°$, and angle $BCA = 70°$, find angles $a$, $c$, and $e$.

NOTE. $O$ is the center of the circumscribed circle.

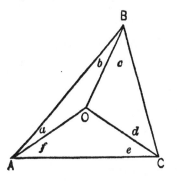

**25.** It takes A two hours longer than B to travel 24 miles, but if A would double his pace, he would walk it in two hours less than B. Find their rates of walking.

# CHAPTER XII*

## 'GRAPHIC REPRESENTATION OF FUNCTIONS AND EQUATIONS

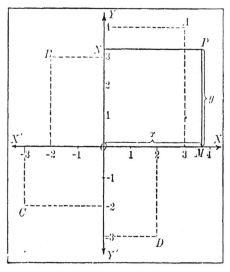

**184. Location of a point.** If two fixed straight lines $XX'$ and $YY'$ meet in $O$ at right angles, and $PM \perp XX'$, and $PN \perp YY'$, then the position of point $P$ is determined if the lengths of $PM$ and $PN$ are given.

**185. Coördinates.** The lines $PM$ and $PN$ are called the **coördinates** of point $P$. $PN$, or its equal $OM$, is the **abscissa**; and $PM$, or its equal $ON$, is the **ordinate** of point $P$. The abscissa is usually denoted by $x$, the ordinate by $y$.

The line $XX'$ is called the **x-axis,** $YY'$ the **y-axis,** and point $O$ the **origin.** Abscissas measured to the *right of the origin,* and ordinates *above the x-axis* are considered **positive ;** hence coördinates lying in opposite directions are **negative.**

**186.** The point whose abscissa is $x$, and whose ordinate is $y$, is usually denoted by $(x, y)$. Thus the points $A$, $B$, $C$, and $D$ are respectively represented by $(3, 4)$, $(-2, 3)$, $(-3, -2)$, and $(2, -3)$.

---

\* This chapter may be omitted on a first reading.

The process of locating a point whose coördinates are given is called **plotting the point.**

NOTE. Graphic constructions are greatly facilitated by the use of cross-section paper, *i.e.* paper ruled with two sets of equidistant and parallel lines intersecting at right angles. (See diagram on page 151.)

## EXERCISE 71

**1.** Plot the points: $(4, 3)$, $(4, -2)$, $(-4, 2)$, $(-3, -3)$.

**2.** Plot the points: $(-4, 2\frac{1}{2})$, $(-5, 1)$, $(4, 0)$, $(-2, 0)$.

**3.** Plot the points: $(0, 3)$, $(4, 0)$, $(0, 0)$, $(0, -2)$.

**4.** Draw the triangle whose vertices are respectively $(4, 1)$, $(-1, 3)$, and $(1, -2)$.

**5.** Plot the points $(6, 4)$ and $(-4, -4)$, and measure their distance.

**6.** What is the distance of the point $(3, 4)$ from the origin?

**7.** Draw the quadrilateral whose vertices are respectively $(4, 1)$, $(-1, 4)$, $(-4, -1)$, $(1, -4)$.

**8.** Where do all points lie whose ordinates equal 4?

**9.** Where do all points lie whose abscissas equal zero?

**10.** Where do all points lie whose ordinates equal zero?

**11.** What is the locus of $(x, y)$ if $y = 3$?

**12.** If a point lies in the *x*-axis, which of its coördinates is known?

**13.** What are the coördinates of the origin?

**187. Graphs.** If two variable quantities are so related that changes of the one bring about definite changes of the other, the mutual dependence of the two quantities may be represented either by a table or by a diagram.

Thus the following tables represent the average temperature of New York City from January 1 to December 1, and the volumes of a certain amount of gas subjected to pressures from 1 pound to 8 pounds.

| Date | Average Temperature |
|---|---|
| January 1 . . | − 1° C. |
| February 1 . | − 1° C. |
| March 1 . . | 1° C. |
| April 1 . . . | 6° C. |
| May 1 . . . | 12° C. |
| June 1 . . . | 18° C. |
| July 1 . . . | 22° C. |
| August 1 . . | 23° C. |
| September 1 . | 20° C. |
| October 1 . . | 16° C. |
| November 1 . | 10° C. |
| December 1 . | 3° C. |

| Pressure | Volume |
|---|---|
| 1 lb. | 12 c.cm. |
| 2 lb. | 6 c.cm. |
| 3 lb. | 4 c.cm. |
| 4 lb. | 3 c.cm. |
| 5 lb. | 2.4 c.cm. |
| 6 lb. | 2 c.cm. |
| 7 lb. | 1.7 c.cm. |
| 8 lb. | 1.5 c.cm. |

**188.** The same data, however, may be represented graphically by making each number in one column the abscissa, and the corresponding number in the adjacent column the ordinate of a point. Thus the first table produces 12 points, $A, B, C, D,$ etc., each representing a temperature at a certain date.

By representing in like manner the average temperatures for every value of the time, we obtain an uninterrupted sequence of points, or the curved line $ABCN$, the so-called **graph** of the temperature.

To find from the diagram the temperature on June 1, we measure the ordinate of $F$. Thus the average temperature on May 15 may be found to be 15°; on April 20, 10°; on Jan. 15, $-1\frac{1}{2}°$.

A graphic representation does not allow the same accuracy of results as a numerical table, but it indicates in a given space a great many more facts than a table, and it impresses upon the eye all the peculiarities of the changes better and quicker than any numerical compilations.

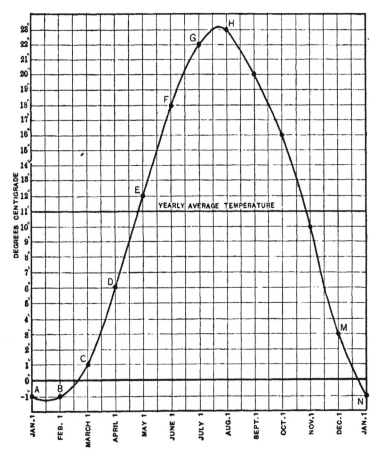

Graphs are possibly the most widely used devices of applied mathematics. The scientist uses them to compile the data found from experiments, and to deduce general laws therefrom. The engineer, the physician, the merchant, uses them. Daily papers represent economical facts graphically, as the prices and production of commodities, the rise and fall of wages, etc. Whenever a clear, concise representation of a number of numerical data is required, the graph is applied.

## EXERCISE 72

From the diagram find approximate answers to the following questions:

1. Determine the average temperature of New York City on (*a*) May 1, (*b*) July 15, (*c*) January 15, (*d*) November 20.

**2.** At what date or dates is the average temperature of New York (*a*) 6° C., (*b*) 16° C., (*c*) −1° C., (*d*) 9° C.?

**3.** At what date is the average temperature highest? What is the highest average temperature?

**4.** At what date is the average temperature lowest? What is the lowest average temperature?

**5.** During what months is the average temperature of New York above 18° C.?

**6.** When is the average temperature below 0° C. (freezing point)?

**7.** From what date to what date does the temperature increase (on the average)?

**8.** When is the temperature equal to the yearly average of 11° C.?

**9.** What is the average temperature from Sept. 1 to Oct. 1?

**10.** How much, on the average, does the temperature increase from June 1 to July 1?

**11.** During what month does the temperature increase most rapidly?

**12.** During what month does the temperature decrease most rapidly?

**13.** During what month does the temperature change least?

**14.** Which month is the coldest of the year?

**15.** Which month is the hottest of the year?

**16.** How much warmer on the average is it on July 1 than on May 1?

**17.** If we would denote the time during which the temperature is above the yearly average of 11° as the warm season, from what date to what date would it extend?

**18.** When is the average temperature the same as on April 1 ?

Note. Use the graphs of the following examples for the solution of concrete numerical examples, in a similar manner as the temperature graph was applied in examples 1–18.

**19.** From the table on page 150 draw a graph representing the volumes of a certain body of gas under varying pressures.

**20.** Construct a diagram containing the graphs of the mean temperatures of the following three cities (in degrees Fahrenheit):

| | JAN. 1 | FEB. 1 | MARCH 1 | APRIL 1 | MAY 1 | JUNE 1 | JULY 1 | AUG. 1 | SEPT. 1 | OCT. 1 | NOV. 1 | DEC. 1 | AVERAGE |
|---|---|---|---|---|---|---|---|---|---|---|---|---|---|
| San Francisco | 50 | 52 | 54 | 55 | 57 | 58 | 58 | 59 | 60 | 59 | 56 | 51 | 56 |
| Tampa | 59 | 66 | 66 | 72 | 76 | 80 | 82 | 81 | 80 | 73 | 65 | 63 | 72 |
| Bismarck | 4 | 10 | 23 | 42 | 54 | 64 | 70 | 68 | 57 | 44 | 26 | 15 | 40 |

**21.** Represent graphically the populations (in hundred thousands) of the following states :

| STATE | 1800 | 1810 | 1820 | 1830 | 1840 | 1850 | 1860 | 1870 | 1880 | 1890 | 1900 |
|---|---|---|---|---|---|---|---|---|---|---|---|
| Illinois | | | .5 | 1.6 | 4.8 | 8.5 | 17.1 | 25.3 | 30.8 | 38.3 | 48.2 |
| Maryland | 3.4 | 3.8 | 4.0 | 4.5 | 4.7 | 5.8 | 6.9 | 7.8 | 9 3 | 10.4 | 11.9 |
| New York | 6.9 | 9.6 | 13.7 | 19.2 | 24.3 | 31.0 | 38.8 | 43.8 | 50.8 | 60.0 | 72.7 |
| Virginia | 8.8 | 9.7 | 10.7 | 12.1 | 12.4 | 14.2 | 16.0 | 12.3 | 15.1 | 16.6 | 18.5 |

**22.** One meter equals 1.09 yards. Draw a graph for the transformation of meters into yards.

**23.** Draw a temperature chart of a patient.

| Hour . . . | 6 p.m. | 7 p.m. | 8 p.m. | 9 p.m. | 10 p.m. | 11 p.m. | m'dn't |
|---|---|---|---|---|---|---|---|
| Temperature | 100.5° | 101° | 101.5° | 103.2° | 102.5° | 102° | 101.4° |

**24.** If $C$ is the circumference of a circle whose radius is $R$, then $C = 2 \pi R$. (Assume $\pi = \frac{22}{7}$.) Represent graphically the circumferences of all circles from $R = 0$ to $R = 8$ inches.

**25.** Represent graphically the weight of iron from 0 to 20 cubic centimeters, if 1 cubic centimeter of iron weighs 7.5 grams.

**26.** Represent graphically the cost of butter from 0 to 5 pounds if 1 pound cost $.50.

**27.** Represent graphically the distances traveled by a train in 3 hours at a rate of 20 miles per hour.

**28.** A dealer in bicycles gains $2 on every wheel he sells. If the daily average expenses for rent, gas, etc., amount to $8, represent his daily gain (or loss), if he sells 0, 1, 2 ... 10 wheels a day.

**29.** The cost of manufacturing a certain book consists of the initial cost of $800 for making the plates, and $.50 per copy for printing, binding, etc. Show graphically the cost of the books from 1 to 1200 copies. (Let 100 copies = about $\frac{1}{2}$ inch.) On the same diagram represent the selling price of the books, if each copy sells for $1.50.

## REPRESENTATION OF FUNCTIONS OF ONE VARIABLE

**189.** An expression involving one or several letters is called a **function** of these letters.

$x^2 - x + 7$ is a function of $x$.

$2 xy - y^2 + 3 y^3$ is a function of $x$ and $y$.

**190.** If the value of a quantity changes, the value of a function of this quantity will change; *e.g.* if $x$ assumes successively the values 1, 2, 3, 4, $x^2 - x + 7$ will respectively assume the values 7, 9, 13, 19. If $x$ increases gradually from 1 to 2, $x^2 - x + 7$ will change gradually from 7 to 9.

**191.** ⸱A **variable** is a quantity whose value changes in the same discussion.

A **constant** is a quantity whose value does not change in the same discussion.

In the example of the preceding article, $x$ is supposed to change, hence it is a variable, while 7 is a constant.

**192. Graph of a function.** The values of a function for the various values of $x$ may be given in the form of a numerical table. Thus the table on page 164 gives the values of the functions $x^2$, $x^3$, and $\sqrt{x}$, for $x = 1, 2, 3 \cdots 50$. The values of functions may, however, be also represented by a graph.

*E.g.* to construct the graph of $x^2$, construct a series of points whose abscissas represent $x$, and whose ordinates are the corresponding values of $x^2$; *i.e.* construct the points $(-3, 9)$, $(-2, 4)$, $(-1, 1)$, $(0, 0)$, $(1, 1)$, $(2, 4)$, and $(3, 9)$, and join the points in order.

If a more exact diagram is required, plot points which lie between those constructed above, as $(\frac{1}{2}, \frac{1}{4})$, $(1\frac{1}{2}, 2\frac{1}{4})$, etc.

| $x$ | $x^2$ |
|-----|-------|
| $-3$ | 9 |
| $-2$ | 4 |
| $-1$ | 1 |
| 0 | 0 |
| 1 | 1 |
| 2 | 4 |
| 3 | 9 |

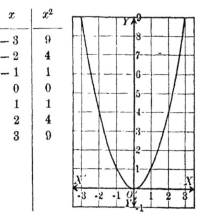

**Ex. 1.** Draw the graph of $x^2 + 2x - 4$ from $x = -4$, to $x = 4$.

To obtain the values of the functions for the various values of $x$, the following arrangement may be found convenient:

| $x =$ | $-4$ | $-3$ | $-2$ | $-1$ | 0 | 1 | 2 | 3 | 4 |
|-------|------|------|------|------|---|---|---|---|---|
| $x^2 =$ | 16 | 9 | 4 | 1 | 0 | 1 | 4 | 9 | 16 |
| $x^2 + 2x =$ | 8 | 3 | 0 | $-1$ | 0 | 3 | 8 | 15 | 11 |
| $y$, or $x^2 + 2x - 4 =$ | 4 | $-1$ | $-4$ | $-5$ | $-4$ | $-1$ | 4 | 11 | 20 |

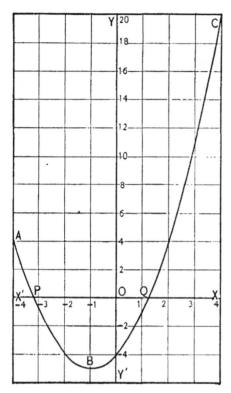

Locating the points $(-4, 4)$, $(-3, -1)$, $(-2, 4)$ ... $(4, 20)$, and joining in order produces the graph $ABC$. (To avoid very large ordinates, the scale unit of the ordinates is taken smaller than that of the $x$.)

For brevity, the function is frequently represented by a single letter, as $y$. Thus in the above example, $y = x^2 + 2x - 4$; and if $x = 2\frac{1}{2}$, $y = 7\frac{1}{4}$; if $x = -1\frac{1}{2}$, $y = -4\frac{3}{4}$, etc.

**193.** A function of the first degree is an integral rational function involving only the first power of the variable.

Thus $4x + 7$, or $ax + b + c$ are functions of the first degree.

**194.** It can be proved that the **graph of a function of the first degree** is a straight line, hence two points are sufficient for the construction of these graphs.

Ex. 2.  Draw the graph of
$$y = 2x - 3.$$

If          $x = 0$, $y = -3$.
If          $x = 4$, $y = 5$.

Locating $(0, -3)$ and $(4, 5)$, **and join-**ing by a straight line produces the required graph.

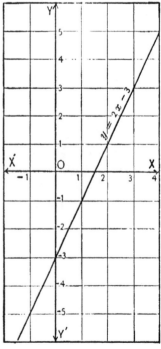

## EXERCISE 73

Draw the graphs of the following functions:

1. $x + 2$.       4. $2x + 1$.       7. $2 - 3x$.       10. $x^2 - 1$.

2. $x - 1$.       5. $3x - 2$.       8. $1 - x$.       11. $x^2 + x$.

3. $2x - 1$.       6. $\frac{3}{2}x$.       9. $\frac{1}{4}x^2$.       12. $4x - x^2$.

13. $x^2 - 4x + 4$.       16. $x^2 - x + 1$.       19. $y = 2x - 3$.

14. $x^2 - 3x - 8$.       17. $6 + x - x^2$.       20. $y = x^2 - 4$.

15. $x^2 + x - 2$.       18. $2 - x - x^2$.

21. Draw the graph of $x^2$ from $x = -4$ to $x = 4$, and from the diagram find:

(a) $(3.5)^2$;       (b) $(-1.5)^2$;       (c) $(-2.8)^2$;       (d) $(-1\frac{3}{4})^2$;

(e) $\sqrt{6.25}$;       (f) $\sqrt{12.25}$;       (g) $\sqrt{5}$;       (h) $\sqrt{10.5}$.

22. Draw the graph of $x^2 - 4x + 2$ from $x = 1$ to $x = 4$, and from the diagram determine:

(a) The values of the function if $x = -\frac{1}{2}, 1\frac{1}{2}, 2\frac{1}{2}$.

(b) The values of $x$, if $x^2 - 4x + 2$ equals $-2, 1, 1\frac{1}{2}$.

(c) The smallest value of the function.

(d) The value of $x$ that produces the smallest value of the function.

(e) The values of $x$ that make $x^2 - 4x + 2 = 0$.

(f) The roots of the equation $x^2 - 4x + 2 = 0$.

(g) The roots of the equation $x^2 - 4x + 2 = -1$.

(h) The roots of the equation $x^2 - 4x + 2 = 2$.

23. Draw the graph of $y = 2 + 2x - x^2$ from $x = -2$ to $x = 4$, and from the diagram determine:

(a) The values of $y$; *i.e.* the function, if $x = \frac{1}{2}, -1\frac{1}{2}, 2\frac{1}{4}$.

(b) The values of $x$, if $y = -2$.

(c) The values of $x$, if the function equals zero.

(d) The roots of the equation $2 + 2x - x^2 = 0$.

(e) The roots of the equation $2 + 2x - x^2 = 1$.

**24.** Degrees of the Fahrenheit ($F.$) scale are expressed in degrees of the Centigrade ($C.$) scale by the formula

$$C = \tfrac{5}{9}(F - 32).$$

(*a*) Draw the graph of

$$C = \tfrac{5}{9}(F - 32)$$

from $\qquad\qquad F = -4$

to $\qquad\qquad F = 41.$

(*b*) From the diagram find the number of degrees of centi-grade equal to $-1°$ F., $9°$ F., $14°$ F., $32°$ F.

(*c*) Change to Fahrenheit readings: $-10°$ C., $0°$ C., $1°$ C.

**25.** A body moving with a uniform velocity of 3 yards per second moves in $t$ seconds a distance $d = 3\,t.$

Represent this formula graphically.

**26.** If two variables $x$ and $y$ are directly proportional, then $y = cx$, where $c$ is a constant.

Show that the graph of two variables that are directly pro-portional is a straight line passing through the origin (assume for $c$ any convenient number).

**27.** If two variables $x$ and $y$ are inversely proportional, then

$$y = \frac{c}{x}, \text{ where } c \text{ is a constant.}$$

Draw the locus of this equation if $c = 12.$

## GRAPHIC SOLUTION OF EQUATIONS INVOLVING ONE UNKNOWN QUANTITY

**195.** Since we can graphically determine the values of $x$ that make a function of $x$ equal to zero, it is evidently possible to find graphically the real roots of an equation. Thus to find what values of $x$ make the function $x^2 + 2x - 4 = 0$ (see § 192), we have to measure the abscissas of the intersection of the graph with the $x$-axis, *i.e.* the abscissas of $P$ and $Q$.

Therefore $x = 1.24$ or $x = -3.24.$

**196**. To solve the equation $x^2 + 2x - 4 = -1$, determine the points where the function is $-1$. If cross-section paper is used, the points may be found by inspection, otherwise draw through $(0, -1)$ a line parallel to the $x$-axis, and determine the abscissas of the points of intersection with the graph, viz. $-2$ and $1$.

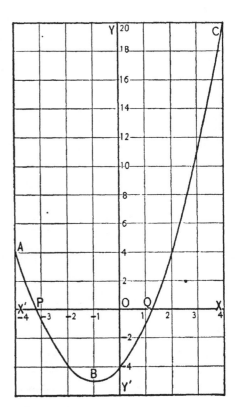

**197**. An equation of the the form $ax^2 + bx + c = 0$, where $a$, $b$, and $c$ represent known quantities, is called a *quadratic equation*. Such equations in general have two roots.

**EXERCISE 74**

Solve graphically the following equations:

1. $4x - 7 = 0$.
2. $2x + 5 = 0$.
3. $6 - x = 0$.
4. $8 - 3x = 0$.
5. $8 - 3x = 2$.
6. $8 - 3x = -1$.
7. $8 - 3x = 11$.
8. $5x - 2 = 4$.
9. $x^2 - x - 6 = 0$.

10. $x^2 - x - 5 = 0$.
11. $x^2 - 2x - 7 = 0$.
12. (a) $x^2 - 6x + 9 = 0$.
    (b) $x^2 - 6x + 9 = 8$.
    (c) $x^2 - 6x + 9 = 4$.
13. (a) $x^2 - 4x - 6 = 0$.
    (b) $x^2 - 4x - 6 = -1$.
14. (a) $x^2 + 2x - 6 = 0$.
    (b) $x^2 + 2x - 6 = 2$.

## GRAPHIC SOLUTION OF EQUATIONS INVOLVING TWO UNKNOWN QUANTITIES

**198. Graph of equations involving two unknown quantities.**
Since we can represent graphically equations of the form $y =$ function of $x$ (§ 192), we can construct the **graph** or **locus** of any equation involving two unknown quantities, that can be reduced to the above form.

Thus to represent $\dfrac{x^2 + x}{y - 5} = 2$ graphically, solve for $y$, *i.e.*

$$y = \frac{x^2 + x + 10}{2},$$

and construct $\dfrac{x^2 + x + 10}{2}$ graphically.

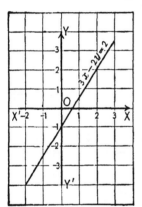

**Ex. 1.** Represent graphically $3x - 2y = 2$.

Solving for $y$,    $y = \dfrac{3x - 2}{2}$.

Hence if   $x = -2,\, y = -4$ ;
if       $x = 2,\, y = 2$.

Locating the points $(-2, -4)$ and $(2, 2)$, and joining by a straight line, produces the required locus.

**199.** If the given equation is of the first degree, we can usually locate two points without solving the equation for $y$. Thus in the preceding example:

$$3x - 2y = 2.$$

If      $x = 0,\, y = -1$.
If      $y = 0,\, x = \tfrac{2}{3}$.

Hence we may join $(0, -1)$ and $(\tfrac{2}{3}, 0)$.

**Ex. 2.** Draw the locus of $4x + 3y = 12$.

If $x = 0,\, y = 4$ ; if $y = 0,\, x = 3$.

Hence, locate points $(0, 4)$ and $(3, 0)$, and join them by straight line $AB$. $AB$ is the required graph.

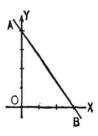

**NOTE.** Equations of the first degree are called *linear* equations, because their graphs are straight lines.

**200.** The coördinates of every point of the graph satisfy the given equation, and every set of real values of $x$ and $y$ satisfying the given equation is represented by a point in the locus.

**201. Graphical solution of a linear system.**

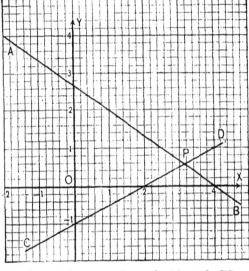

To find the roots of the system.

$$2\,x + 3\,y = 8, \quad (1)$$
$$x - 2\,y = 2. \quad (2)$$

By the method of the preceding article construct the graphs $AB$ and $CD$ of (1) and (2) respectively. The coördinates of every point in $AB$ satisfy the equation (1), but only one point in $AB$ also satisfies equation (2), viz. $P$, the point of intersection of $AB$ and $CD$.

By measuring the coördinate of $P$, we obtain the roots, $x = 3.15,\ y = .57$.

**202.** The roots of two simultaneous equations are represented by the coördinates of the point (or points) at which their graphs intersect.

**203.** Since two straight lines which are not coincident nor parallel have only one point of intersection, simultaneous linear equations have only one pair of roots.

**Ex. 3.** Solve graphically the equations:

$$\begin{cases} 4\,x + 2\,y = 8, & (1) \\ x - y + 1 = 0. & (2) \end{cases}$$

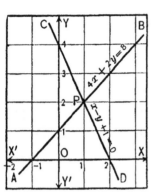

Using the method of the preceding paragraph, construct $AB$ the locus of (1), and $CD$ the locus of (2).

Measuring the coördinates of $P$, the point of intersection, we obtain

$$x = 1, \ y = 2.$$

**Ex. 4.** Solve graphically the following system :

$$\begin{cases} x^2 + y^2 = 25, & (1) \\ 3\,x - 2\,y = -6. & (2) \end{cases}$$

Solving (1) for $y$, $y = \sqrt{25 - x^2}$.

Therefore, if $x$ equals $-5, -4, -3, -2, -1, 0, 1, 2, 3, 4, 5$, $y$ equals respectively $0, \pm 3, \pm 4, \pm 4.5, \pm 4.9, \pm 5, \pm 4.9, \pm 4.5, \pm 4, \pm 3, 0$.

Locating the points $(-5, 0)$, $(-4, +3)$, $(-4, -3)$, etc., and joining, we obtain the graph (a circle) $ABC$ of the equation $x^2 + y^2 = 25$.

Locating two points of equation (2), e.g. $(-2, 0)$ and $(0, 3)$, and joining by a straight line, we obtain $DE$, the graph of $3\,x - 2\,y = -6$.

Since the two graphs meet in two points $P$ and $Q$, there are two pairs of roots. By measuring the coördinates of $P$ and $Q$ we find :

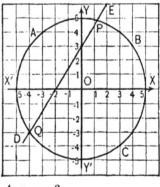

$$x = 1\tfrac{1}{2}, \ y = 4\tfrac{2}{3}, \text{ or } x = -4, \ y = -3.$$

**204. Inconsistent equations.** The equations

$$\begin{cases} 2\,x + y - 2 = 0, & (1) \\ 2\,x + y - 4 = 0, & (2) \end{cases}$$

cannot be satisfied by the same values of $x$ and $y$, i.e. they are *inconsistent*. This is clearly shown by the graphs of (1) and (2), which consist of a pair of parallel lines. There can be no point of intersection, and hence no roots.

In general, parallel graphs indicate inconsistent equations.

**205.** Dependent equations, as

$$\frac{x}{2}+\frac{y}{3}=1$$

and $\qquad 3x+2y=6$

have identical graphs, and, *vice versa*, identical graphs indicate dependent equations.

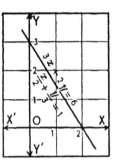

**EXERCISE 75**

Construct the loci of the following equations:

1. $x+y=6$.  3. $2x-3y=6$.  5. $y=-4$.  7. $y=x^2+2x-8$.

2. $2x-y=6$.  4. $x-y=0$.  6. $y=x^2+5$.  8. $x^2-4x+y=0$.

Draw the graphs of the following systems, and solve each system, if possible. If there are no solutions, state reasons.

9. $\begin{cases} 2x+4y=8, \\ 4x-2y=6. \end{cases}$

10. $\begin{cases} x+y=5, \\ x-y=-1. \end{cases}$

11. $\begin{cases} 4x+2y=10, \\ x-y=4. \end{cases}$

12. $\begin{cases} 2x+3y=6, \\ 2x-y=2. \end{cases}$

13. $\begin{cases} 2x+y+4=0, \\ 2x-3y=4. \end{cases}$

14. $\begin{cases} 3x-2y=6, \\ 5x+4y=21. \end{cases}$

15. $\begin{cases} 2x-3y=9, \\ 3x+y=8. \end{cases}$

16. $\begin{cases} 4x-3y=5, \\ 5x-2y=15. \end{cases}$

17. $\begin{cases} 2x+5y=1, \\ 6x+7y=3. \end{cases}$

18. $\begin{cases} \frac{x}{2}+\frac{y}{3}=1, \\ 3x+2y=12. \end{cases}$

19. $\begin{cases} 4x-5y=1, \\ 5x+2y=26. \end{cases}$

20. $\begin{cases} x+2y=7, \\ 2x+4y=5. \end{cases}$

21. $\begin{cases} 6x+4y=19, \\ 4x+10y=31. \end{cases}$

22. $\begin{cases} x+1=2y, \\ 5x-2y=16. \end{cases}$

23. $\begin{cases} 5x-2y=11, \\ 2x+3y=12. \end{cases}$

24. $\begin{cases} 6\,x + 4\,y = 13, \\ 4\,x - 14\,y = -8. \end{cases}$
     26. $\begin{cases} 5\,x = 11 + 2\,y, \\ 4\,x = 3(6 - y). \end{cases}$

25. $\begin{cases} 2\,x = y + 4, \\ 2\,y = 4\,x - 8. \end{cases}$
     27. $\begin{cases} 2\,x + 10\,y - 29 = 0, \\ 2(x + 3\,y) = 19. \end{cases}$

28. $\begin{cases} 3\,x + 2\,y = 6, \\ \dfrac{x}{2} = 1 - \dfrac{y}{3}. \end{cases}$
   29. $\begin{cases} x^2 - 2\,x - 3 = y, \\ 2\,x + y = 1. \end{cases}$
   30. $\begin{cases} y = x^2, \\ y - x - 6 = 0. \end{cases}$

**31.** Show that the same values of $x$ and $y$ cannot satisfy the three equations:

$$\begin{cases} x - y = 5, \\ 2\,x + y = 2, \\ x + 5\,y = -5. \end{cases}$$

TABLE OF SQUARES, CUBES, AND SQUARE ROOTS

| $x$ | $x^2$ | $x^3$ | $\sqrt{x}$ | $x$ | $x^2$ | $x^3$ | $\sqrt{x}$ |
|---|---|---|---|---|---|---|---|
| 1 | 1 | 1 | 1.000 | 2 6 | 6 76 | 17 576 | 5.099 |
| 2 | 4 | 8 | 1.414 | 2 7 | 7 29 | 19 683 | 5.196 |
| 3 | 9 | 27 | 1.732 | 2 8 | 7 84 | 21 952 | 5.292 |
| 4 | 16 | 64 | 2.000 | 2 9 | 8 41 | 24 389 | 5.385 |
| 5 | 25 | 125 | 2.236 | 3 0 | 9 00 | 27 000 | 5.477 |
| 6 | 36 | 216 | 2.449 | 3 1 | 9 61 | 29 791 | 5.568 |
| 7 | 49 | 343 | 2.646 | 3 2 | 10 24 | 32 768 | 5.657 |
| 8 | 64 | 512 | 2.828 | 3 3 | 10 89 | 35 937 | 5.745 |
| 9 | 81 | 729 | 3.000 | 3 4 | 11 56 | 39 304 | 5.831 |
| 1 0 | 1 00 | 1 000 | 3.162 | 3 5 | 12 25 | 42 875 | 5.916 |
| 1 1 | 1 21 | 1 331 | 3.317 | 3 6 | 12 96 | 46 656 | 6.000 |
| 1 2 | 1 44 | 1 728 | 3.464 | 3 7 | 13 69 | 50 653 | 6.083 |
| 1 3 | 1 69 | 2 197 | 3.606 | 3 8 | 14 44 | 54 872 | 6.164 |
| 1 4 | 1 96 | 2 744 | 3.742 | 3 9 | 15 21 | 59 319 | 6.245 |
| 1 5 | 2 25 | 3 375 | 3.873 | 4 0 | 16 00 | 64 000 | 6.325 |
| 1 6 | 2 56 | 4 096 | 4.000 | 4 1 | 16 81 | 68 921 | 6.403 |
| 1 7 | 2 89 | 4 913 | 4.123 | 4 2 | 17 64 | 74 088 | 6.481 |
| 1 8 | 3 24 | 5 832 | 4.243 | 4 3 | 18 49 | 79 507 | 6.557 |
| 1 9 | 3 61 | 6 859 | 4.359 | 4 4 | 19 36 | 85 184 | 6.633 |
| 2 0 | 4 00 | 8 000 | 4.472 | 4 5 | 20 25 | 91 125 | 6.708 |
| 2 1 | 4 41 | 9 261 | 4.583 | 4 6 | 21 16 | 97 336 | 6.782 |
| 2 2 | 4 84 | 10 648 | 4.690 | 4 7 | 22 09 | 103 823 | 6.856 |
| 2 3 | 5 29 | 12 167 | 4.796 | 4 8 | 23 04 | 110 592 | 6.928 |
| 2 4 | 5 76 | 13 824 | 4.899 | 4 9 | 24 01 | 117 649 | 7.000 |
| 2 5 | 6 25 | 15 625 | 5.000 | 5 0 | 25 00 | 125 000 | 7.071 |

# CHAPTER XIII

## INVOLUTION

**206. Involution** is the operation of raising a quantity to a positive integral power.

To find $(3a^5b)^n$ is a problem of involution. Since a power is a special kind of product, involution may be effected by repeated multiplication.

**207. Law of Signs.** According to § 50,

$$+ a \cdot + a \cdot + a = + a^3.$$
$$- a \cdot - a \cdot = + a^2.$$
$$- a \cdot - a \cdot - a = - a^3, \text{ etc.}$$

Obviously it follows that

1. *All powers of a positive quantity are positive.*
2. *All even powers of a negative quantity are positive.*
3. *All odd powers of a negative quantity are negative.*

$(- a)^6$ is positive, $(- ab^2)^9$ is negative.

## INVOLUTION OF MONOMIALS

**208.** According to § 52,

1. $(a^2)^3 = a^2 \cdot a^2 \cdot a^2 = a^{2+2+2} = a^6.$

2. $(b^5)^4 = b^5 \cdot b^5 \cdot b^5 \cdot b^5 = b^{5+5+5+5} = b^{20}.$

3. $(a^n)^m = a^n \cdot a^n \cdots \text{ to } m \text{ factors}$
   $$= a^{n+n+n+n} \text{ to } m \text{ terms}$$
   $$= a^{mn}.$$

4. $(- 3 a^2b^3)^4 = (- 3 a^2b^3) \cdot (- 3 a^2b^3) \cdot (- 3 a^2b^3) \cdot (- 3 a^2b^3)$
   $$= 81 a^8b^{12}.$$

5. $\left( - \dfrac{2\, m^2}{3\, n^5} \right)^3 = - \dfrac{(2\, m^2)^3}{(3\, n^5)^3} = - \dfrac{8\, m^6}{27\, n^{15}}.$

*To find the exponent of the power of a power, multiply the given exponents.*

*To raise a product to a given power, raise each of its factors to the required power.*

*To raise a fraction to a power, raise its terms to the required power.*

### EXERCISE 76

Perform the operations indicated:

1. $(x^3)^5$.
2. $(-x^2)^4$.
3. $(-a^2)^5$.
4. $(ab^2)^{11}$.

5. $(a^3b^2c^4)^5$.
6. $(-2\,mn^4)^3$.
7. $(-5\,a)^3$.
8. $(-2\,x^2y^2z^2)^6$.
9. $(-x^2y^3)^9$.
10. $(-2\,x^7y^2)^4$.
11. $(3\,xy^2z^3)^4$.
12. $(-a^4b^9c^{10})^{15}$.
13. $(-11\,a^2b^3c)^2$
14. $(-3\,a^{10}x^{20}y)^3$.
15. $(\tfrac{1}{2}\,m^3n^3p^3)^3$.
16. $(-\tfrac{2}{3}\,x^2y^2z)^2$.

17. $\left(\dfrac{2\,m}{3}\right)^4$.
18. $\left(-\dfrac{2\,m}{3}\right)^3$.
19. $\left(-\dfrac{2\,x}{3\,y}\right)^4$.
20. $\left(\dfrac{11\,x^2y^2z^3}{5\,a^4}\right)^2$.
21. $\left(\dfrac{a^7b^8c^9}{x^{10}y^{11}}\right)^{10}$.
22. $\left(\dfrac{-2\,mn}{3\,x^2y}\right)^4$.
23. $\left(\dfrac{4\,a^2b^2c^2}{3\,xy}\right)^3$.

24. $\left(-\dfrac{7\,x^{10}}{5\,a^4b^3}\right)^3$.
25. $\left(\dfrac{-1}{5\,x^3y^3z^3}\right)^3$.
26. $\left(\dfrac{-1}{a^{11}}\right)^{11}$.
27. $\left(-\dfrac{4}{a^4b^9}\right)^3$.
28. $\left(\dfrac{-1}{a^{10}b^{11}}\right)^4$.
29. $(a^m)^4$.
30. $(-a^m)^6$.
31. $\left(\dfrac{2^3}{3^2}\right)^2$.

## INVOLUTION OF BINOMIALS

**209.** The **square of a binomial** was discussed in § 63.

$$(a+b)^2 = a^2 + 2\,ab + b^2.$$

**210.** The **cube of a binomial** we obtain by multiplying $(a+b)^2$ by $a+b$.

$$(a+b)^3 = a^3 + 3\,a^2b + 3\,ab^2 + b^3,$$

and
$$(a-b)^3 = a^3 - 3\,a^2b + 3\,ab^2 - b^3.$$

**Ex. 1.** Find the cube of $2x + 3y$.

$$(2x + 3y)^3 = (2x)^3 + 3(2x)^2(3y) + 3(2x)(3y)^2 + (3y)^3$$
$$= 8x^3 + 36x^2y + 54xy^2 + 27y^3.$$

**Ex. 2.** Find the cube of $3x^2 - y^n$.

$$(3x^2 - y^n)^3 = (3x^2)^3 - 3(3x^2)^2(y^n) + 3(3x^2)(y^n)^2 - (y^n)^3$$
$$= 27x^6 - 27x^4y^n + 9x^2y^{2n} - y^{3n}.$$

**EXERCISE 77**

Perform the operations indicated:

| | | |
|---|---|---|
| 1. $(a + b)^3$. | 7. $(5 - x)^3$. | 13. $(3a + 2b)^3$. |
| 2. $(x - y)^3$. | 8. $(1 + 2x)^3$. | 14. $(5m + 2n)^3$. |
| 3. $(a + 1)^3$. | 9. $(3a - 1)^3$. | 15. $(3a - 5b)^3$. |
| 4. $(m - 2)^3$. | 10. $(1 + 4x)^3$. | 16. $(3a^2b^2 - 1)^3$. |
| 5. $(m + n)^3$. | 11. $(7a^2 - 1)^3$. | 17. $(am + bn)^3$. |
| 6. $(a + 7)^3$. | 12. $(1 + 5a^3)^3$. | 18. $(4x^2 - 5y^2)^3$. |

Find the cube root of:

19. $a^3 + 3a^2b + 3ab^2 + b^3$.      22. $1 + 3m + 3m^2 + m^3$.

20. $x^3 - 3x^2y + 3xy^2 - y^3$.      23. $8b^3 - 12b^2 + 6b - 1$.

21. $a^3 - 3a^2 + 3a - 1$.

**211.** The **higher powers of binomials**, frequently called ex-pansions, are obtained by multiplication, as follows:

$$(a + b)^2 = a^2 + 2ab + b^2.$$
$$(a + b)^3 = a^3 + 3a^2b + 3ab^2 + b^3.$$
$$(a + b)^4 = a^4 + 4a^3b + 6a^2b^2 + 4ab^3 + b^4.$$
$$(a + b)^5 = a^5 + 5a^4b + 10a^3b^2 + 10a^2b^3 + 5ab^4 + b^5, \text{ etc.}$$

An examination of these results shows that:

1. *The number of terms is 1 greater than the exponent of the binomial.*

2. *The exponent of a in the first term is the same as the exponent of the binomial, and decreases in each succeeding term by 1.*

3. *The exponent of b is 1 in the second term of the result, and increases by 1 in each succeeding term.*

4. *The coefficient of the first term is 1.*

5. *The coefficient of the second term equals the exponent of the binomial.*

6. *The coefficient of any term of the power multiplied by the exponent of a, and the result divided by 1 plus the exponent of b, is the coefficient of the next term.*

Ex. 1. Expand $(x + y)^5$.

$$(x+y)^5 = x^5 + 5x^4y + \frac{5\cdot 4}{1\cdot 2}x^3y^2 + \frac{5\cdot 4\cdot 3}{1\cdot 2\cdot 3}x^2y^3 + \frac{5\cdot 4\cdot 3\cdot 2}{1\cdot 2\cdot 3\cdot 4}xy^4 + \frac{5\cdot 4\cdot 3\cdot 2\cdot 1}{1\cdot 2\cdot 3\cdot 4\cdot 5}y^5$$
$$= x^5 + 5x^4y + 10x^3y^2 + 10x^2y^3 + 5xy^4 + y^5.$$

Ex. 2. Expand $(x - y)^5$.

$$(x-y)^5 = x^5 + 5x^4(-y) + 10x^3(-y)^2 + 10x^2(-y)^3 + 5x(-y)^4 + (-y)^5$$
$$= x^5 - 5x^4y + 10x^3y^2 - 10x^2y^3 + 5xy^4 - y^5.$$

**212.** The signs of the last answer are alternately plus and minus, since the even powers of $-y$ are positive, and the odd powers negative.

Ex. 3. Expand $(2x^2 - 3y^3)^4$.

$$(2x^2 - 3y^3)^4 = (2x^2)^4 - 4(2x^2)^3(3y^3) + 6(2x^2)^2(3y^3)^2$$
$$- 4(2x^2)(3y^3)^3 + (3y^3)^4$$
$$= 16x^8 - 96x^6y^3 + 216x^4y^6 - 216x^2y^9 + 81y^{12}.$$

Expand:

### EXERCISE 78

1. $(p+q)^4$.
2. $(m-n)^4$.
3. $(x+1)^4$.
4. $(1+y)^4$.
5. $(m-1)^4$.
6. $(1-ab)^4$.

7. $(1+x^2)^4$.
8. $(a-b)^5$.
9. $(c+d)^5$.
10. $(m-n)^6$.
11. $(a+b)^7$.
12. $(a-1)^5$.

13. $(2+a)^5$.
14. $(m^3+1)^4$.
15. $(1-m^2n^2)^3$.
16. $(m^3-1)^5$.
17. $(m+n)^8$.
18. $(mn+c)^8$.

19. $(mnp-1)^5$.
20. $(2m^2+1)^5$.
21. $(3a^2+5)^4$.
22. $(2x^2-5)^4$.
23. $(2a-5c)^4$.
24. $(1+2x)^4$.

25. $(1+a^2b^2)^5$.

# CHAPTER XIV

## EVOLUTION

**213.** **Evolution** is the operation of finding a root of a quantity; it is the inverse of involution.

$$\sqrt[n]{a} = x \text{ means } x^n = a.$$

$$\sqrt[3]{-27} = y \text{ means } y^3 = -27, \text{ or } y = -3.$$

$$\sqrt[5]{b^{20}} = x \text{ means } x^5 = b^{20}, \text{ or } x = b^4.$$

**214.** It follows from the law of signs in evolution that:

**1.** *Any even root of a positive quantity may be either positive or negative.*

**2.** *Every odd root of a quantity has the same sign as the quantity.*

$\sqrt{9} = +3,$ or $-3$ (usually written $\pm 3$); for $(+3)^2$ and $(-3)^2$ equal 9.

$\sqrt[3]{-27} = -3,$ for $(-3)^3 = -27.$

$\sqrt[4]{a^4} = \pm a,$ for $(+a)^4 = a^4,$ and $(-a)^4 = a^4.$

$\sqrt[5]{32} = 2,$ etc.

**215.** Since even powers can never be negative, it is evidently impossible to express an even root of a negative quantity by the usual system of numbers. Such roots are called **imaginary numbers**, and all other numbers are, for distinction, called **real** numbers.

Thus $\sqrt{-1}$ is an imaginary number, which can be simplified no further.

160

## EVOLUTION OF MONOMIALS

The following examples are solved by the definition of a root:

**Ex. 1.** $\sqrt[4]{a^{12}} = \pm a^3$, for $(\pm a^3)^4 = a^{12}$.

**Ex. 2.** $\sqrt[n]{a^{mn}} = a^m$, for $(a^m)^n = a^{mn}$.

**Ex. 3.** $\sqrt[3]{8\,a^6b^3c^{12}} = 2\,a^2bc^4$, for $(2\,a^2bc^4)^3 = 8\,a^6b^3c^{12}$.

**Ex. 4.** $\sqrt[4]{\dfrac{81\,a^{12}b^4}{c^4}} = \pm \dfrac{3\,a^3b}{c}$, for $\left(\pm \dfrac{3\,a^3b}{c}\right)^4 = \dfrac{81\,a^{12}b^4}{c^4}$.

**Ex. 5.** $\sqrt[5]{-\dfrac{32\,a^{15}}{243\,b^{10}c^{20}}} = -\dfrac{2\,a^3}{3\,b^2c^4}$, for $\left(-\dfrac{2\,a^3}{3\,b^2c^4}\right)^5 = -\dfrac{32\,a^{15}}{243\,b^{10}c^{20}}$.

**216.** *To extract the root of a power, divide the exponent by the index.*

*A root of a product equals the product of the roots of the factors.*

*To extract a root of a fraction, extract the roots of the numerator and denominator.*

**Ex. 6.** $\sqrt{18 \cdot 14 \cdot 63 \cdot 25} = \sqrt{2 \cdot 3^2 \cdot 2 \cdot 7 \cdot 7 \cdot 3^2 \cdot 5^2} = \sqrt{2^2 \cdot 3^4 \cdot 5^2 \cdot 7^2}$
$$= 2 \cdot 3^2 \cdot 5 \cdot 7 = 630.$$

**Ex. 7.** $\sqrt{18225} = \sqrt{25 \cdot 729} = \sqrt{25 \cdot 9 \cdot 81} = 5 \cdot 3 \cdot 9 = 135.$

**Ex. 8.** Find $(\sqrt{19472})^2$.

Since by definition $(\sqrt[n]{a})^n = a$, we have $(\sqrt{19472})^2 = 19472$.

**Ex. 9.** $(\sqrt{199})^2 + (\sqrt[3]{-198})^3 - (\sqrt[4]{200})^4 - (\sqrt[5]{-201})^5$
$$= 199 + (-198) - 200 - (-201) = 2.$$

### EXERCISE 79

1. $\sqrt[3]{2^3}$.    3. $\sqrt[3]{-5^3}$.    5. $\sqrt{5^2 \cdot 7^2}$.      7. $\sqrt{25 \cdot 9 \cdot 16}$.

2. $\sqrt{3^4}$.    4. $\sqrt[6]{2^{12}}$.    6. $\sqrt[3]{2^3 \cdot 3^3 \cdot 5^3}$.    8. $\sqrt[3]{-125 \cdot 64}$.

9. $\sqrt{36 \cdot 9 \cdot 100 \cdot a^2}$.        10. $\sqrt[4]{2^4 \cdot 9^4 \cdot 5^4}$.

11. $\sqrt{x^4}$.

12. $\sqrt[3]{m^{12}}$.

13. $\sqrt{5^4 \cdot a^2}$.

14. $\sqrt[3]{7^6 \cdot a^3}$.

15. $\sqrt{100\,x^2}$.

16. $\sqrt[3]{-1000}$.

17. $\sqrt[3]{27\,a^6}$.

18. $\sqrt[4]{16 \cdot x^4 y^8}$.

19. $\sqrt[10]{a^{20} b^{30}}$.

20. $\sqrt[4]{16\,m^4 n^8}$.

21. $\sqrt[5]{\dfrac{-32\,a^5}{b^5}}$.

22. $\sqrt[4]{\dfrac{a^{16}}{81\,b^4}}$.

23. $\sqrt[3]{\dfrac{27 \cdot 64 \cdot 125}{1000}}$.

24. $\sqrt[7]{\dfrac{a^{14} b^{49}}{c^{70}}}$.

25. $\sqrt[5]{(-x)^5}$.

26. $\sqrt{(x+y)^2}$.

27. $\sqrt[3]{8\,(a+b)^3}$.

28. $\sqrt{a^2 + 2\,ab + b^2}$.

29. $\sqrt{8 \cdot 75 \cdot 98 \cdot 3}$.

30. $\sqrt{20 \cdot 45 \cdot 9}$.

31. $\sqrt{5184}$.

32. $\sqrt{9216}$.

33. $(\sqrt{11})^2 + (\sqrt{19})^2 - (\sqrt{200})^2 + (\sqrt{240})^2$.

34. $(\sqrt{15})^2 \times (\sqrt{17})^2 + (\sqrt{15})^2 \times (\sqrt{3})^2$.

35. $(\sqrt{2441})^2 - (\sqrt{2401})^2$.

36. $(\sqrt{124})^2 - (\sqrt[3]{120})^3 + (\sqrt[7]{19})^7$.

## EVOLUTION OF POLYNOMIALS AND ARITHMETICAL NUMBERS

**217. A trinomial is a perfect square** if one of its terms is equal to twice the product of the square roots of the other terms. (§ 116.) In such a case the square root can be found by inspection.

Ex. 1. Find the square root of $x^6 - 6\,x^3 y^2 + 9\,y^4$.

$$x^6 - 6\,x^3 y^2 + 9\,y^4 = (x^3 - 3\,y^2)^2. \quad (\text{§ 116.})$$

Hence $\sqrt{x^6 - 6\,x^3 y^2 + 9\,y^4} = \pm\,(x^3 - 3\,y^2)$.

### EXERCISE 80

Extract the square roots of the following expressions:

1. $a^2 + 2\,a + 1$.

2. $1 - 2\,y + y^2$.

3. $x^2 - 4\,xy + 4\,y^2$.

4. $9\,x^2 + 6\,xy + y^2$.

5. $x^4 + 1 - 2\,x^2$.

6. $1 + 64\,x^4 + 16\,x^2$.

7. $4\,a^2 - 44\,ab + 121\,b^2.$　　10. $a^4 + b^8 - 2\,a^2b^4.$

8. $4\,a^2 + b^2 + 4\,ab.$　　11. $49\,a^8 - 42\,a^4b^4 + 9\,b^8.$

9. $m^2n^2 - 14\,mnp + 49\,p^2.$　　12. $16\,a^4 - 72\,a^2b^2 + 81\,b^4.$

13. $x^2 + y^2 + z^2 + 2\,xy + 2\,xz + 2\,yz.$

14. $x^2 + y^2 + 1 + 2\,xy + 2\,x + 2\,y.$

15. $a^2 + b^2 + c^2 - 2\,ab + 2\,ac - 2\,bc.$

16. $a^2 + b^2 + c^2 + 2\,ab - 2\,ac - 2\,bc.$

---

**218.** In order to find a general method for extracting the **square root of a polynomial**, let us consider the relation of $a + b$ to its square, $a^2 + 2\,ab + b^2.$

The first term $a$ of the root is the square root of the first term $a^2.$

The second term of the root can be obtained by dividing the second term $2\,ab$ by the double of $a$, the so-called trial divisor;

$$\frac{2\,ab}{2\,a} = b.$$

$a + b$ is the root if the given expression is a perfect square. In most cases, however, it is not known whether the given expression is a perfect square, and we have then to consider that $2\,ab + b^2 = b(2\,a + b)$, *i.e.* the sum of trial divisor $2\,a$, and $b$, multiplied by $b$ must give the last two terms of the square.

The work may be arranged as follows:

$$a^2 + 2\,ab + b^2\,\lfloor\,a + b$$
$$a^2$$
$$2\,a + b\,\lvert\,2\,ab + b^2$$
$$\lfloor\,2\,ab + b^2$$

**Ex. 1.** Extract the square root of $16\,x^4 - 24\,x^2y^3 + 9\,y^6$.

$$16\,x^4 - 24\,x^2y^3 + 9\,y^6 \;\underline{|\,4\,x^2 - 3\,y^3}$$
$$16\,x^4$$
$$8\,x^2 - 3\,y^3\,\underline{|\;-24\,x^2y^3 + 9\,y^6}$$
$$\underline{-24\,x^2y^3 + 9\,y^6}$$

*Explanation.* Arrange the expression according to descending powers of $x$. The square root of $16\,x^4$ is $4\,x^2$, the first term of the root. Subtracting the square of $4\,x^2$ from the trinomial gives the remainder $-24\,x^2y^3 + 9\,y^6$. By doubling $4\,x^2$, we obtain $8\,x^2$, the trial divisor. Dividing the first term of the remainder, $-24\,x^2y^3$, by the trial divisor $8\,x^2$, we obtain the next term of the root $-3\,y^3$, which has to be added to the trial divisor. Multiply the complete divisor $8\,x^2 - 3\,y^3$ by $-3\,y^3$, and subtract the product from the remainder. As there is no remainder, $4\,x^2 - 3\,y^3$ is the required square foot.

**219.** The process of the preceding article can be extended to polynomials of more than three terms. We find the first two terms of the root by the method used in Ex. 1, and consider their sum one term, the first term of the answer. Hence the double of this term is the new trial divisor; by division we find the next term of the root, and so forth.

**Ex. 2.** Extract the square root of
$$16\,a^4 - 24\,a^3 + 4 - 12\,a + 25\,a^2.$$

Arranging according to descending powers of $a$.

$$\underline{|\,4\,a^2 - 3\,a + 2}$$
$$16\,a^4 - 24\,a^3 + 25\,a^2 - 12\,a + 4$$

Square of $4\,a^2$. $\qquad 16\,a^4$

First remainder. $\qquad\quad -24\,a^3 + 25\,a^2 - 12\,a + 4$

First trial divisor, $8\,a^2$.

First complete divisor, $8\,a^2 - 3\,a$. $\quad -24\,a^3 + 9\,a^2$

Second remainder. $\qquad\qquad +16\,a^2 - 12\,a + 4$

Second trial divisor, $8\,a^2 - 6\,a$.

Second complete divisor, $8\,a^2 - 6\,a + 2$. $\quad +16\,a^2 - 12\,a + 4$

As there is no remainder, the required root is $\pm(4\,a^2 - 3\,a + 2)$.

### EXERCISE 81

Extract the square roots of the following expressions:

1. $x^4 - 6 x^3 + 13 x^2 - 12 x + 4.$
2. $3 a^2 - 2 a + a^4 - 2 a^3 + 1.$
3. $x^4 + 2 x^3 + 1 - x^2 - 2 x.$
4. $16 a^4 + 24 a^3 + 81 a^2 + 54 a + 81.$
5. $25 m^4 - 20 m^3 + 34 m^2 - 12 m + 9.$
6. $4 - 12 ab + 37 a^2b^2 - 42 a^3b^3 + 49 a^4b^4.$
7. $25 x^4 + 40 x^3y + 46 x^2y^2 + 24 xy^3 + 9 y^4.$
8. $16 x^6 + 73 x^4 + 40 x^5 + 36 x^2 + 60 x^3.$
9. $1 + 2 x + 3 x^2 + 2 x^3 + x^4.$
10. $1 + 4 x + 10 x^2 + 20 x^3 + 25 x^4 + 24 x^5 + 16 x^6.$
11. $36 a^6 + 60 a^5 + 73 a^4 + 40 a^3 + 16 a^2.$
12. $36 x^6 + 36 x^5y + 69 x^4y^2 + 30 x^3y^3 + 25 x^2y^4.$
13. $4 m^6 + 12 m^5 + 9 m^4 - 20 m^3 - 30 m^2 + 25.$
14. $49 a^4 - 42 a^3b + 37 a^2b^2 - 12 ab^3 + 4 b^4.$
15. $x^6 + 4 x^5 + 20 x^2 - 16 x + 16.$
16. $13 x^4 + 13 x^2 + 4 x^6 - 14 x^3 + 4 - 4 x - 12 x^5.$
17. $x^4 + y^4 + 2 x^3y - 2 xy^3 - x^2y^2.$
18. $729 + 162 a^2 - 6 a^5 + a^6 - 54 a^3 + 9 a^4.$
19. $60 a^{10} + 73 a^8 + 36 a^{12} + 40 a^6 + 16 a^4.$
20. $46 a^4 + 44 a^3 + 25 a^2 + 12 a + 4 + 25 a^6 + 40 a^5.$
21. $\dfrac{z^2}{25} + \dfrac{yz}{10} + \dfrac{y^2}{16} + \dfrac{2 xz}{15} + \dfrac{xy}{6} + \dfrac{x^2}{9}.$
22. $16 - \dfrac{24}{x} + \dfrac{25}{x^2} - \dfrac{12}{x^3} + \dfrac{4}{x^4}.$
23. $1 + \dfrac{4}{x} + \dfrac{10}{x^2} + \dfrac{20}{x^3} + \dfrac{25}{x^4} + \dfrac{24}{x^5} + \dfrac{16}{x^6}.$
24. $\dfrac{36}{x^2} + \dfrac{36}{x} + 69 + 30 x + 25 x^2.$

**220.** The **square root of arithmetical numbers** can be found by a method very similar to the one used for algebraic expressions.

Since the square root of 100 is 10; of 10,000 is 100; of 1,000,000 is 1000, etc., the integral part of the square root of a number less than 100 has one figure, of a number between 100 and 10,000, two figures, etc. Hence if we divide the digits of the number into groups, beginning at the units, and each group contains two digits (except the last, which may contain one or two), then the number of groups is equal to the number of digits in the square root, and the square root of the greatest square in the first group is the first digit in the root. Thus the square root of $96'04'$ consists of two digits, the first of which is 9; the square root of $21'06'81$ has three digits, the first of which is 4.

## Ex. 1. Find the square root of 7744.

From the preceding explanation it follows that the root has two digits, the first of which is 8. Hence the root is 80 plus an unknown number, and we may apply the method used in algebraic process.

A comparison of the algebraical and arithmetical method given below will show the identity of the methods.

$$
\begin{array}{ll}
\underline{a^2 + 2\,ab + b^2}\,\lfloor a+b \quad & 7744\,\lfloor 80 + 8 \\
\quad a^2 & \quad 6400 \\
\hline
2a+b\,\lfloor 2\,ab + b^2 & 160 + 8 = 168\,\lceil 1344 \\
\qquad\lfloor 2\,ab + b^2 & \qquad\qquad\qquad\lfloor 1344
\end{array}
$$

*Explanation.* Since $a = 80$, $a^2 = 6400$, and the first remainder is 1344. The trial divisor $2\,a = 160$. Therefore $b = 8$, and the complete divisor is 168.

As $8 \times 168 = 1344$, the square root of 7744 equals 88.

## Ex. 2. Find the square root of 524,176.

$$
\begin{array}{rl}
 & \quad\ \ a \quad\ \ b \quad\ c \\
 & 52'41'76\,\lfloor 700 + 20 + 4 = 724 \\
a^2 = & \quad 49\ 00\ 00 \\
\hline
2\,a + b = 1400 + 20 = 1420\,\lceil & 3\ 41\ 76 \\
 & 2\ 84\ 00 \\
\hline
2(a+b) + c = 1440 + \ \ 4 = 1444\,\lceil & 57\ 76 \\
 & 57\ 76
\end{array}
$$

**221.** In marking off groups in a number which has decimal places, we must begin at the decimal point, and if the right-hand group contains only one digit, annex a cipher.

Thus the groups in .0961 are ′.09′61. The groups of 16724.1 are 1′67′24.10.

Ex. **3.** Find the square root of 6.7 to three decimal places.

$$
\begin{array}{r}
6'.70 \ \lfloor 2.588 \\
4 \\
\hline
45 | \ 2 \ 70 \\
\underline{2 \ 25} \\
508 | 4500 \\
\underline{4064} \\
5168 | 43600 \\
\underline{41344} \\
2256
\end{array}
$$

**222. Roots of common fractions** are extracted either by dividing the root of the numerator by the root of the denominator, or by transforming the common fraction into a decimal.

$$\sqrt{\tfrac{4}{9}} = \pm \tfrac{2}{3}; \quad \sqrt{\tfrac{2}{5}} = \sqrt{.4}.$$

### EXERCISE 82

Extract the square roots of :

| | | |
|---|---|---|
| 1. 5625. | 11. 95,481. | 21. 4,153,444. |
| 2. 4096. | 12. 61,009. | 22. 57,198,969. |
| 3. 3249. | 13. 582,169. | 23. 25,836,889. |
| 4. 5041. | 14. 956,484. | 24. 43,046,721. |
| 5. 7056. | 15. 8.0089. | 25. 236,144,689. |
| 6. 9604. | 16. 42.25. | 26. $42\tfrac{1}{4}$. |
| 7. 9801. | 17. 72.25. | 27. $1\tfrac{11}{25}$. |
| 8. 14,161. | 18. .8836. | 28. $\tfrac{625}{4096}$. |
| 9. 10,201. | 19. .001369. | |
| 10. 56,169. | 20. 1,555,009. | |

Find to three decimal places the square roots of the following numbers:

**29.** 5.      **31.** .22.      **33.** 1.01.      **35.** $\frac{5}{12}$.

**30.** 13.      **32.** 1.53.      **34.** $\frac{7}{8}$.      **36.** $\frac{1}{17}$.

**37.** Find the side of a square whose area equals 56.58 square feet.

**38.** Find the side of a square whose area equals 96 square yards.

**39.** Find the radius of a circle whose area equals 48.4 square feet. (Area of a circle equals $\pi R^2$, when $R =$ radius and $\pi = 3.1416$.)

**40.** Find the mean proportional between 2 and 11.

# CHAPTER XV

## QUADRATIC EQUATIONS INVOLVING ONE UNKNOWN QUANTITY

**223.** A **quadratic equation,** or equation of the second degree, is an integral rational equation that contains the square of the unknown number, but no higher power; *e.g.* $x^2 - 4x = 7$, $6y^2 = 17$, $ax^2 + bx + c = 0$.

**224.** A **complete,** or affected, **quadratic equation** is one which contains both the square and the first power of the unknown quantity.

**225.** A **pure,** or incomplete, **quadratic equation** contains only the square of the unknown quantity.

$ax^2 + bx + c = 0$ is a complete quadratic equation.
$ax^2 = m$ is a pure quadratic equation.

**226.** The **absolute term** of an equation is the term which does not contain any unknown quantities.

In $4x^2 - 7x + 12 = 0$ the absolute term is 12.

### PURE QUADRATIC EQUATIONS

**227.** A **pure quadratic** is solved by reducing it to the form $x^2 = a$, and extracting the square root of both members.

**Ex. 1.** Solve $13x^2 - 19 = 7x^2 + 5$.

Transposing, etc., $\qquad 6x^2 = 24.$
Dividing, $\qquad\qquad x^2 = 4.$

Extracting the square root of each member,
$$x = +2 \text{ or } x = -2.$$
This answer is frequently written $x = \pm 2$.

*Check.* $\qquad 13(\pm 2)^2 - 19 = 33; \ 7(\pm 2)^2 + 5 = 33.$

Ex. 2.   Solve $\dfrac{-ax}{a+x} = \dfrac{x+4a}{x-4a}$.

Clearing of fractions, $ax - x^2 - 4a^2 + 4ax = ax + 4a^2 + x^2 + 4ax$.

Transposing and combining,   $\qquad -2x^2 = 8a^2$.

Dividing by $-2$,   $\qquad\qquad\qquad x^2 = -4a^2$.

Extracting the square root,   $\qquad\quad x = \pm\sqrt{-4a^2}$,

or   $\qquad\qquad\qquad\qquad\qquad\quad x = \pm\sqrt{4a^2}\cdot\sqrt{-1}$.

Therefore,   $\qquad\qquad\qquad\qquad x = \pm 2a\sqrt{-1}$.

### EXERCISE 83

Solve the following equations:

1.  $x^2 - 7 = 162$.

2.  $x^2 + 1 = 1.25$.

3.  $19x^2 + 9 = 5500$.

4.  $16x^2 - 393 = 7$.

5.  $15x^2 - 5 = 9x^2 + 19$.

6.  $(x + \tfrac{1}{3})(x - \tfrac{1}{3}) = \tfrac{8}{9}$.

7.  $(x + 2)^2 + (x - 2)^2 = 26$.

8.  $(x - 2)(x - 3) + (x + 6)(x - 1) = 32$.

9.  $5x(x - 2) = -10(x - 1)$.

11.  $\dfrac{x+4}{x-4} + \dfrac{x-4}{x+4} = 9\tfrac{1}{3}$.

10.  $\dfrac{x}{x-3} + \dfrac{x}{x+3} = 4$.

12.  $\dfrac{5}{2x^2} - \dfrac{4}{3} = \dfrac{7}{4x^2}$.

13.  $\dfrac{4(x^2 - 5)}{3} - \dfrac{1}{12} = 20 + \dfrac{3(25 - x^2) + 10}{4}$.

14.  $\dfrac{1+x}{1-x} = \dfrac{x+25}{x-25}$.

15.  $(4 + x)(5 - 7x) + 23 = (9 + 2x)(2 - 3x)$.

16.  $\dfrac{12}{x-5} + \dfrac{4}{x+5} = 9 + \dfrac{8(2x - 3)}{x^2 - 25}$.

17.  $ax^2 = b$.

18.  $\dfrac{a^2 - x^2}{x^2 - b^2} = \dfrac{a}{b}$.

19.  $\dfrac{a^2 - x^2}{b^2} + \dfrac{b^2 - x^2}{a^2} + \dfrac{2x^2}{ab} = 2$.

20.  $\dfrac{a+x}{a-x} = \dfrac{x+b}{x-b}$.

21. $x = \dfrac{a + bx}{b + cx}$.     22. $\dfrac{x + a}{x - a} + \dfrac{x - a}{x + a} - 2 = \dfrac{4 a^2}{2 a + 1}$.

23. If $a^2 + b^2 = c^2$, find $a$ in terms of $b$ and $c$.

24. If $s = \dfrac{g}{2} t^2$, solve for $t$.

25. If $s = \pi r^2$, solve for $r$.     26. If $s = 4 \pi r^2$, solve for $r$.

27. If $2 a^2 + 2 b^2 = 4 m^2 + c^2$, solve for $m$.

28. If $E = \dfrac{mv^2}{2}$, solve for $v$.     29. If $G = \dfrac{m \cdot m'}{d^2}$, solve for $d$.

### EXERCISE 84

**1.** Find a positive number which is equal to its reciprocal (§ 144).

**2.** A number multiplied by its fifth part equals 45. Find the number.

**3.** The ratio of two numbers is $2 : 3$, and their product is 150. Find the numbers. (See § 168.)

**4.** Three numbers are to each other as $1 : 2 : 3$, and the sum of their squares is 56. Find the numbers.

**5.** The sides of two square fields are as $3 : 5$, and they contain together 306 square feet. Find the side of each field.

**6.** The sides of two square fields are as $7 : 2$, and the first exceeds the second by 405 square yards. Find the side of each field.

---

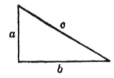

**228.** A *right triangle* is a triangle, one of whose angles is a right angle. The side opposite the right angle is called the *hypotenuse* ($c$ in the diagram). If the hypotenuse contains $c$ units of length, and the two other sides respectively $a$ and $b$ units, then $\qquad c^2 = a^2 + b^2$.

Since such a triangle may be considered one half of a rectangle, its area contains $\dfrac{ab}{2}$ square units.

**7.** The hypotenuse of a right triangle is 35 inches, and the other two sides are as $3:4$. Find the sides.

**8.** The hypotenuse of a right triangle is to one side as $13:12$, and the third side is 15 inches. Find the unknown sides and the area.

**9.** The hypotenuse of a right triangle is 2, and the other two sides are equal. Find these sides.

**10.** The area of a right triangle is 24, and the two smaller sides are as $3:4$. Find these sides.

**11.** A body falling from a state of rest, passes in $t$ seconds over a space $s = \frac{1}{2}gt^2$. Assuming $g = 32$ feet, in how many seconds will a body fall (*a*) 64 feet, (*b*) 100 feet?

**12.** The area $S$ of a circle whose radius equals $r$ is found by the formula $S = \pi r^2$. Find the radius of circle whose area $S$ equals (*a*) 154 square inches, (*b*) 44 square feet. (Assume $\pi = \frac{22}{7}$.)

**13.** Two circles together contain 3850 square feet, and their radii are as $3:4$. Find the radii.

**14.** If the radius of a sphere is $r$, its surface $S = 4\pi r^2$. Find the radius of a sphere whose surface equals 440 square yards. (Assume $\pi = \frac{22}{7}$.)

## COMPLETE QUADRATIC EQUATIONS

**229. Method of completing the square.** The following example illustrates the method of solving a complete quadratic equation by completing the square.

Solve $x^2 - 7x + 10 = 0$.

Transposing, $x^2 - 7x = -10$.

The left member can be made a complete square by adding another term. To find this term, let us compare $x^2 - 7x$ with the perfect square $x^2 - 2mx + m^2$. Evidently 7 takes the place of $2m$, or $m = \frac{7}{2}$. Hence to make $x^2 - 7x$ a complete square we have to add $(\frac{7}{2})^2$, which corresponds to $m^2$.

Adding $(\frac{7}{2})^2$ to each member,

$$x^2 - 7x + (\tfrac{7}{2})^2 = \tfrac{49}{4} - 10,$$

Or $$(x - \tfrac{7}{2})^2 = \tfrac{9}{4}.$$

Extracting square roots, $x - \frac{7}{2} = \pm \frac{3}{2}.$

Hence $$x = \tfrac{7}{2} \pm \tfrac{3}{2}.$$

Therefore $$x = 5 \text{ or } x = 2.$$

*Check.* $\qquad 5^2 - 7 \cdot 5 + 10 = 0,\ 2^2 - 7 \cdot 2 + 10 = 0.$

**Ex. 1.** Solve $9x^2 - 2 = 15x + 4.$

Transposing, $\qquad 9x^2 - 15x = 6.$

Dividing by 9, $\qquad x^2 - \frac{5}{3}x = \frac{2}{3}.$

Completing the square (*i.e.* adding $(\frac{5}{6})^2$ to each member),

$$x^2 - \tfrac{5}{3}x + (\tfrac{5}{6})^2 = \tfrac{2}{3} + \tfrac{25}{36}.$$

Simplifying, $\qquad (x - \tfrac{5}{6})^2 = \tfrac{49}{36}.$

Extracting square roots, $\quad x - \frac{5}{6} = \pm \frac{7}{6}.$

Transposing, $\qquad x = \frac{5}{6} \pm \frac{7}{6}.$

Therefore, $\qquad x = 2, \text{ or } -\frac{1}{3}.$

**230.** Hence to solve a complete quadratic :

*Reduce the equation to the form* $x^2 + px = q.$ *Complete the square by adding the square of one half the coefficient of* $x.$ *Extract the square root and solve the equation of the first degree thus formed.*

**Ex. 2.** Solve $\dfrac{x-1}{a} + \dfrac{2a+1}{x} = 2.$

Clearing of fractions, $\ x^2 - x + 2a^2 + a = 2ax.$

Transposing, $\qquad x^2 - x - 2ax = -2a^2 - a.$

Uniting, $\qquad x^2 - x(1 + 2a) = -2a^2 - a.$

Completing the square,

$$x^2 - x(1 + 2\,a) + \left(\frac{1 + 2\,a}{2}\right)^2 = -2\,a^2 - a + \frac{1 + 4\,a + 4\,a^2}{4}.$$

Simplifying, $\qquad \left(x - \frac{1 + 2\,a}{2}\right)^2 = \frac{1 - 4\,a^2}{4}.$

Extracting square root, $\quad x - \dfrac{1 + 2\,a}{2} = \pm\dfrac{1}{2}\sqrt{1 - 4\,a^2}.$

Transposing, $\qquad\qquad\qquad x = \dfrac{1 + 2\,a}{2} \pm \dfrac{1}{2}\sqrt{1 - 4\,a^2}.$

Therefore $\qquad\qquad\qquad\quad x = \dfrac{1 + 2\,a \pm \sqrt{1 - 4\,a^2}}{2}.$

## EXERCISE 85

Solve:

1. $x^2 - 8\,x = 20.$

2. $x^2 + 2\,x = 35.$

3. $x^2 + 6\,x = 27.$

4. $x^2 - 7 = 6\,x.$

5. $x^2 - 7\,x = 60.$

6. $x^2 + 28 = 11\,x.$

7. $x^2 - 2\,x - 2 = 0.$

8. $15 = 12\,x - x^2.$

9. $70 + x^2 = -17\,x.$

10. $x^2 - 7\,x + 3\frac{1}{4} = 0.$

11. $4\,x^2 - 23\,x = 72.$

12. $3\,x^2 - 2\,x = 65.$

13. $3\,x^2 + x = 4.$

14. $x = 1 - x^2.$

15. $4\,x^2 + 16\,x = 33.$

16. $9\,x^2 - 18\,x = 7.$

17. $2\,x^2 - 11\,x = 30.$

18. $x + \dfrac{11}{x} = 12.$

19. $\dfrac{x^2 - 21}{x} = 4.$

20. $\dfrac{x^2}{3} + 5 = 4\,x.$

21. $12 + \dfrac{77}{x} = -\dfrac{30}{x^2}.$

22. $\dfrac{84}{x + 5} = x.$

23. $(x - 11)(x + 3) = 0.$

24. $11\,x - 16 = \dfrac{42}{3\,x + 1}.$

25. $(x - 3)(x - 2) = 3(5\,x - 26).$

26. $x^2 + \frac{1}{2} = \frac{11}{2}(x + 2).$

27. $(3\,x - 4)(x - 2) = 5.$

28. $(x + 1)^2 + x^2 = 25.$

29. $(x + 1)^2 + (x + 2)^2 = 61.$

**30.** $(2x - 5)^2 - (x - 6)^2 = 80.$

**31.** $(x + 3)^2 + (x - 2)^2 = 37.$

**32.** $4x - \dfrac{36 - x}{x} = 46.$

**33.** $\dfrac{x}{x + 60} = \dfrac{7}{3x - 5}.$

**34.** $\dfrac{x + 3}{x - 1} = x.$

**35.** $\dfrac{x - 7}{x + 9} = \dfrac{1}{x}.$

**36.** $\dfrac{x + 11}{x + 3} = \dfrac{2x + 1}{x + 5}.$

**37.** $\dfrac{x + 1}{x + 4} = \dfrac{2x - 1}{x + 8}.$

**38.** $\dfrac{40}{x - 5} + \dfrac{27}{x} = 13.$

**39.** $\dfrac{8x}{x + 2} - 6 = \dfrac{2x}{30}.$

**40.** $x + \dfrac{1}{x} = \dfrac{17}{4}.$

**41.** $\dfrac{x}{9 - x} + \dfrac{9 - x}{x} = 2\frac{1}{2}.$

**42.** $\dfrac{48}{x + 3} = \dfrac{165}{x + 10} - 5.$

**43.** $\dfrac{10}{x - 3} - \dfrac{12}{x + 1} = 3.$

**44.** $\dfrac{5x - 1}{9} + \dfrac{3x - 1}{5} = \dfrac{2}{x} + x - 1.$

**45.** $\dfrac{6x + 4}{5} - \dfrac{15 - 2x}{x - 3} = \dfrac{7(x - 1)}{5}.$

**46.** $\dfrac{7}{2x - 3} + \dfrac{5}{x - 1} = 12.$

**47.** $\dfrac{4(x + 1)}{x + 2} + \dfrac{4(x - 3)}{x - 4} = 11.$

**48.** $x^2 - 3ax = 4a^2.$

**49.** $x^2 + 7m^2 = 8mx.$

**50.** $x^2 - (a + b)x + ab = 0.$

**231. Solution by formula.** Every quadratic equation can be reduced to the general form,

$$ax^2 + bx + c = 0.$$

Solving this equation by the method of the preceding article, we obtain

$$x = \frac{-b \pm \sqrt{b^2 - 4ac}}{2a}.$$

The roots of any quadratic equation may be obtained by substituting the values of $a$, $b$, and $c$ in the general answer.

**Ex. 1.** Solve $5\,x^2 = 26\,x - 5$.

Transposing, $\qquad 5\,x^2 - 26\,x + 5 = 0$.

Hence $\qquad\qquad a = 5,\ b = -26,\ c = 5$.

Therefore $\qquad x = \dfrac{+\,26 \pm \sqrt{(-\,26)^2 - 4 \cdot 5 \cdot 5}}{10}$

$$= \frac{26 \pm 24}{10} = 5 \text{ or } \frac{1}{5}.$$

**Ex. 2.** Solve $px^2 - p^2x - x = -p$.

Reducing to general form, $\qquad px^2 - (p^2 + 1)x + p = 0$.

Hence $\qquad\qquad a = p,\ b = -\,(p^2 + 1),\ c = p$.

Therefore $\qquad x = \dfrac{p^2 + 1 \pm \sqrt{(p^2 + 1)^2 - 4\,p^2}}{2\,p}$

$$= \frac{p^2 + 1 \pm (p^2 - 1)}{2\,p} = p \text{ or } \frac{1}{p}.$$

### EXERCISE 86

Solve by the above formula :

1. $2\,x^2 - 5\,x + 2 = 0$.

2. $3\,x^2 - 11\,x + 10 = 0$.

3. $2\,x^2 - 11\,x + 15 = 0$.

4. $2\,x^2 + 3\,x - 35 = 0$.

5. $2\,x^2 + 13\,x + 15 = 0$.

6. $30\,x^2 + x = 1$.

7. $x^2 = 64 - 12\,x$.

8. $11\,x = x^2 + 10$.

9. $x^2 = 17\,x - 60$.

10. $3\,x^2 - 22\,x + 35 = 0$.

11. $21 = 44\,x - 15\,x^2$.

12. $25\,x^2 = 30\,x - 2$.

13. $6\,x^2 + x = 15$.

14. $7\,x^2 = 12 - 25\,x$.

15. $6\,x^2 + 5\,x = 56$.

16. $7\,x^2 + 9\,x = 90$.

17. $x^2 - 6\,mx = 7\,m^2$.

18. $x^2 + 12\,nx = 64\,n^2$.

19. $x^2 + 2\,x + 4 = 0$.

20. $nx^2 + (n^2 + 1)x + n = 0$.

21. $x^2 - (a + b)x + a + b - 1 = 0$.

Find the roots of the following equations to two decimal places :

22. $x^2 = 1 - x.$

23. $3 x^2 + x = 7.$

24. $x^2 - 8 x = 14.$

25. $x = \dfrac{4 - 2 x}{x}.$

26. $x(x - 4) = -2.$

27. $2 x = 28 - 3 x^2.$

28. $7 x - 1 = \dfrac{1}{7 x}.$

**232. Solution by factoring.** Let it be required to solve the equation:
$$5 x^2 + 5 = 26 x;$$

or, transposing all terms to one member,
$$5 x^2 - 26 x + 5 = 0.$$

Resolving into factors,
$$(5 x - 1)(x - 5) = 0.$$

Now, if either of the factors $5 x - 1$, or $x - 5$ is zero, the product is zero. Therefore the equation will be satisfied if $x$ has such a value that either

$$5 x - 1 = 0, \tag{1}$$

or
$$x - 5 = 0. \tag{2}$$

Solving (1) and (2), we obtain the roots
$$x = \tfrac{1}{5} \text{ or } x = 5.$$

**233.** Evidently this method can be applied to equations of any degree, if one member of the equation is zero and the other member can be factored.

Ex. 1.   Solve $x^3 = \dfrac{7 x^2 + 15 x}{2}.$

Clearing for fractions,      $2 x^3 = 7 x^2 + 15 x.$

Transposing,          $2 x^3 - 7 x^2 - 15 x = 0.$

Factoring,          $x(2 x + 3)(x - 5) = 0.$

Therefore      $x = 0, 2 x + 3 = 0, \text{ or } x - 5 = 0.$

Hence the equation has three roots, 0, $-\tfrac{3}{2}$, and 5.

**Ex. 2.** Solve $x^3 - 3x^2 - 4x + 12 = 0$.

Factoring, $\qquad x^2(x - 3) - 4(x - 3) = 0$.

$\qquad\qquad\qquad (x^2 - 4)(x - 3) = 0$.

Or $\qquad\qquad (x - 2)(x + 2)(x - 3) = 0$.

Hence the roots are 2, $-$ 2, 3.

**234.** If both members of an equation are divided by an expression involving the unknown quantity, the resulting equation contains fewer roots than the original one. In order to obtain all roots of the original equation, such a common divisor must be made equal to zero, and the equation thus formed be solved. *E.g.* let it be required to solve

$$x^2 - 9 = 5(x - 3).$$

If we divide both members by $x - 3$, we obtain $x + 3 = 5$ or $x = 2$. But evidently the value $x = 3$ obtained from the equation $x - 3 = 0$ is also a root, for $x^2 - 9 - 5(x - 3) = 0$, or $(x - 3)(x + 3 - 5) = 0$. Therefore $x = 3$ or $x = 2$.

**Ex. 3.** Form an equation whose roots are 4 and $-$ 6.

The equation is evidently $(x - 4)(x - (- 6)) = 0$.

*I.e.* $\qquad\qquad\qquad (x - 4)(x + 6) = 0$.

Or $\qquad\qquad\qquad x^2 + 2x - 24 = 0$.

Solve by factoring :

1. $x^2 + 7x + 6 = 0$.

2. $x^2 + 21 = 10x$.

3. $x^2 - 8x = - 12$.

4. $x^2 - 10x = 24$.

5. $x^2 + 10x = - 24$.

6. $x^2 + 10x = 24$.

7. $x^2 - 10x = - 24$.

8. $x(x + 8) = - 7$.

9. $2x^2 + 9x - 5 = 0$.

10. $3x^2 - 25x + 28 = 0$.

11. $x^3 + 9x^2 + 20x = 0$.

12. $4x^3 + 18x^2 + 8x = 0$.

13. $3y^2 - y - 5 = 0$.

14. $3y^3 = y(11y - 6)$.

15. $y(y - 2) = 7(y - 2)$.

16. $(5t^2 + t)t = 18t$.

17. $u(3u^2 + 7u) = 6u.$          19. $u(u^2 - u) = 6u.$

18. $u^2 + u = 2.$          20. $x^2 - a^2 = (x - a)b.$

21. $(x + 1)(x - 3) = (x + 1)(3 - x).$

22. $(2x - 3)(x + 2) = (x + 3)(x + 2).$      23. $x^3 - x^2 - 2x = 0.$

24. $(y - 1)(y - 2)(y + 3) = 6.$

25. $(u - 1)(u^2 - 4u + 2) = -2.$

26. $x^2 = (a + b + c)x.$          27. $y(y^2 - 1) = 6(y - 1).$

Solve:

28. $(3x - 11)(2x - 10) = x^2 - 9.$

29. $\dfrac{x}{4} + \dfrac{25}{x} = 3.$

39. $\dfrac{6}{x - 6} - \dfrac{5}{x - 4} = \dfrac{6}{x - 3}.$

30. $x + \dfrac{1}{2} = \dfrac{1}{2x}.$

40. $\dfrac{1}{x - 1} + \dfrac{2}{x - 2} = \dfrac{3}{x - 3}.$

31. $\dfrac{x}{100} - \dfrac{21}{25x} = \dfrac{1}{4}.$

41. $\dfrac{1}{x - 2} + \dfrac{2}{x - 1} = \dfrac{6}{x}.$

32. $\dfrac{15}{x} - \dfrac{72 - 6x}{2x^2} = 2.$

42. $\dfrac{3x + 4}{x + 1} = 2x + \dfrac{3}{x + 1}.$

33. $\dfrac{x}{100} + \dfrac{21}{25x} = -\dfrac{1}{4}.$

43. $ax^2 + bx = cx.$

34. $\dfrac{x}{x + 1} + \dfrac{x + 1}{x} = 2\tfrac{1}{6}.$

44. $\dfrac{a}{x - b} + \dfrac{b}{x - a} = 2.$

35. $\dfrac{5}{x + 3} + \dfrac{6}{x + 4} = 2.$

45. $\dfrac{y + a}{y - a} + \dfrac{y - a}{y + a} = 2\tfrac{1}{2}.$

36. $\dfrac{4}{3x - 5} - \dfrac{5}{2x + 1} = 3.$

46. $\dfrac{a}{x - a} + \dfrac{b}{x - b} = 1.$

47. $x^2 - (b - a)x - ab = 0.$

37. $\dfrac{3}{x - 5} + \dfrac{2x}{x - 3} = 5.$

48. $\dfrac{a + x}{b + x} + \dfrac{b + x}{a + x} = 2\tfrac{1}{2}.$

38. $\dfrac{2x - 6}{x + 1} + \dfrac{24}{x^2 - 1} = \dfrac{x + 2}{x - 1}.$

49. $x^3 + 5x^2 - 4x - 20 = 0.$

50. $x^3 - 3x^2 - 9x + 27 = 0.$

Form the equations whose roots are:

**51.** 3, 1.  **53.** −2, −5.  **55.** 1, −2, 3.  **57.** 1, 2, 3.

**52.** 3, −4.  **54.** 0, 9.  **56.** −2, 3, 0.  **58.** 2, 0, −2.

## PROBLEMS INVOLVING QUADRATICS

**235.** Problems involving quadratics have in general two answers, but frequently the conditions of the problem exclude negative or fractional answers, and consequently many problems of this type have only one solution.

### EXERCISE 88

**1.** A number increased by three times its reciprocal equals $6\frac{1}{2}$. Find the number.

**2.** Divide 60 into two parts whose product is 875.

**3.** The difference of two numbers is 4, and the difference of their reciprocals is $\frac{1}{3}$. Find the numbers.

**4.** Find two numbers whose product is 288, and whose sum is 36.

**5.** The sum of the squares of two consecutive numbers is 85. What are the numbers?

**6.** The product of two consecutive numbers is 210. Find the numbers.

**7.** Find a number which exceeds its square by $\frac{2}{9}$.

**8.** Find two numbers whose difference is 6, and whose product is 40.

**9.** Twenty-nine times a number exceeds the square of the number by 190. Find the number.

**10.** The sides of a rectangle differ by 9 inches, and its area equals 190 square inches. Find the sides.

**11.** A rectangular field has an area of 8400 square feet and a perimeter of 380 feet. Find the dimensions of the field.

**12.** The length *AB* of a rectangle, *ABCD*, exceeds its width

*AD* by 119 feet, and the line *BD* joining two opposite vertices (called "diagonal") equals 221 feet. Find *AB* and *AD*.

**13.** The diagonal of a rectangle is to the length of the rectangle as 5 : 4, and the area of the figure is 96 square inches. Find the sides of the rectangle.

**14.** A man sold a watch for $24, and lost as many per cent as the watch cost dollars. Find the cost of the watch.

**15.** A man sold a watch for $21, and lost as many per cent as the watch cost dollars. Find the cost of the watch.

**16.** A man sold a horse for $144, and gained as many per cent as the horse cost dollars. Find the cost of the horse.

**17.** Two steamers ply between the same two ports, a distance of 420 miles. One steamer travels half a mile faster than the other, and is two hours less on the journey. At what rates do the steamers travel?

**18.** If a train had traveled 10 miles an hour faster, it would have needed two hours less to travel 120 miles. Find the rate of the train.

**19.** Two vessels, one of which sails two miles per hour faster than the other, start together on voyages of 1152 and 720 miles respectively, and the slower reaches its destination one day before the other. How many miles per hour did the faster vessel sail?

**20.** A man bought a certain number of apples for $2.10. If he had paid 2 ¢ more for each apple, he would have received 12 apples less for the same money. What did he pay for each apple?

**21.** A man bought a certain number of horses for $1200. If he had paid $20 less for each horse, he would have received two horses more for the same money. What did he pay for each horse?

**22.** On the prolongation of a line $AC$, 23 inches long, a point $B$ is taken, so that the rectangle, constructed with $AB$ and $CB$ as sides, contains 78 square inches. Find $AB$ and $CB$.

**23.** A rectangular grass plot, 30 feet long and 20 feet wide, is surrounded by a walk of uniform width. If the area of the walk is equal to the area of the plot, how wide is the walk?

**24.** A circular basin is surrounded by a path 5 feet wide, and the area of the path is $\frac{7}{9}$ of the area of the basin. Find the radius of the basin. (Area of a circle $= \pi r^2$.)

**25.** A needs 8 days more than B to do a certain piece of work, and working together, the two men can do it in 3 days. In how many days can B do the work?

**26.** Find the side of an equilateral triangle whose altitude equals 3 inches.

**27.** The number of eggs which can be bought for $1 is equal to the number of cents which 4 eggs cost. How many eggs can be bought for $1?

## EQUATIONS IN THE QUADRATIC FORM

**236.** An equation is said to be in the **quadratic form** if it contains only two unknown terms, and the unknown factor of one of these terms is the square of the unknown factor of the other, as

$$ax^{2n} + bx^n + c = 0, \quad x^6 - 3x^3 = 7, \quad (x^2 - 1)^2 - 4(x^2 - 1) = 9.$$

**237.** Equations in the quadratic form can be solved by the methods used for quadratics.

**Ex. 1.** Solve $\quad x^6 - 9x^3 + 8 = 0.$

By formula, $\qquad x^3 = \dfrac{9 \pm \sqrt{81 - 32}}{2}$

$\qquad\qquad\qquad = \dfrac{9 \pm 7}{2} = 8,$ or 1.

Therefore $\qquad x = \sqrt[3]{8} = 2,$ or $x = \sqrt[3]{1} = 1.$

**238.** In more complex examples it is advantageous to substitute a letter for an expression involving $x$.

Ex. **2.**   Solve $\left(\dfrac{x+4}{x}\right)^2 + 15 = \dfrac{8(x+4)}{x}$.

Let                                    $\dfrac{x+4}{x} = y$.

Then                                 $y^2 + 15 = 8y$,

or                                     $y^2 - 8y + 15 = 0$,

or                                     $(y-5)(y-3) = 0$.

Hence                               $y = 5$, or $y = 3$.

I.e.          $\dfrac{x+4}{x} = 5$,   or $\dfrac{x+4}{x} = 3$.

Solving,      $x + 4 = 5x$,        $x + 4 = 3x$.

                   $4x = 4$,            $2x = 4$.

                   $x = 1$,            $x = 2$.

### EXERCISE 89

Solve the following equations:

**1.** $x^4 - 10x^2 + 9 = 0$.   **3.** $x^4 - 5x^2 = -4$.   **5.** $x^4 - 8 = 2x^2$.

**2.** $x^4 + 36 = 13x^2$.   **4.** $x^4 - 21x^2 = 100$.   **6.** $x^6 - 7x^3 - 8 = 0$.

**7.** $3x^4 - 44x^2 + 121 = 0$.   **9.** $4x^4 - 37x^2 = -9$.

**8.** $16x^4 - 40a^2x^2 + 9a^4 = 0$.   **10.** $(x^2 + x)^2 - 18(x^2 + x) + 72 = 0$.

**11.** $(x^2 - x)^2 - 8(x^2 - x) + 12 = 0$.

**12.** $(x^2 + 3x)^2 - 2(x^2 + 3x) = 8$.

**13.** $\dfrac{x^2+1}{x+1} + 2\left(\dfrac{x+1}{x^2+1}\right) = 3$.   **15.** $\dfrac{x^2+1}{x^2-1} + \dfrac{x^2-1}{x^2+1} = 3\tfrac{1}{3}$.

**14.** $\dfrac{x}{x^2+1} + \dfrac{x^2+1}{x} = 2\tfrac{1}{2}$.   **16.** $x^2 - 6x + 12 = \dfrac{-28}{x^2-6x+1}$.

**17.** $(x^2 - 4x + 5)(x^2 - 4x + 2) + 2 = 0$.

**18.** $(x^2 + x + 1)^2 - 4(x^2 + x - 1) - 5 = 0$.

**19.** $(x^2 + 3x - 2)^2 - 4(x^2 + 3x - 2) = 32$.

**20.** $\dfrac{x^2+x+1}{x^2+x-1} + \dfrac{x^2+x+3}{x^2+x-3} + 2 = 0$.

## CHARACTER OF THE ROOTS

**239.** The quadratic equation $ax^2 + bx + c = 0$ has two roots,

$$\frac{-b + \sqrt{b^2 - 4ac}}{2a} \quad \text{and} \quad \frac{-b - \sqrt{b^2 - 4ac}}{2a} \qquad (\S 231.)$$

Hence it follows:

1. *If $b^2 - 4ac$ is a positive or equal to zero, the roots are real.*
   *If $b^2 - 4ac$ is negative, the roots are imaginary.*
2. *If $b^2 - 4ac$ is a perfect square, the roots are rational.*
   *If $b^2 - 4ac$ is not a perfect square, the roots are irrational.*
3. *If $b^2 - 4ac$ is zero, the roots are equal.*
   *If $b^2 - 4ac$ is not zero, the roots are unequal.*

**240.** The expression $b^2 - 4ac$ is called the **discriminant** of the equation $ax^2 + bx + c = 0$.

**Ex. 1.** Determine the character of the roots of the equation $3x^2 - 2x - 5 = 0$.

The discriminant $= (-2)^2 - 4 \cdot 3 \cdot (-5) = 64$.
Hence the roots are real, rational, and unequal.

**Ex. 2.** Determine the character of the roots of the equation $4x^2 - 12x + 9 = 0$.

Since $(-12)^2 - 4 \cdot 4 \cdot 9 = 0$, the roots are real, rational, and equal.

**241. Relations between roots and coefficients.** If the roots of the equation $ax^2 + bx + c = 0$ are denoted by $r_1$ and $r_2$, then

$$r_1 = \frac{-b + \sqrt{b^2 - 4ac}}{2a},$$

$$r_2 = \frac{-b - \sqrt{b^2 - 4ac}}{2a}.$$

Hence $\qquad r_1 + r_2 = -\dfrac{b}{a}$,

and $\qquad r_1 r_2 = \dfrac{(-b)^2 - (b^2 - 4ac)}{4a^2}$.

Or $\qquad r_1 r_2 = \dfrac{c}{a}$.

o

If the given equation is written in the form $x^2 + \dfrac{b}{a}x + \dfrac{c}{a} = 0$, these results may be expressed as follows:

*If the coefficient of $x^2$ in a quadratic equation is unity,*

(a) *The sum of the roots is equal to the coefficient of $x$ with the sign changed.*

(b) *The product of the roots is equal to the absolute term.*

*E.g.* the sum of the roots of $4x^2 + 5x - 3 = 0$ is $-\frac{5}{4}$, their product is $-\frac{3}{4}$.

### EXERCISE 89 *a*

Determine without solution the character of the roots of the following equations:

1. $x^2 - 11x + 18 = 0.$

2. $5x^2 - 26x + 5 = 0.$

3. $2x^2 + 6x + 3 = 0.$

4. $x^2 + 10x + 4520 = 0.$

5. $x^2 - 12x + 36 = 0.$

6. $3x^2 + 4x + 240 = 0.$

7. $9x^2 - 24x + 16 = 0.$

8. $5x^2 + 2 = x.$

9. $x^2 - 7 = 5x.$

10. $12 - x = x^2.$

11. $\dfrac{x-3}{x-8} = \dfrac{1}{x}.$

12. $10x = 25x^2 + 1.$

In each of the following equations determine by inspection the sum and the product of the roots:

13. $x^2 - 5x + 2 = 0.$

14. $x^2 - 9x - 3 = 0.$

15. $2x^2 - 4x - 5 = 0.$

16. $3x^2 + 9x - 2 = 0.$

17. $x^2 - mx + p = 0.$

18. $5x^2 - x + 1 = 0.$

Solve the following equations and check the answers by forming the sum and the product of the roots:

19. $x^2 - 19x + 60 = 0.$

20. $x^2 + 2x - 2 = 0.$

21. $x^2 + 2x - 15 = 0.$

22. $x^2 - 4x - 12 = 0.$

23. $x^2 + 2x + 2 = 0.$

24. $x^2 - 7x + 3 = 0.$

# CHAPTER XVI

## THE THEORY OF EXPONENTS

**242.** The following **four fundamental laws for positive integral exponents** have been developed in preceding chapters:

I. $a^m \cdot a^n = a^{m+n}$.

II. $a^m \div a^n = a^{m-n}$, provided $m > n$.*

III. $(a^m)^n = a^{mn}$.

IV. $(ab)^m = a^m \cdot b^m$.

The first of these laws is the direct consequence of the definition of power, while the second and third are consequences of the first.

### FRACTIONAL AND NEGATIVE EXPONENTS

**243. Fractional and negative exponents,** such as $2^{\frac{1}{3}}$, $4^{-3}$, have no meaning according to the original definition of power, and we may choose for such symbols any definition that is convenient for other work.

It is, however, very important that all exponents should be governed by the same laws; hence, instead of giving a formal definition of fractional and negative exponents, we let these quantities be what they must be if the exponent law of multiplication is generally true.

**244.** We assume, therefore, that $a^m \cdot a^n = a^{m+n}$, for all values of $m$ and $n$. Then the law of involution, $(a^m)^n = a^{mn}$, must be

---

* The symbol $>$ means "is greater than"; similarly $<$ means "is smaller than."

true for positive integral values of $n$, since the raising to a positive integral power is only a repeated multiplication.

Assuming these two laws, we try to discover the meaning of $8^{\frac{1}{3}}$, $a^0$, $4^{-2}$, $a^{\frac{m}{n}}$, etc. In every case we let the unknown quantity equal $x$, and apply to both members of the equation that operation which makes the negative, fractional, or zero exponent disappear.

**245. To find the meaning of a fractional exponent;** *e.g.* $a^{\frac{1}{3}}$.

Let $\qquad\qquad x = a^{\frac{1}{3}}.$

The operation which makes the fractional exponent disappear is evidently the raising of both members to the third power.

Hence $\qquad\qquad x^3 = (a^{\frac{1}{3}})^3.$

Or $\qquad\qquad x^3 = a.$

Therefore $\qquad\qquad x = \sqrt[3]{a}.$

Similarly, we find $\qquad a^{\frac{p}{q}} = \sqrt[q]{a^p}.$

*Hence we define $a^{\frac{p}{q}}$ to be the qth root of $a^p$.*

Find the values of:

1. $9^{\frac{1}{2}}$.
2. $4^{\frac{1}{2}}$.
3. $16^{\frac{1}{4}}$.
4. $8^{\frac{1}{3}}$.
5. $27^{\frac{1}{3}}$.
6. $64^{\frac{1}{3}}$.
7. $32^{\frac{1}{5}}$.

8. $8^{\frac{2}{3}}$.
9. $4^{\frac{3}{2}}$.
10. $27^{\frac{2}{3}}$.
11. $125^{\frac{1}{3}}$.
12. $32^{\frac{2}{5}}$.
13. $1^{\frac{11}{12}}$.
14. $(-8)^{\frac{1}{3}}$.

15. $16^{\frac{3}{4}}$.
16. $0^{\frac{7}{17}}$.
17. $(\frac{1}{8})^{\frac{1}{3}}$.
18. $(\frac{1}{4})^{\frac{3}{2}}$.
19. $(\frac{1}{8})^{\frac{2}{3}}$.
20. $(m^2 - 2mn + n^2)^{\frac{1}{2}}$.
21. $(x^3 + 3x^2y + 3xy^2 + y^3)^{\frac{1}{3}}$.

Write the following expressions as radicals:

22. $m^{\frac{7}{3}}$.
23. $x^{\frac{1}{4}}$.
24. $a^{\frac{2}{3}}$.
25. $a^{\frac{3}{2}}$.
26. $(xy)^{\frac{1}{5}}$.
27. $3^{\frac{1}{4}}$.
28. $(bcd)^{\frac{3}{4}}$.
29. $a^{\frac{m}{3}}$.
30. $b^{\frac{1}{2}}x^{\frac{1}{4}}$.
31. $m^{\frac{a}{b}}$.

Express with fractional exponents:

**32.** $\sqrt[3]{x^2}$.     **33.** $\sqrt[5]{abc}$.     **34.** $\sqrt[8]{x^3}$.     **35.** $\sqrt[3]{m}$.

**36.** $\sqrt[16]{a^8}$.          **37.** $\sqrt[4]{xy}$.          **38.** $\sqrt[6]{m} \cdot \sqrt[5]{n}$.

Solve the following equations:

**39.** $2^x = 4$.     **41.** $x^{\frac{1}{2}} = 3$.     **43.** $3^x = 27$.     **45.** $5 x^{\frac{1}{4}} = 10$.

**40.** $4^x = 2$.     **42.** $x^{\frac{1}{3}} = 2$.     **44.** $27^x = 3$.     **46.** $7 x^{\frac{1}{2}} = 49$.

Find the values of:

**47.** $4^{\frac{1}{2}} + 9^{\frac{1}{2}} + 16^{\frac{1}{2}} + 25^{\frac{1}{2}} + 36^{\frac{1}{2}}$.

**48.** $1^{\frac{1}{3}} + 8^{\frac{1}{3}} + 27^{\frac{1}{3}} + 64^{\frac{1}{3}} + 0^{\frac{1}{3}}$.

**49.** $64^{\frac{1}{3}} + 9^{\frac{1}{2}} + 16^{\frac{1}{4}} + (-32)^{\frac{1}{5}}$.

**50.** $(\frac{1}{4})^{\frac{5}{2}} + (\frac{1}{27})^{\frac{1}{3}} - (\frac{1}{81})^{\frac{1}{4}}$.

**246.** **To find the meaning of zero exponent,** *e.g.* $a^0$.

Let $\qquad\qquad x = a^0$.

The operation which makes the zero exponent disappear is evidently a multiplication by any power of $a$, *e.g.* $a^2$.

$$a^2 x = a^{0+2} = a^2.$$

$$x = \frac{a^2}{a^2} = 1.$$

Or $\qquad\qquad a^0 = 1$.

Therefore *the zero power of any number is equal to unity.*

NOTE. If, however, the base is zero, $\dfrac{a^2}{a^2}$ is indeterminate; hence $0^0$ is indeterminate.

**247.** To find the meaning of a negative exponent, *e.g.* $a^{-n}$.

Let $$x = a^{-n}.$$

Multiplying both members by $a^n$, $a^n x = a^0$.

Or $$a^n x = 1.$$

Hence $$x = \frac{1}{a^n}.$$

Therefore $$a^{-n} = \frac{1}{a^n}.$$

**248.** *Factors may be transferred from the numerator to the denominator of a fraction, or vice versa, by changing the sign of the exponent.*

$$1. \quad a^{-n} = \frac{1}{a^n}.$$

$$2. \quad \frac{1}{a^{-n}} = \frac{1}{\frac{1}{a^n}} = a^n.$$

NOTE. The fact that $a^0 = 1$ sometimes appears peculiar to beginners. It loses its singularity if we consider the following equations, in which each is obtained from the preceding one by dividing both members by $a$.

$$a^3 = 1 \cdot a \cdot a \cdot a$$

$$a^2 = 1 \cdot a \cdot a$$

$$a^1 = 1 \cdot a$$

$$a^0 = 1$$

$$a^{-1} = \frac{1}{a}$$

$$a^{-2} = \frac{1}{a^2}, \text{ etc.}$$

## EXERCISE 91

Find the values of:

1. $7^{-2}$.

2. $5^{-1}$.

3. $3^{-3}$.

4. $1^{-9}$.

5. $199^0$.

6. $2^{-5}$.

7. $(\tfrac{3}{17})^0$.

19. $625^{-\frac{1}{4}}$.

8. $15^{-2}$.

9. $217^{-1}$.

10. $\dfrac{1}{4^{-1}}$.

11. $\dfrac{1}{2^{-3}}$.

12. $(\tfrac{2}{3})^{-1}$.

13. $(\tfrac{3}{4})^{-2}$.

14. $4^{-\frac{1}{2}}$.

15. $8^{-\frac{2}{3}}$.

16. $64^{-\frac{1}{2}}$.

17. $(\tfrac{1}{2})^{-5}$.

18. $\dfrac{1}{7^{-2}}$.

20. $12^0 - 5^{-1} - 25^{\frac{1}{2}}$.

Express with positive exponents:

21. $x^{-5}$.

22. $6\,x^{-2}$.

23. $9\,a^{-2}b^{-1}c^5$.

24. $7^{-1}a^2b^2$.

25. $\dfrac{3\,x^2b^{-3}\cdot c^2}{x^{-3}}$.

26. $\dfrac{5^{-2}\cdot a^{-2}}{b^{-3}}$.

27. $\dfrac{6\,a^2b^{-3}c^{-4}}{5^{-1}\,xyz^{-2}}$.

28. $\dfrac{5\,a^{-7}b^{-7}}{6^{-2}\,ax^{-4}}$.

Write without denominators:

29. $\dfrac{6\,a}{b}$.

30. $\dfrac{5\,a^2b^2}{c^2}$.

31. $\dfrac{a^3b^3c^3}{x^3y^3}$.

32. $\dfrac{abx}{y^{-2}}$.

33. $\dfrac{m^{-2}x^{\frac{1}{2}}}{n^{-2}}$.

34. $\dfrac{1}{7\,mnp}$.

Write with radical signs and positive exponents:

35. $m^{\frac{1}{2}}$.

36. $m^{-\frac{1}{2}}$.

37. $3\,x^{\frac{2}{3}}$.

38. $3\,x^{-\frac{1}{3}}$.

39. $2\,m^{-\frac{1}{5}}$.

40. $(2\,m)^{-\frac{1}{3}}$.

41. $\dfrac{1}{2\,x^{-\frac{1}{2}}}$.

42. $m^{\frac{n}{2}}$.

43. $n^{-\frac{a}{3}}$.

44. $\dfrac{a^{\frac{1}{3}}}{6\,b^{-\frac{1}{2}}c^{-\frac{1}{4}}}$.

45. $\dfrac{1}{m^{-\frac{1}{n}}}$.

Solve the equations:

46. $x^{-1} = \frac{1}{5}$.

47. $x^{-2} = \frac{1}{4}$.

48. $2^x = \frac{1}{2}$.

49. $3^x = \frac{1}{9}$.

50. $17^x = 1$.

51. $x^{\frac{1}{3}} = 5$.

52. $5\,x^{\frac{1}{4}} = 10$.

53. $10^x = .001$.

54. $10^x = .1$.

55. $5^x = \frac{1}{125}$.

Find the values of:

56. $2^0 + 3^{-1} + 9^{\frac{1}{2}} - (\frac{1}{7})^{-1} - (\frac{1}{9})^{\frac{1}{2}}$.

57. $3^2 - 11^0 - (\frac{1}{4})^{-\frac{1}{2}} + 0^7 + 9^{\frac{3}{2}}$.

58. $4^{-\frac{1}{2}} + 1^{-\frac{1}{2}} - (\frac{1}{2})^1 + 21^0 - 9^{\frac{1}{2}}$.

59. $(81)^{\frac{3}{4}} + (3\frac{3}{8})^{\frac{1}{3}} - (5\frac{1}{16})^{\frac{1}{4}} - 32^{.2}$.

60. $49^{.5} + 16^{.25} - 81^{.75} + (a - b)^0$.

61. $(.343)^{\frac{1}{3}} + (.25)^{\frac{1}{2}} - (.008)^{\frac{1}{3}} + \dfrac{1}{2^{-1}} + \dfrac{1}{3^{-1}}$.

## USE OF NEGATIVE AND FRACTIONAL EXPONENTS

**249.** It can be demonstrated that the last three laws for any exponents are consequences of the first law, and we shall hence assume that all four laws are generally true. It then follows that:

*Fractional and negative exponents may be treated by the same methods as positive integral exponents.*

**250.** Examples relating to roots can be reduced to examples containing fractional exponents.

**Ex. 1.** $(a^{\frac{1}{2}}b^{-\frac{1}{4}})^{\frac{1}{3}} \div (a^{\frac{2}{3}}b^{-\frac{3}{2}})^{\frac{1}{3}} = a^{\frac{1}{6}}b^{-\frac{1}{12}} \div a^{\frac{2}{9}}b^{-\frac{1}{2}} = a^{-\frac{1}{18}}b^{\frac{5}{12}}$

$$= \frac{b^{\frac{1}{2}}}{a^{\frac{1}{6}}} = \sqrt[6]{\frac{b^2}{a}}.$$

**Ex. 2.** $\left(\dfrac{2\sqrt{a}\,\sqrt[3]{b}}{3\sqrt[6]{a}\,\sqrt{b}}\right)^6 = \left(\dfrac{2\,a^{\frac{1}{2}}b^{\frac{1}{3}}}{3\,a^{\frac{1}{6}}b^{\frac{1}{2}}}\right)^6 = \dfrac{64\,a^3b^2}{729\,ab^3} = \dfrac{64\,a^2}{729\,b}.$

**251.** Expressions containing radicals should be simplified as follows:

(a) **Write all radical signs as fractional exponents.**

(b) **Perform the operation indicated.**

(c) **Remove the negative exponents.**

(d) **If required, remove the fractional exponents.**

NOTE. *Negative exponents should not be removed until all operations of multiplication, division, etc., are performed.*

## EXERCISE 92

Simplify:

1. $2^3 \cdot 2^4 \cdot 2^{-5} \cdot 2$.

2. $3^8 \cdot 3^5 \cdot 3^9 \cdot 3^{-10} \cdot 3^{-11}$.

3. $7^2 \cdot 7^9 \cdot 7^{-5} \cdot 7^{-6}$.

4. $2^5 \cdot 2^6 \cdot 2^7 \cdot 2^{-8} \cdot 2^{-9} \cdot 2^{-3}$.

5. $a^{-3} \cdot a^{-4} \cdot a^8$.

6. $x^5 \cdot 3 x^{-4} \cdot 2 x^2 \cdot 2 x^{-1}$.

7. $6 a^{-4} \cdot 5 a^5$.

8. $5^{\frac{1}{2}} \cdot 5^{\frac{1}{3}} \cdot 5^{\frac{1}{6}}$.

9. $7^{\frac{1}{2}} \cdot 7^{\frac{1}{3}} \cdot 7^{\frac{1}{8}} \cdot 7^{\frac{11}{24}}$.

10. $x^{\frac{3}{4}} \cdot x^{-\frac{1}{3}} \cdot x^{\frac{1}{4}}$.

11. $\sqrt{5} \cdot \sqrt[3]{5} \cdot \sqrt[4]{5}$.

12. $9^5 \div 9^{\frac{3}{2}}$.

13. $5^{-2} \div 5$.

14. $3^{-4} \div 3^{-5}$.

15. $a^9 \div a^{-4}$.

16. $14 a^2 \div 7 a^{\frac{3}{2}}$.

17. $(4 x^{\frac{2}{3}} - 3 x^{\frac{1}{2}}) \div 6 x^{-2}$.

18. $(\sqrt{a})^4$.

19. $\sqrt{11} \sqrt[3]{11} \cdot \sqrt[6]{11}$.

20. $\sqrt[6]{x^7} \div \sqrt[6]{x^5}$.

21. $\sqrt[3]{5} \cdot \sqrt[2]{5} \div \sqrt[6]{5}$.

22. $3 \div \sqrt{3}$.

23. $7 \div \sqrt[3]{7}$.

24. $(a^{\frac{3}{4}})^{\frac{2}{3}}$.

25. $(\sqrt[5]{x^2})^{10}$.

26. $(\sqrt[3]{m^2})^{\frac{3}{2}}$.

27. $\sqrt[4]{x^3} \cdot \sqrt[6]{x} \cdot \sqrt[12]{x}$.

28. $\sqrt[5]{m} \sqrt[10]{m^3} \sqrt{m}$.

29. $\dfrac{\sqrt[5]{m} \cdot \sqrt{m}}{\sqrt[10]{m^7}}$.

30. $4 n^{-\frac{2}{3}} \div 2 n^{\frac{4}{5}}$.

31. $\dfrac{\sqrt{x^3} \cdot \sqrt[3]{y^3} \cdot \sqrt[3]{z}}{\sqrt{x} \sqrt{y} \sqrt[6]{z}}$.

**32.** $\dfrac{\sqrt[4]{m}\ \sqrt[36]{m}}{\sqrt[6]{m^5}\cdot\sqrt[9]{m^4}}.$

**37.** $\sqrt[4]{\dfrac{x^2\cdot y^2}{a^6\cdot b^4}}\cdot\sqrt{\dfrac{a^3b^2}{xy}}.$

**33.** $\sqrt[4]{x^{\frac{1}{2}}}.$

**34.** $\sqrt[4]{\sqrt[3]{a}}.$

**38.** $\dfrac{\sqrt[3]{a^2x}}{\sqrt[4]{ax^2}}\cdot\dfrac{\sqrt[4]{a^2x}}{\sqrt[3]{ax^2}}.$

**35.** $\left(\dfrac{64\,x^{-3}}{125\,y^{-3}}\right)^{-\frac{1}{3}}.$

**39.** $\left(\dfrac{a^{-2}b}{a^2b^{-3}}\right)^{-2}\cdot\left(\dfrac{ab}{a^{-3}b^{-2}}\right)^{-3}.$

**36.** $\sqrt{\left(\dfrac{a^2b^2}{9\,a^4}\right)^3}.$

**40.** $\sqrt{\dfrac{\sqrt{m}}{\sqrt{n}}}\cdot\sqrt{\dfrac{\sqrt{n}}{\sqrt{p}}}\cdot\sqrt{\dfrac{\sqrt{p}}{\sqrt{m}}}.$

**252.** If we wish to arrange terms according to descending powers of $x$, we have to remember that the term which does not contain $x$ may be considered as a term containing $x^0$. The powers of $x$ arranged are:

$$x^3,\ x^2,\ x,\ x^0,\ x^{-1},\ x^{-2},\ x^{-3}.$$

**Ex. 1.** Multiply $3\,x^{-1}+x-5$ by $2\,x-1$.

Arrange in descending powers of $x$.

Check.   If $x=1$

$$x-5+3\,x^{-1}$$
$$\underline{\phantom{xxxxx}2\,x-1}$$
$$2\,x^2-10\,x+\ 6$$
$$\underline{-\ \ \ x+\ \ 5-3\,x^{-1}}$$
$$2\,x^2-11\,x+11-3\,x^{-1}$$

$=-1$

$=+1$

$\underline{\underline{=-1}}$

**Ex. 2.** Divide

$\sqrt[3]{a^4}-4\sqrt[3]{b^4}-6\,a\sqrt[3]{b}+9\sqrt[3]{a^2b^2}$ by $2\sqrt[3]{b^2}-3\sqrt[3]{ab}+\sqrt[3]{a^2}$.

$$\begin{array}{l}a^{\frac{4}{3}}-6\,ab^{\frac{1}{3}}+9\,a^{\frac{2}{3}}b^{\frac{2}{3}}-4\,b^{\frac{4}{3}}\ \Big|\ a^{\frac{2}{3}}-3\,a^{\frac{1}{3}}b^{\frac{1}{3}}+2\,b^{\frac{2}{3}}\\[4pt]\underline{a^{\frac{4}{3}}-3\,ab^{\frac{1}{3}}+2\,a^{\frac{2}{3}}b^{\frac{2}{3}}}\ \ \Big|\overline{\ a^{\frac{2}{3}}-3\,a^{\frac{1}{3}}b^{\frac{1}{3}}-2\,b^{\frac{2}{3}}}\\[4pt]\quad-3\,ab^{\frac{1}{3}}+7\,a^{\frac{2}{3}}b^{\frac{2}{3}}-4\,b^{\frac{4}{3}}\\[4pt]\quad\underline{-3\,ab^{\frac{1}{3}}+9\,a^{\frac{2}{3}}b^{\frac{2}{3}}-6\,a^{\frac{1}{3}}b}\\[4pt]\qquad\qquad-2\,a^{\frac{2}{3}}b^{\frac{2}{3}}+6\,a^{\frac{1}{3}}b-4\,b^{\frac{4}{3}}\\[4pt]\qquad\qquad\underline{-2\,a^{\frac{2}{3}}b^{\frac{2}{3}}+6\,a^{\frac{1}{3}}b-4\,b^{\frac{4}{3}}}\end{array}$$

Perform the operations indicated:

**1.** $(a + \sqrt{b})(a - 2\sqrt{b})$.

**2.** $(\sqrt{x} + 2\sqrt{y})(\sqrt{x} - 3\sqrt{y})$.

**3.** $(8 r^{-1} - 3 s)(2 r^{-1} + 5 s)$.

**4.** $(7 r - 8\sqrt{r} + 5)(9\sqrt{r} - 7)$.

**5.** $(a^{-2} + a^2 + 1)(a^{-2} + a^2 - 1)$.

**6.** $(\sqrt{x} + \sqrt{y} + 1)(\sqrt{x} + \sqrt{y} - 1)$.

**7.** $(\sqrt[3]{x^2} + \sqrt[3]{x} + 1)(\sqrt[3]{x} - 1)$.

**8.** $(a^m + a^{-m} - 1)(a^m + a^{-m} + 1)$.

**9.** $(4 x + 1 - \sqrt{x})(x - 3\sqrt{x} + 9)$.

**10.** $(x + y) \div (\sqrt[3]{x} + \sqrt[3]{y})$.

**11.** $(4 a^{-3} - 24 a^{-1} - 9 - 3 a^{-2}) \div (a^{-1} - 3)$.

**12.** $(13\sqrt{p} + 1 + 47 p + 35\sqrt{p^3}) \div (5\sqrt{p} + 1)$.

**13.** $(\sqrt{a^3} + a\sqrt{b} - b\sqrt{a} - \sqrt{b^3}) \div (\sqrt{a} - \sqrt{b})$.

**14.** $(3 a^{-5} + 16 a^{-4}b^{-1} - 33 a^{-3}b^{-2} + 14 a^{-2}b^{-3}) \div (a^{-2} + 7 a^{-1}b^{-1})$.

**15.** $(a - b) \div (\sqrt[3]{a^2} + \sqrt[3]{ab} + \sqrt[3]{b^2})$.

**16.** $(a - b + 2\sqrt{bc} - c) \div (\sqrt{a} + \sqrt{b} - \sqrt{c})$.

**17.** $\sqrt{x^2 - 6 x + 13 - 12 x^{-1} + 4 x^{-2}}$.

**18.** $\sqrt{x^{-2} - 2 x + x^2 - 2 x^{-1} + 3}$.

**19.** $\sqrt{25 x^2 - 20 x^{-1} + 34 - 12 x + 9 x^2}$.

**20.** $\sqrt{x^{\frac{4}{3}} - 6 x + 13 x^{\frac{2}{3}} - 12 x^{\frac{1}{3}} + 4}$.

**21.** $\sqrt{1 + 2\sqrt{x} + 3 x + 2\sqrt{x^3} + x^2}$.

**22.** $(1 + 4\sqrt[3]{x} + 10\sqrt[3]{x^2} + 20 x + 25\sqrt[3]{x^4} + 24\sqrt[3]{x^5} + 16 x^2)^{\frac{1}{2}}$.

**23.** $(1 + \sqrt{2})\sqrt{2}$.

**24.** $(2 + \sqrt{2})(\sqrt{2} - 2)$.

**25.** $(5 + \sqrt{3})(5 - 2\sqrt{3})$.

**26.** $(1 - 3\sqrt{5})(2 + \sqrt{5})$.

**27.** $(\sqrt{11} - \sqrt{2})(\sqrt{11} - 3\sqrt{2})$

Find by inspection:

28. $(x^{\frac{1}{2}} + 3)(x^{\frac{1}{2}} + 2)$.

29. $(x^{\frac{1}{2}} + 3)(x^{\frac{1}{2}} - 5)$.

30. $(x^{\frac{1}{2}} + y^{\frac{1}{2}})(x^{\frac{1}{2}} - y^{\frac{1}{2}})$.

31. $(5^{\frac{1}{2}} - 3^{\frac{1}{2}})(5^{\frac{1}{2}} + 3^{\frac{1}{2}})$.

32. $(3^{\frac{1}{2}} - 2^{\frac{1}{2}})^2$.

33. $(x^{\frac{1}{2}} + 5)(x^{\frac{1}{2}} - 5)$.

34. $(x^{\frac{1}{3}} + 1)^2$.

35. $(x^{\frac{1}{2}} - y^{\frac{1}{2}})^3$.

36. $(5^{\frac{1}{2}} - 2^{\frac{1}{2}})^2$.

37. $(11^{\frac{1}{2}} + 2^{\frac{1}{2}})^2$.

38. $\sqrt{x + 2\,x^{\frac{1}{2}}y^{\frac{1}{2}} + y}$.

39. $\sqrt{a + 4\,a^{\frac{1}{2}} + 4}$.

40. $(m - n) \div (m^{\frac{1}{2}} + n^{\frac{1}{2}})$.

# CHAPTER XVII

## RADICALS

**253.** A **radical** is the root of a quantity, indicated by a radical sign.

**254.** The radical is **rational**, if the root can be extracted exactly; **irrational**, if the root cannot be exactly obtained. Irrational quantities are frequently called *surds*.

$$\sqrt{9}, \ (x+y)^{\frac{1}{3}} \text{ are radicals.}$$
$$4^{\frac{1}{2}} = 2, \ \sqrt{(a+b)^2} \text{ are rational.}$$
$$\sqrt{2}, \ \sqrt{4\,a+b} \text{ are irrational.}$$

**255.** The **order** of a surd is indicated by the index of the root.

$\sqrt{a}$ is of the second order, or quadratic.

$\sqrt[3]{2}$ is of the third order, or cubic.

$\sqrt[4]{c}$ is of the fourth order, or biquadratic.

**256.** A **mixed surd** is the product of a rational factor and a surd factor; as $3\sqrt{a}$, $x\sqrt{3}$. The rational factor of a mixed surd is called the **coefficient** of the surd.

An **entire surd** is one whose coefficient is unity; as $\sqrt{a}$, $\sqrt[3]{x^2 + y^2}$.

**257.** **Similar surds** are surds which contain the same irrational factor.

$3\sqrt{2}$ and $5\,a\sqrt{2}$ are similar.

$3\sqrt{2}$ and $3\sqrt{3}$ are dissimilar.

**258. Conventional restriction of the signs of roots.**

All even roots may be positive or negative,

e.g.                   $\sqrt{4} = +2$ or $-2$.

Hence          $5\sqrt{4} + 2\sqrt{4} = 5(\pm 2) + 2(\pm 2),$

which results in four values, viz. $14, 6, -14,$ or $-6$. To avoid this ambiguity, it is customary in elementary algebra to restrict the sign of a root to the prefixed sign.

Thus          $5\sqrt{4} + 2\sqrt{4} = 7\sqrt{4} = 14.$

$$5\sqrt{2} - \sqrt{2} = 4\sqrt{2}.$$

If the object of an example, however, is merely an evolution, the complete answer is usually given; thus

$$\sqrt{x^2 - 4x + 4} = \pm (x - 2).$$

**259.** *Since radicals can be written as powers with fractional exponents, all examples relating to radicals may be solved by the methods employed for fractional exponents.*

Thus, to find the $n$th root of a product $ab$ we have

$$(ab)^{\frac{1}{n}} = a^{\frac{1}{n}} b^{\frac{1}{n}} \quad (\S\ 242).$$

*I.e.* to extract the root of a product, multiply the roots of the factors.

## TRANSFORMATION OF RADICALS

**260. Simplification of surds.** A radical is simplified when the expression under the radical sign is integral, and contains no factor whose power is equal to the index.

**Ex. 1.** Simplify $\sqrt{25\,a^4b}$.

$$\sqrt{25\,a^4b} = \sqrt{25\,a^4} \cdot \sqrt{b} = 5\,a^2\sqrt{b}.$$

**Ex. 2.** Simplify $\sqrt[3]{16}$.

$$\sqrt[3]{16} = \sqrt[3]{8} \cdot \sqrt[3]{2} = 2\sqrt[3]{2}.$$

**261. When the quantity under the radical sign is a fraction,** we multiply both numerator and denominator by such a quantity as will make the denominator a perfect power of the same degree as the surd.

**Ex. 3.** Simplify $\sqrt{\tfrac{1}{3}}$.

$$\sqrt{\tfrac{1}{3}} = \sqrt{\tfrac{1}{3} \cdot \tfrac{3}{3}} = \sqrt{\tfrac{3}{9}} = \tfrac{1}{3}\sqrt{3}.$$

**Ex. 4.** Simplify $\sqrt[3]{\dfrac{4\,x^2 y^3}{9\,a^2 b}}$.

$$\sqrt[3]{\frac{4\,x^2 y^3}{9\,a^2 b}} = \sqrt[3]{\frac{4\,x^2 y^3}{9\,a^2 b} \cdot \frac{3\,ab^2}{3\,ab^2}} = \sqrt[3]{\frac{12\,ab^2 x^2 y^3}{27\,a^3 b^3}} = \frac{y}{3\,ab}\sqrt[3]{12\,ab^2 x^2}.$$

### EXERCISE 94

Simplify :

1. $\sqrt{27}$.

2. $\sqrt{45}$.

3. $\sqrt{32}$.

4. $\sqrt{28}$.

5. $\sqrt{24}$.

6. $\sqrt{243}$.

7. $\sqrt{363}$.

8. $\sqrt{x^3}$.

9. $\sqrt{a^7}$.

10. $\sqrt{a^3 b^3}$.

11. $\sqrt{28\,a^2}$.

12. $\sqrt{320\,a^2 b^2}$.

13. $\sqrt{8\,a^3 b^2}$.

14. $5\sqrt{80\,b^2}$.

15. $5\sqrt{40\,a^2 c}$.

16. $7\sqrt{48\,a^2 x^3}$.

17. $\tfrac{5}{6}\sqrt{45\,c^3}$.

18. $\sqrt[3]{16}$.

19. $\sqrt[3]{54}$.

20. $\sqrt[3]{1000\,a^4 b}$.

21. $\sqrt[3]{250\,a^7 b^6}$.

22. $\sqrt[3]{48\,a^3 b^3}$.

23. $\sqrt[3]{-108\,x^6 y^7}$.

24. $\sqrt{2(a+b)^2}$.

25. $\sqrt{3\,a^2 + 6\,ab + 3\,b^2}$.

26. $(125\,x^2 y^3)^{\frac{1}{3}}$.

27. $\sqrt{\tfrac{1}{2}}$.

28. $\sqrt{\tfrac{1}{5}}$.

29. $\sqrt{\dfrac{3\,x}{25}}$.

30. $\sqrt{\dfrac{5\,a^2 b^2}{7}}$.

31. $\sqrt{\dfrac{2\,a^2 b^2}{c}}$.

32. $a\sqrt{\dfrac{x}{a}}$.

33. $m\sqrt{\dfrac{n^2}{m}}$.

34. $t^2\sqrt{\dfrac{2\,a^2}{t^3}}$.

35. $\sqrt[3]{\tfrac{1}{2}}$.

36. $\sqrt[3]{\dfrac{3\,a^3}{4}}$.

**37.** $\sqrt[3]{\dfrac{9\,a^3b^4}{8\,c^2}}.$

**39.** $(x+y)\sqrt{\dfrac{x-y}{x+y}}.$

**38.** $\dfrac{2\,m}{n}\sqrt[3]{\dfrac{27\,n^4}{2\,m}}.$

**40.** $\dfrac{10}{a}\sqrt[3]{\dfrac{a^3b^4}{500}}.$

**262.** An imaginary surd can be simplified in precisely the same manner as a real surd; thus,

$$\sqrt{-16}=4\sqrt{-1}, \quad \sqrt{-\dfrac{m^2n^4}{36}}=\dfrac{mn^2}{6}\sqrt{-1}, \quad \sqrt{-\dfrac{9\,a^2}{8}}=\dfrac{3\,a}{4}\sqrt{-2}.$$

Simplify:

**41.** $\sqrt{-9}.$

**43.** $\sqrt{\dfrac{-225\,a^2b^2}{c^2}}.$

**45.** $\sqrt{-18\,a^2b^7}.$

**42.** $\sqrt{-16\,a^2}.$

**44.** $2\sqrt{-\dfrac{25\,x^3y^3}{4}}.$

**46.** $\dfrac{4}{a^2}\sqrt{-\dfrac{5\,a^5}{16}}.$

Simplify and find to three decimal places the numerical values of:

**47.** $\sqrt{\tfrac{1}{3}}.$*

**49.** $\sqrt{\tfrac{2}{5}}.$

**48.** $\sqrt{\tfrac{1}{2}}.$

**50.** $\sqrt{\tfrac{3}{10}}.$

**51.** $\sqrt{\dfrac{3.5}{10}}.$

**263.** **Reduction of a surd to an entire surd.**

**Ex.** Express $4\,a\sqrt{b}$ as an entire surd.

$$4\,a\sqrt{b}=\sqrt{16\,a^2}\sqrt{b}=\sqrt{16\,a^2b}.$$

**EXERCISE 95**

Express as entire surds:

**1.** $4\sqrt{5}.$　　**3.** $2\sqrt[3]{11}.$　　**5.** $\tfrac{1}{2}\sqrt[3]{12}.$　　**7.** $\tfrac{4}{5}\sqrt{35\,a}.$

**2.** $3\sqrt{7}.$　　**4.** $3\sqrt[3]{5}.$　　**6.** $a\sqrt{b}.$　　**8.** $\tfrac{3}{2}\sqrt[3]{12\,ab}.$

**9.** $\dfrac{a}{b}\sqrt{\dfrac{b^3}{a^3}}.$

**10.** $\dfrac{2\,m}{3\,n}\sqrt[3]{\dfrac{27\,n^2}{4}}.$

**11.** $\dfrac{xy}{z}\sqrt[3]{\dfrac{4\,z^4}{x^3y^3}}.$

* See table of square roots on page 164.

**264. Transformation of surds to surds of different order.**

**Ex. 1.** Transform $\sqrt[4]{a^3b^2}$ into a surd of the 20th order.

$$\sqrt[4]{a^3b^2} = a^{\frac{3}{4}}b^{\frac{2}{4}} = a^{\frac{15}{20}}b^{\frac{10}{20}} = \sqrt[20]{a^{15}b^{10}}.$$

**Ex. 2.** Transform $\sqrt{2}$, $\sqrt[3]{3}$, and $\sqrt[4]{5}$ into surds of the same lowest order.

$$\sqrt{2} = 2^{\frac{1}{2}} = 2^{\frac{6}{12}} = \sqrt[12]{64}.$$
$$\sqrt[3]{3} = 3^{\frac{1}{3}} = 3^{\frac{4}{12}} = \sqrt[12]{81}.$$
$$\sqrt[4]{5} = 5^{\frac{1}{4}} = 5^{\frac{3}{12}} = \sqrt[12]{125}.$$

**Ex. 3.** Reduce the order of the surd $\sqrt[16]{a^{12}}$.

$$\sqrt[16]{a^{12}} = a^{\frac{12}{16}} = a^{\frac{3}{4}} = \sqrt[4]{a^3}.$$

*Exponent and index bear the same relation as numerator and denominator of a fraction ; and hence both may be multiplied by the same number, or both divided by the same number, without changing the value of the radical.*

**EXERCISE 96**

Reduce to surds of the 6th order:

1. $\sqrt{x}$. 2. $\sqrt[3]{mn}$. 3. $\sqrt{\dfrac{xy}{z}}$. 4. $\sqrt[12]{a^3}$. 5. $\sqrt{\dfrac{2\,xy}{3}}$. 6. $mn$.

Reduce to surds of the 12th order:

7. $\sqrt{2\,a}$. 9. $\sqrt[3]{a^4b^2c}$. 11. $\sqrt[6]{a^{-5}b}$. 13. $\sqrt[24]{a^2b^2c^2}$.

8. $\sqrt[3]{m^2n}$. 10. $\sqrt[4]{3\,ax}$. 12. $\sqrt[6]{5\,x^9z^7}$. 14. $a$.

Express as surds of lowest order with integral exponents and indices:

15. $\sqrt[6]{a^2}$. 16. $\sqrt[4]{a^2b^2c^2}$. 17. $\sqrt[4]{4\,a^2}$. 18. $\sqrt[6]{a^9b^{12}c^3}$.

19. $\sqrt[5]{\dfrac{a^{25}}{b^{10}}}$. 21. $\sqrt[30]{2^{10}m^{20}}$. 23. $\sqrt[11]{\dfrac{(a+b)^{22}}{c^{44}}}$.

20. $\sqrt[14]{\dfrac{m^7n^{14}}{2^7}}$. 22. $\sqrt[4]{81\,m^2n^4}$. 24. $\sqrt[6]{512}$.

P

Express as surds of the same lowest order:

25. $\sqrt{3}$, $\sqrt[3]{2}$.  29. $2^{\frac{1}{2}}$, $3^{\frac{1}{3}}$.  32. $\sqrt{a^3}$, $\sqrt[6]{a}$, $\sqrt[10]{ab}$.

26. $\sqrt[4]{2}$, $\sqrt[3]{3}$.  30. $\sqrt{2}$, $\sqrt[3]{3}$, $\sqrt[6]{5}$.  33. $\sqrt{3}$, $\sqrt[3]{3}$, $\sqrt[4]{4}$.

27. $\sqrt[4]{3}$, $\sqrt[3]{2}$.  31. $\sqrt[3]{3}$, $\sqrt[4]{5}$, $\sqrt[6]{7}$.  34. $\sqrt[3]{2}$, $\sqrt[6]{4}$, $\sqrt[4]{20}$.

28. $\sqrt[5]{7}$, $\sqrt{2}$.

Arrange in order of magnitude:

35. $\sqrt[3]{3}$, $\sqrt{2}$.  37. $\sqrt[3]{7}$, $\sqrt{5}$.  39. $5\sqrt{2}$, $4\sqrt[3]{4}$.

36. $\sqrt[3]{4}$, $\sqrt[4]{5}$.  38. $\sqrt{5}$, $\sqrt[3]{11}$, $\sqrt[6]{126}$.  40. $\sqrt[3]{2}$, $\sqrt[5]{3}$, $\sqrt[15]{30}$.

## ADDITION AND SUBTRACTION OF RADICALS

**265.** *To add or subtract surds, reduce them to their simplest form. If the resulting surds are similar, add them like similar terms (i.e. add their coefficients); if dissimilar, connect them by their proper signs.*

Ex. 1. Simplify $\sqrt{\frac{1}{2}} + 3\sqrt{18} - 2\sqrt{50}$.

$$\sqrt{\tfrac{1}{2}} + 3\sqrt{18} - 2\sqrt{50} = \tfrac{1}{2}\sqrt{2} + 9\sqrt{2} - 10\sqrt{2} = -\tfrac{1}{2}\sqrt{2}.$$

Ex. 2. Simplify $\sqrt[3]{a^3x} - 3\sqrt[3]{\dfrac{b^6}{x^2}} + \sqrt[3]{\dfrac{27x}{y^3}}$.

$$\sqrt[3]{a^3x} - 3\sqrt[3]{\frac{b^6}{x^2}} + \sqrt[3]{\frac{27x}{y^3}} = a\sqrt[3]{x} - \frac{3b^2}{x}\sqrt[3]{x} + \frac{3}{y}\sqrt[3]{x} = \left(a - \frac{3b^2}{x} + \frac{3}{y}\right)\sqrt[3]{x}.$$

Ex. 3. Simplify $\sqrt{\frac{9}{2}} - \sqrt[3]{\frac{1}{4}} + \sqrt{72} - 4\sqrt{\frac{1}{3}} + \sqrt[3]{16}$.

$$\sqrt{\tfrac{9}{2}} - \sqrt[3]{\tfrac{1}{4}} + \sqrt{72} - 4\sqrt{\tfrac{1}{3}} + \sqrt[3]{16} = \tfrac{3}{2}\sqrt{2} - \tfrac{1}{2}\sqrt[3]{2} + 6\sqrt{2} - \tfrac{4}{3}\sqrt{3} + 2\sqrt[3]{2}$$

$$= \tfrac{15}{2}\sqrt{2} + \tfrac{3}{2}\sqrt[3]{2} - \tfrac{4}{3}\sqrt{3}.$$

**EXERCISE 97**

Simplify the following expressions:

1. $\sqrt{24} + \sqrt{54} - \sqrt{6}$.

2. $2\sqrt{8} - 7\sqrt{18} + 5\sqrt{72} - \sqrt{50}$.

3. $\sqrt{12} + 2\sqrt{27} + 3\sqrt{75} - 9\sqrt{48}$.

4. $\sqrt{18} + \sqrt{32} - \sqrt{128} + \sqrt{2}$.

5. $\sqrt{175} - \sqrt{28} + \sqrt{63} - 4\sqrt{7}$.

6. $\sqrt{\frac{1}{2}} + \sqrt{8} - \sqrt{\frac{9}{2}} + \sqrt{50}$.

7. $4\sqrt{80} - 5\sqrt{45} - 3\sqrt{20} + 6\sqrt{5}$.

8. $8\sqrt{18} + 2\sqrt{32} + 3\sqrt{8} - 35\sqrt{2}$.

9. $8\sqrt{\frac{3}{4}} - \frac{1}{2}\sqrt{12} + 4\sqrt{27} - 2\sqrt{\frac{3}{16}}$.

10. $2\sqrt{\frac{5}{3}} + \sqrt{60} - \sqrt{15} + \sqrt{\frac{3}{5}}$.

11. $7\sqrt[3]{54} + 3\sqrt[3]{16} + \sqrt[3]{2} - 5\sqrt[3]{128}$.

12. $\sqrt[3]{81} - 2\sqrt[3]{24} + \sqrt{28} + 2\sqrt{63}$.

13. $\sqrt{45\,c^3} - \sqrt{80\,c^3} + \sqrt{5\,a^2c} + c\sqrt{5\,c}$.

14. $3\,ab\sqrt{ab} + 3\,a\sqrt{ab^3} - 3\sqrt{a^3b^3} - ab\sqrt{4\,ab}$.

15. $\sqrt{\dfrac{m}{3}} + \sqrt{3\,m} + \sqrt{27m} - \sqrt{\dfrac{100\,m}{3}}$.

16. $\dfrac{1}{x}\sqrt{21\,x} - 3\sqrt{\dfrac{3}{7\,x}} + 4\sqrt{\dfrac{7}{3\,x}} + \dfrac{1}{21\,x}\sqrt{84\,x}$.

17. $\frac{31}{40}\sqrt{10} - \frac{1}{2}\sqrt{\frac{5}{8}} - 3\sqrt{\frac{5}{2}} + \sqrt{1\frac{2}{8}\frac{5}{}}$.

18. $\sqrt{1\frac{1}{3}} + \sqrt{16\frac{1}{3}} + \sqrt{21\frac{1}{3}}$.

19. $\sqrt{(a+b)^2x} + \sqrt{(a-b)^2x} - \sqrt{a^2x}$.

20. $\sqrt{4 + 4\,x^2} + \sqrt{9 + 9\,x^2} - \sqrt{16 + 16\,x^2}$.

21. $\sqrt{\dfrac{1}{x}} - \sqrt{\dfrac{1}{x^3}} + \sqrt{\dfrac{1}{x^5}}.$

22. $\sqrt{\dfrac{11\,b}{a}} - \sqrt{\dfrac{11\,a}{b}} + \sqrt{11\,ab}.$

23. $\sqrt{8\,a^2} + \sqrt[3]{8\,a^2} + \sqrt{18\,a^2} - \sqrt[3]{27\,a^2}.$

24. $\dfrac{\sqrt{98\,ab}}{16} - \sqrt{\dfrac{ab}{2}} + \sqrt{\dfrac{ab}{8}} - \sqrt{\dfrac{ab}{32}} - \sqrt{\dfrac{ab}{128}}.$

25. $\sqrt[3]{\dfrac{27\,a^5x}{2\,b}} - \sqrt[3]{\dfrac{a^2x}{2\,b}} - 3\sqrt[3]{\dfrac{a^5x}{2\,b}}.$

## MULTIPLICATION OF RADICALS

**266.** *Surds of the same order are multiplied by multiplying the product of the coefficients by the product of the irrational factors, for* $a\sqrt[n]{x} \cdot b\sqrt[n]{y} = ab\sqrt[n]{xy}.$

*Dissimilar surds are reduced to surds of the same order, and then multiplied.*

**Ex. 1.** Multiply $3\sqrt[3]{25\,y^2}$ by $5\sqrt[3]{50\,y^2}.$

$$3\sqrt[3]{25\,y^2} \cdot 5\sqrt[3]{50\,y^2} = 15\sqrt[3]{5^2 \cdot 2 \cdot 5^2 \cdot y^4} = 75\,y\sqrt[3]{10\,y}.$$

**Ex. 2.** Multiply $\sqrt{2}$ by $3\sqrt[3]{4}.$

$$\sqrt{2} \cdot 3\sqrt[3]{4} = \sqrt[6]{2^3} \cdot 3\sqrt[6]{4^2} = \sqrt[6]{2^3} \cdot 3\sqrt[6]{2^4} = 3\sqrt[6]{2^7} = 6\sqrt[6]{2}.$$

**Ex. 3.** Multiply $5\sqrt{7} - 2\sqrt{5}$ by $3\sqrt{7} + 10\sqrt{5}.$

$$
\begin{array}{r}
5\sqrt{7} - \ 2\sqrt{5} \\
3\sqrt{7} + 10\sqrt{5} \\
\hline
105 - \ 6\sqrt{35} \\
+\ 50\sqrt{35} - 100 \\
\hline
105 + 44\sqrt{35} - 100 = 5 + 44\sqrt{35}.
\end{array}
$$

## EXERCISE 98

1. $\sqrt{3} \cdot \sqrt{12}$.

2. $\sqrt{2} \cdot \sqrt{50}$.

3. $\sqrt{3} \cdot \sqrt{6}$.

4. $\sqrt{5} \cdot \sqrt{10}$.

5. $\sqrt{7} \cdot \sqrt{42}$.

6. $\sqrt{10} \cdot \sqrt{15}$.

7. $\sqrt{14} \cdot \sqrt{35}$.

8. $\sqrt{20} \cdot \sqrt{30}$.

9. $\sqrt[3]{4} \cdot \sqrt[3]{2}$.

10. $\sqrt[3]{3} \cdot \sqrt[3]{9}$.

11. $\sqrt[3]{18} \cdot \sqrt[3]{3}$.

12. $\sqrt[3]{5} \cdot \sqrt[3]{50}$.

13. $\sqrt{a} \cdot \sqrt{x}$.

14. $\sqrt{a} \cdot \sqrt{5\,a}$.

15. $\sqrt{y} \cdot \sqrt{8\,y^2}$.

16. $a\sqrt{x} \cdot b\sqrt{4\,x^3}$.

17. $\sqrt{2\,a} \cdot \sqrt{8\,a^3}$.

18. $\sqrt{14\,c^3} \cdot \sqrt{70\,c^5}$.

19. $\sqrt[n]{a} \cdot \sqrt[n]{b} \cdot \sqrt[n]{c}$.

20. $\sqrt{a} \cdot \sqrt{\dfrac{3\,x}{4}}$.

21. $\sqrt{mn} \cdot \sqrt{\dfrac{n}{4\,m}}$.

22. $\sqrt{\dfrac{7\,a}{40}} \cdot \sqrt{\dfrac{21\,a}{10}}$.

23. $\sqrt{\dfrac{5\,a}{6\,b}} \cdot \sqrt{\dfrac{10\,a}{3\,b}}$.

24. $\sqrt[5]{2\,ab^3} \cdot \sqrt[5]{16\,a^9 b^{11}}$.

25. $(\sqrt{2} + \sqrt{3} + \sqrt{4})\sqrt{3}$.

26. $(3\sqrt{3} + 4\sqrt{5})\sqrt{5}$.

27. $(5\sqrt{2} - 2\sqrt{3} - 6\sqrt{5})\sqrt{3}$.

28. $(3 + \sqrt{5})(2 - \sqrt{5})$.

29. $(a + \sqrt{b})(a - \sqrt{b})$.

30. $(\sqrt{m} - \sqrt{n})(\sqrt{m} + \sqrt{n})$.

31. $(\sqrt{5\,x} + \sqrt{2\,y})(\sqrt{5\,x} - \sqrt{2\,y})$.

32. $(4\sqrt{2\,m} - 3\sqrt{3\,n})(4\sqrt{2\,m} + 3\sqrt{3\,n})$.

33. $(\sqrt{m+1} - \sqrt{m})(\sqrt{m+1} + \sqrt{m})$.

34. $(\sqrt{a} - \sqrt{a-b})(\sqrt{a} + \sqrt{a-b})$.

35. $\left(\sqrt{\dfrac{m+n}{2}} - \sqrt{\dfrac{m-n}{2}}\right)\left(\sqrt{\dfrac{m+n}{2}} + \sqrt{\dfrac{m-n}{2}}\right)$.

36. $(6\sqrt{2} - 3\sqrt{3})(6\sqrt{2} + 3\sqrt{3})$.

37. $(5\sqrt{5} - 8\sqrt{2})(5\sqrt{5} + 8\sqrt{2})$.

38. $(\sqrt{m} - \sqrt{n})^2$.

39. $(\sqrt{3} - \sqrt{2})^2$.

40. $(\sqrt{5} + 1)^2$.

41. $(2 - \sqrt{3})^2$.

**42.** $(3\sqrt{5} - 5\sqrt{3})^2$.      **43.** $\left(\sqrt{\dfrac{a}{b}} - \sqrt{\dfrac{b}{a}}\right)^2$.

**44.** $(3\sqrt{3} - 2\sqrt{5})(2\sqrt{3} + \sqrt{5})$.

**45.** $(2\sqrt{3} - \sqrt{5})(\sqrt{3} + 2\sqrt{5})$.

**46.** $(5\sqrt{7} - 2\sqrt{2})(2\sqrt{7} - 7\sqrt{2})$.

**47.** $(5\sqrt{2} + \sqrt{10})(2\sqrt{5} - 1)$.

**48.** $(1 + \sqrt{2} - \sqrt{3})(1 + \sqrt{2} + \sqrt{3})$.

**49.** $(3\sqrt{5} - 2\sqrt{3})(2\sqrt{3} - \sqrt{3})$.

**50.** $(\sqrt{5} + \sqrt{3} + \sqrt{2})(\sqrt{5} - \sqrt{3})$.

**51.** $(\sqrt{6} + \sqrt{2} + \sqrt{8})(\sqrt{6} + \sqrt{2} - \sqrt{8})$.

**52.** $\sqrt{a} \cdot \sqrt[3]{a}$.      **53.** $\sqrt[3]{a} \cdot \sqrt[6]{a^5}$.

## DIVISION OF RADICALS

**267.** *Monomial surds of the same order may be divided by multiplying the quotient of the coefficients by the quotient of the surd factors.* E.g. $a\sqrt{b} \div x\sqrt{y} = \dfrac{a}{x}\sqrt{\dfrac{b}{y}}$.

Since surds of different orders can be reduced to surds of the same order, all monomial surds may be divided by this method.

Ex. 1. $4\sqrt{48} \div 3\sqrt{6} = \frac{4}{3}\sqrt{8} = \frac{8}{3}\sqrt{2}$.

Ex. 2. $(\sqrt{50} + 3\sqrt{12}) \div \sqrt{2} = \sqrt{25} + 3\sqrt{6} = 5 + 3\sqrt{6}$.

**268.** If, however, the quotient of the surds is a fraction, it is more convenient to multiply dividend and divisor by a factor which makes the divisor rational.

This method, called **rationalizing the divisor**, is illustrated by the following examples:

**Ex. 1.** Divide $\sqrt{11}$ by $\sqrt{7}$.

In order to make the divisor ($\sqrt{7}$) rational, we have to multiply by $\sqrt{7}$.

$$\frac{\sqrt{11}}{\sqrt{7}} = \frac{\sqrt{11}}{\sqrt{7}} \cdot \frac{\sqrt{7}}{\sqrt{7}} = \frac{\sqrt{77}}{7} = \tfrac{1}{7}\sqrt{77}.$$

**Ex. 2.** Divide $4\sqrt[3]{3\,a}$ by $3\sqrt[3]{2\,b^2}$.

The rationalizing factor is evidently $\sqrt[3]{4\,b}$; hence,

$$\frac{4\sqrt[3]{3\,a}}{3\sqrt[3]{2\,b^2}} = \frac{4\sqrt[3]{3\,a}}{3\sqrt[3]{2\,b^2}} \cdot \frac{\sqrt[3]{4\,b}}{\sqrt[3]{4\,b}} = \frac{4\sqrt[3]{12\,ab}}{3 \cdot 2\,b} = \frac{2\sqrt[3]{12\,ab}}{3\,b}.$$

**Ex. 3.** Divide $12\sqrt{3} + 4\sqrt{5}$ by $\sqrt{8}$.

Since $\sqrt{8} = 2\sqrt{2}$, the rationalizing factor is $\sqrt{2}$,

$$\frac{12\sqrt{3} + 4\sqrt{5}}{\sqrt{8}} = \frac{12\sqrt{3} + 4\sqrt{5}}{\sqrt{8}} \cdot \frac{\sqrt{2}}{\sqrt{2}} = \frac{12\sqrt{6} + 4\sqrt{10}}{4} = 3\sqrt{6} + \sqrt{10}.$$

**269.** To show that expressions with rational denominators are simpler than those with irrational denominators, arithmetical problems afford the best illustrations. To find, *e.g.*, $\dfrac{1}{\sqrt{3}}$ by the usual arithmetical method, we have

$$\frac{1}{\sqrt{3}} = \frac{1}{1.73205}.$$

But if we simplify $\quad \dfrac{1}{\sqrt{3}} = \dfrac{\sqrt{3}}{3} = \dfrac{1.73205}{3}.$

Either quotient equals .57735. Evidently, however, the division by 3 is much easier to perform than the division by 1.73205. Hence in arithmetical work it is always best to rationalize the denominators before dividing.

Simplify:

1. $\dfrac{\sqrt{14}}{\sqrt{7}}$.

2. $\dfrac{\sqrt{3\,n}}{\sqrt{n}}$.

3. $\dfrac{\sqrt{21}}{\sqrt{7}}$.

4. $\dfrac{\sqrt{48}}{\sqrt{12}}$.

5. $\dfrac{\sqrt{11\,n}}{\sqrt{11}}$.

6. $\dfrac{\sqrt[3]{54}}{\sqrt[3]{2}}$.

7. $\dfrac{\sqrt{84\,xy}}{\sqrt{7\,xy}}$.

8. $\sqrt{\tfrac{5}{4}} \div \sqrt{\tfrac{4}{5}}$.

9. $\dfrac{1}{\sqrt{2}}$.

10. $\dfrac{3}{\sqrt{3}}$.

11. $\dfrac{a}{\sqrt{a}}$.

12. $\dfrac{p}{\sqrt[3]{p}}$.

13. $\dfrac{11\,n}{\sqrt{11}}$.

14. $\dfrac{5}{2\sqrt{5}}$.

15. $\dfrac{3}{2\sqrt{3}}$.

16. $\dfrac{a+b}{\sqrt{a+b}}$.

17. $\dfrac{5\,a}{\sqrt{8}}$.

18. $\dfrac{6}{\sqrt{50}}$.

Given $\sqrt{2} = 1.4142$, $\sqrt{3} = 1.7320$, and $\sqrt{5} = 2.2361$, find to four decimal places the numerical values of:

19. $\dfrac{1}{\sqrt{2}}$.

20. $\dfrac{3}{\sqrt{3}}$.

21. $\dfrac{1}{\sqrt{8}}$.

22. $\dfrac{10}{\sqrt{5}}$.

23. $\dfrac{3}{\sqrt{8}}$.

24. $\dfrac{8}{\sqrt{48}}$.

25. $\dfrac{2}{\sqrt{50}}$.

**270.** Two binomial quadratic surds are said to be **conjugate,** if they differ only in the sign which connects their terms.

$\sqrt{a} + \sqrt{b}$ and $\sqrt{a} - \sqrt{b}$ are conjugate surds.

**271.** *The product of two conjugate binomial surds is rational.*

$$(\sqrt{a} + \sqrt{b})(\sqrt{a} - \sqrt{b}) = a - b.$$

**272.** To rationalize the denominator of a fraction whose denominator is a binomial quadratic surd, multiply numerator and denominator by the conjugate surd of the denominator.

**Ex. 1.** Simplify $\dfrac{2\sqrt{3}-\sqrt{2}}{\sqrt{3}-\sqrt{2}}$.

$$\frac{2\sqrt{3}-\sqrt{2}}{\sqrt{3}-\sqrt{2}}=\frac{2\sqrt{3}-\sqrt{2}}{\sqrt{3}-\sqrt{2}}\cdot\frac{\sqrt{3}+\sqrt{2}}{\sqrt{3}+\sqrt{2}}=\frac{4+\sqrt{6}}{3-2}=4+\sqrt{6}.$$

**Ex. 2.** Simplify $\dfrac{x-\sqrt{x^2-1}}{x+\sqrt{x^2-1}}$.

$$\frac{x-\sqrt{x^2-1}}{x+\sqrt{x^2-1}}=\frac{x-\sqrt{x^2-1}}{x+\sqrt{x^2-1}}\cdot\frac{x-\sqrt{x^2-1}}{x-\sqrt{x^2-1}}=\frac{x^2-2\,x\sqrt{x^2-1}+x^2-1}{x^2-(x^2-1)}$$

$$=2\,x^2-1-2\,x\sqrt{x^2-1}.$$

**Ex. 3.** Find the numerical value of:

$$\frac{\sqrt{2}+2}{2\sqrt{2}-1}.$$

$$\frac{\sqrt{2}+2}{2\sqrt{2}-1}=\frac{\sqrt{2}+2}{2\sqrt{2}-1}\cdot\frac{2\sqrt{2}+1}{2\sqrt{2}+1}=\frac{6+5\sqrt{2}}{7}=\frac{13.07105}{7}=1.8673.$$

**EXERCISE 100**

Rationalize the denominators of:

1. $\dfrac{1}{2+\sqrt{3}}$.

2. $\dfrac{3}{1+\sqrt{2}}$.

3. $\dfrac{12}{5-\sqrt{21}}$.

4. $\dfrac{7}{\sqrt{8}-2}$.

5. $\dfrac{\sqrt{3}}{2\sqrt{5}-3\sqrt{2}}$.

6. $\dfrac{1}{\sqrt{2}+\sqrt{3}}$.

7. $\dfrac{1}{\sqrt{7}-\sqrt{2}}$.

8. $\dfrac{2\sqrt{3}}{2-\sqrt{3}}$.

9. $\dfrac{14}{8-5\sqrt{2}}$.

10. $\dfrac{12}{7-3\sqrt{5}}$.

11. $\dfrac{1+\sqrt{2}}{2-\sqrt{2}}$.

12. $\dfrac{5-7\sqrt{3}}{1+\sqrt{3}}$.

13. $\dfrac{6 - 3\sqrt{5}}{\sqrt{5} - 1}$.

16. $\dfrac{6\sqrt{7} - 3\sqrt{3}}{\sqrt{5} - 2}$.

19. $\dfrac{1}{1 - \sqrt{x}}$.

14 $\dfrac{\sqrt{3} + \sqrt{2}}{\sqrt{3} - \sqrt{2}}$.

17. $\dfrac{5\sqrt{7} - 7\sqrt{5}}{\sqrt{5} - \sqrt{7}}$.

20. $\dfrac{\sqrt{x} - \sqrt{y}}{\sqrt{x} + \sqrt{y}}$.

15. $\dfrac{3\sqrt{5} - 2\sqrt{2}}{2\sqrt{5} - \sqrt{18}}$.

18. $\dfrac{m}{m - \sqrt{m}}$.

21. $\dfrac{a\sqrt{b} - b\sqrt{a}}{\sqrt{a} - \sqrt{b}}$.

22. $\dfrac{\sqrt{a}}{\sqrt{b} + \sqrt{c}}$.

Given $\sqrt{2} = 1.4142$, $\sqrt{3} = 1.7320$, and $\sqrt{5} = 2.2361$; find to four places of decimals:

23. $\dfrac{1}{\sqrt{2} - 1}$.

25. $\dfrac{12}{\sqrt{5} - 1}$.

27. $\dfrac{\sqrt{3}}{2 - \sqrt{3}}$.

24. $\dfrac{2}{\sqrt{3} + 1}$.

26. $\dfrac{\sqrt{5}}{1 + \sqrt{5}}$.

28. $\dfrac{\sqrt{18}}{3 - \sqrt{5}}$.

29. $\dfrac{5}{\sqrt{5} + 2}$.

30. $\dfrac{\sqrt{50}}{\sqrt{3} - 2}$.

**31.** Find the third proportional to $1 + \sqrt{2}$ and $3 + 2\sqrt{2}$.

## INVOLUTION AND EVOLUTION OF RADICALS

**273.** By the use of fractional exponents it can easily be shown that $\sqrt[m]{a^n} = (\sqrt[m]{a})^n$.

Hence $\sqrt{25^3} = (\sqrt{25})^3 = 5^3 = 125$.

$\sqrt[3]{8^5} = (\sqrt[3]{8})^5 = 2^5 = 32$.

**274.** In other examples of involution and evolution, introduce fractional exponents:

Ex. **1.** Simplify $\sqrt[3]{\sqrt[2]{x}}$.

$$\sqrt[3]{\sqrt[2]{x}} = (x^{\frac{1}{2}})^{\frac{1}{3}} = x^{\frac{1}{6}} = \sqrt[6]{x}.$$

Ex. **2.** Find the square of $\sqrt[3]{3\,ax^2}$.

$$(\sqrt[3]{3\,ax^2})^2 = \sqrt[3]{(3\,ax^2)^2} = \sqrt[3]{9\,a^2x^4} = x\sqrt[3]{9\,a^2x}.$$

**EXERCISE 101**

Simplify:

1. $(3\sqrt{mn})^2$.

2. $(\sqrt[3]{4\,x^2})^2$.

3. $(\sqrt{2\,x})^3$.

4. $\sqrt{25^3}$.

5. $\sqrt{64^3}$.

6. $\sqrt[3]{8^2}$.

7. $\sqrt[4]{16^3}$.

8. $\sqrt[3]{125^2}$.

9. $(5\,n\sqrt[3]{n})^6$.

10. $\sqrt[3]{\sqrt[2]{a^3b^3c^3}}$.

11. $(\sqrt[3]{\sqrt{a}})^2$.

12. $(\sqrt[3]{\sqrt[4]{5\,n}})^4$.

## SQUARE ROOTS OF QUADRATIC SURDS

**275.** **To find the square root of a binomial square by inspection.** According to § 63,

$$(\sqrt{5}+\sqrt{3})^2 = 5 + 2\sqrt{5\cdot 3} + 3$$
$$= 8 + 2\sqrt{15}.$$

If, on the other hand, we had to find $\sqrt{8 + 2\sqrt{15}}$, the problem would be quite simple if presented in the form $\sqrt{5 + 2\sqrt{3\cdot 5} + 3}$. To reduce it to this form, we must find two numbers whose sum is 8 and whose product is 15, viz. 5 and 3.

**Ex. 1.** Find $\sqrt{12 + 2\sqrt{20}}$.

Find two numbers whose sum is 12 and whose product is 20. These numbers are 10 and 2.

$$\sqrt{12 + 2\sqrt{20}} = \sqrt{10 + 2\sqrt{10 \times 2} + 2}$$
$$= \sqrt{10} + \sqrt{2}.$$

**Ex. 2.** Find $\sqrt{11 - 6\sqrt{2}}$.

Write the binomial so that the coefficient of the irrational term is 2.

$$\sqrt{11 - 6\sqrt{2}} = \sqrt{11 - 2\sqrt{18}}.$$

Find two numbers whose sum is 11, and whose product is 18. The numbers are 9 and 2.

Hence
$$\sqrt{11 - 6\sqrt{2}} = \sqrt{9 - 2\sqrt{2 \cdot 9} + 2}$$
$$= \sqrt{9} - \sqrt{2}$$
$$= 3 - \sqrt{2}.$$

**Ex. 3.** Find $\sqrt{4 + \sqrt{15}}$.

$$\sqrt{4 + \sqrt{15}} = \sqrt{\frac{8 + 2\sqrt{15}}{2}}$$
$$= \frac{\sqrt{5} + \sqrt{3}}{\sqrt{2}}$$
$$= \tfrac{1}{2}(\sqrt{10} + \sqrt{6}).$$

### EXERCISE 102

Extract the square roots of the following binomials:

1. $8 + 2\sqrt{15}$.
2. $3 - 2\sqrt{2}$.
3. $6 - 2\sqrt{5}$.
4. $11 + 2\sqrt{18}$.
5. $5 - \sqrt{24}$.

6. $7 - \sqrt{40}$.
7. $2 + \sqrt{3}$.
8. $4 + 2\sqrt{3}$.
9. $7 - \sqrt{48}$.
10. $3 - \sqrt{5}$.

11. $6 - \sqrt{32}$.
12. $9 + 4\sqrt{5}$.
13. $15 - 4\sqrt{11}$.
14. $8 + \sqrt{55}$.
15. $14 + 8\sqrt{3}$.

16. $a + b + 2\sqrt{ab}$.

17. $2a + b - 2\sqrt{a(a + b)}$.

Simplify the following expressions:

**18.** $\sqrt{13 - 2\sqrt{22}}.$

**21.** $\dfrac{1}{\sqrt{5 - 2\sqrt{6}}} - \dfrac{1}{\sqrt{5 + 2\sqrt{6}}}.$

**19.** $\dfrac{1}{\sqrt{7 - 2\sqrt{6}}}.$

**22.** $\dfrac{1}{\sqrt{7 + \sqrt{48}}} + \dfrac{1}{\sqrt{7 - \sqrt{48}}}.$

**20.** $\dfrac{4}{\sqrt{6 + 2\sqrt{5}}}.$

**23.** $\dfrac{\sqrt{3} - 1}{\sqrt{4 + \sqrt{12}}}.$

## RADICAL EQUATIONS

**276.** **A radical equation** is an equation involving an irrational root of an unknown number.

$$\sqrt{x} = 5, \quad \sqrt[3]{x + 3} = 7, \quad (2x - x^2)^{\frac{1}{3}} = 1, \text{ are radical equations.}$$

**277.** **Radical equations are rationalized,** *i.e.* they are transformed into rational equations, by raising both members to equal powers.

Before performing the involution, it is necessary in most examples to simplify the equation as much as possible, and to transpose the terms so that one radical stands alone in one member.

If all radicals do not disappear through the first involution, the process must be repeated.

**Ex. 1.** Solve $\sqrt{x^2 + 12} - x = 2.$

Transposing $x$, $\qquad\qquad \sqrt{x^2 + 12} = x + 2.$

Squaring both members, $\qquad x^2 + 12 = x^2 + 4x + 4.$

Transposing and uniting, $\qquad -4x = -8.$

Dividing by $-4$, $\qquad\qquad x = 2.$

*Check.* The value $x = 2$ reduces each member to 2.

Ex. 2.  Solve $\sqrt{4x+1} + \sqrt{4x+25} = 12$.

Transpose $\sqrt{4x+1}$,  $\sqrt{4x+25} = 12 - \sqrt{4x+1}$.

Squaring both members,  $4x+25 = 144 - 24\sqrt{4x+1} + 4x + 1$.

Transposing and uniting, $24\sqrt{4x+1} = 120$.

Dividing by 24,  $\sqrt{4x+1} = 5$.

Squaring both members,  $4x + 1 = 25$.

Therefore  $x = 6$.

*Check.*  $\sqrt{24+1} + \sqrt{24+25} = 5 + 7 = 12$.

**278. Extraneous roots.**  Squaring both members of an equation usually introduces a new root.  Thus $x - 2 = 3$ has only one root, viz. 5.

Squaring both members we obtain $x^2 - 4x + 4 = 9$, an equation which has two roots, viz. 5 and $-1$.

The squaring of both members of the given equation introduced the new root $-1$, a so-called *extraneous root*.  Since radical equations require for their solution the squaring of both members, the roots found are not necessarily roots of the given equation; they may be extraneous roots.

**279.**  *The results of the solution of radical equations must be substituted in the given equation to determine whether the roots are true roots or extraneous roots.*

Ex. 3.  Solve $\sqrt{x+1} + \sqrt{x+6} = \sqrt{8x+1}$.

Squaring both members,

$$x + 1 + 2\sqrt{x^2 + 7x + 6} + x + 6 = 8x + 1.$$

Transposing and uniting, $2\sqrt{x^2 + 7x + 6} = 6x - 6$.

Dividing by 2,  $\sqrt{x^2 + 7x + 6} = 3x - 3$.

Squaring both members,  $x^2 + 7x + 6 = 9x^2 - 18x + 9$.

Transposing,  $8x^2 - 25x + 3 = 0$.

Factoring,  $(x - 3)(8x - 1) = 0$.

Therefore  $x = 3$, or $x = \frac{1}{8}$.

*Check.*  If $x = \frac{1}{8}$, the first member $= \frac{3}{4}\sqrt{2} + \frac{7}{4}\sqrt{2} = \frac{5}{2}\sqrt{2}$; the second member $= \sqrt{2}$.

Hence $x = \frac{1}{3}$ does not satisfy the given equation ; it is an extraneous root. If $x = 3$, both members reduce to 5. Hence there is only one root, viz. $x = 3$.

NOTE. If the signs of the roots were not restricted, $x = \frac{1}{3}$ would be a root of the preceding equation, for it satisfies the equation

$$-\sqrt{x+1} + \sqrt{x+6} = \sqrt{8x+1}.$$

Ex. 4. Solve $\sqrt{x+1} + \sqrt{2x+3} = \dfrac{15}{\sqrt{2x+3}}$.

Clearing of fractions, $\sqrt{2x^2 + 5x + 3} + 2x + 3 = 15$.

| | |
|---|---|
| Transposing, | $\sqrt{2x^2 + 5x + 3} = 12 - 2x$. |
| Squaring, | $2x^2 + 5x + 3 = 144 - 48x + 4x^2$. |
| Transposing, | $2x^2 - 53x + 141 = 0$. |
| Factoring, | $(x - 3)(2x - 47) = 0$. |
| Therefore, | $x = 3$, or $x = \frac{47}{2}$. |

*Check.* If $x = 3$, both members reduce to 5.
If $x = \frac{47}{2}$, the left member $= \frac{47}{2}\sqrt{2}$, and the right member $= \frac{3}{2}\sqrt{2}$.
Hence $x = 3$ is the only root.

### EXERCISE 103 *

Solve the following equations :

1. $\sqrt{x} - 2 = 0$.

2. $2\sqrt{x} - 3 = 2$.

3. $\sqrt{x} - b = a$.

4. $2 + \sqrt[3]{x} = 5$.

5. $\sqrt{x+2} - 1 = 2$.

6. $12 - (x-1)^{\frac{1}{2}} = 10$.

7. $1 + \sqrt{x^2 + x - 11} = x$.

8. $2x - \sqrt{4x^2 + 5x - 2} = -1$.

9. $\sqrt{9x^2 - 12x - 51} + 3 = 3x$.

10. $\sqrt{x + 16} = x - 4$.

11. $\sqrt{x+6} + x = 14$.

12. $\sqrt{12 + x} = 2 + \sqrt{x}$.

13. $\sqrt{x+2} + \sqrt{x+14} = 6$.

14. $\sqrt{x+1} = 2 - \sqrt{x+3}$.

15. $3 + \sqrt{x+9} = 2\sqrt{x}$.

16. $\sqrt{6-x} = 4 - \sqrt{2x-1}$.

17. $\sqrt{25+x} + \sqrt{25-x} = 8$.

18. $\sqrt{x+4} + \sqrt{2x-1} = 6$.

19. $\sqrt{13x-1} - \sqrt{2x-1} = 5$.

20. $\sqrt{x+8} - \sqrt{x} = \sqrt{x+3}$.

* Exclude all solutions which do not satisfy the equation or which make the given radicals imaginary.

21. $\sqrt{x+2}+\sqrt{x-3}=\sqrt{3\,x+4}.$

22. $\sqrt{x+4}+\sqrt{2\,x+6}=\sqrt{8\,x+9}.$

23. $\sqrt{3\,x-3}+\sqrt{5\,x-19}=\sqrt{2\,x+8}.$

24. $x+\sqrt{x^2-\sqrt{1-2\,x}}=1.$

25. $(2\sqrt{x}-1)(3\sqrt{x}-1)=0.$

26. $\sqrt{x}+\sqrt{x-9}=\dfrac{36}{\sqrt{x-9}}.$

27. $\sqrt{x}+\sqrt{x-21}=\dfrac{35}{\sqrt{x}}.$     29. $\sqrt{x+7}+\sqrt{x}=\dfrac{28}{\sqrt{x+7}}.$

28. $\sqrt{x}+\sqrt{x-4}=\dfrac{8}{\sqrt{x-4}}.$     30. $\dfrac{\sqrt{x}+16}{\sqrt{x}+4}=\dfrac{\sqrt{x}+32}{\sqrt{x}+12}.$

31. $\dfrac{\sqrt{x}-8}{\sqrt{x}-6}=\dfrac{\sqrt{x}-4}{\sqrt{x}+2}.$

32. $\sqrt{x}+\sqrt{3\,a+x}=\dfrac{9\,a}{\sqrt{3\,a+x}}.$

33. $\sqrt{x+5}\cdot\sqrt{x+12}=12.$     34. $\dfrac{1-x}{1-\sqrt{x}}=\dfrac{4\,x+5}{4}.$

35. $\sqrt{m-y}+\sqrt{n-y}=\dfrac{m}{\sqrt{m-y}}.$

36. $\dfrac{1}{\sqrt{x+1}}+\dfrac{1}{\sqrt{x-1}}=\dfrac{1}{\sqrt{x^2-1}}.$

37. $\left(\dfrac{7-2\,x}{7+2\,x}\right)^{\frac{1}{2}}+\left(\dfrac{7+2\,x}{7-2\,x}\right)^{\frac{1}{2}}=\dfrac{3}{2}\sqrt{2}.$

**280.** Many radical equations may be solved by the method of § 238.

Ex. 1. Solve $x^{-\frac{1}{2}}-33\,x^{-\frac{1}{4}}+32=0.$

Factoring, $\qquad\qquad (x^{-\frac{1}{4}}-32)(x^{-\frac{1}{4}}-1)=0.$

Therefore $\qquad\qquad\qquad x^{-\frac{1}{4}}=32 \text{ or } 1.$

Raising both members to the $-\frac{2}{3}$ power,

$$x = 32^{-\frac{2}{3}} \text{ or } 1^{-\frac{2}{3}} = \tfrac{1}{16} \text{ or } 1.$$

**Ex. 2.** Solve $x^2 - 8\,x - 2\sqrt{x^2 - 8\,x + 40} = -5$.

Adding 40 to both members,  $x^2 - 8\,x + 40 - 2\sqrt{x^2 - 8\,x + 40} = 35.$

Let  $\sqrt{x^2 - 8\,x + 40} = y$, then $x^2 - 8\,x + 40 = y^2$.

Hence  $y^2 - 2\,y = 35.$

$$y^2 - 2\,y - 35 = 0.$$
$$(y - 7)(y + 5) = 0.$$

Therefore  $y = 7, \text{ or } y = -5.$

| | |
|---|---|
| $\sqrt{x^2 - 8\,x + 40} = 7,$ | or $\sqrt{x^2 - 8\,x + 40} = -5.$ |
| $x^2 - 8\,x + 40 = 49,$ | $x^2 - 8\,x + 40 = 25.$ |
| $x^2 - 8\,x - 9 = 0,$ | $x^2 - 8\,x + 15 = 0.$ |
| $(x - 9)(x + 1) = 0,$ | $(x - 5)(x - 3) = 0.$ |
| $x = 9 \text{ or } -1.$ | $x = 5 \text{ or } 3.$ |

Since both members of the equation were squared, some of the roots may be extraneous. Substituting, it will be found that 9 and $-1$ satisfy the equation, while 5 and 3 are extraneous roots.

This can be seen without substituting, for 5 and 3 are the roots of the equation $\sqrt{x^2 - 8\,x + 40} = -5$. But as the square root is restricted to its positive values, it cannot be equal to a negative quantity.

**EXERCISE 104***

Solve the following equations:

1. $x + \sqrt{x} = 6.$

2. $x - 2\sqrt{x} - 3 = 0.$

3. $4\,x - 12\,x^{\frac{1}{2}} = 16.$

4. $45 - 14\sqrt{x} = -x.$

5. $x^{\frac{2}{3}} - 2\,x^{\frac{1}{3}} - 24 = 0.$

6. $\sqrt{x} + \sqrt[4]{x} = 6.$

7. $x + 5 + \sqrt{x + 5} = 12.$

8. $x^2 + 9 + \sqrt{x^2 + 9} = 30.$

9. $x + \sqrt{x + 4} = 8.$

10. $x^2 + \sqrt{x^2 - 5} = 11.$

* Exclude extraneous roots and roots which make the given radicals imaginaries.

Q

**11.** $x^2 - 8x + 40 - 2\sqrt{x^2 - 8x + 40} = 35.$

**12.** $x^2 - 5x + 1 + 5\sqrt{x^2 - 5x + 1} = 50.$

**13.** $x^2 + x - 4\sqrt{x^2 + x + 3} = -6.$

**14.** $5x^2 + 11x - 12\sqrt{5x^2 + 11x - 36} = 36.$

**15.** $2x^2 - 3x + 6\sqrt{2x^2 - 3x + 2} = 14.$

**16.** $x^2 + 5x + 3\sqrt{x^2 + 5x + 7} + 3 = 0.$

**17.** $3x^2 - 4x + 2\sqrt{3x^2 - 4x - 6} - 21 = 0.$

**18.** $7 + 3\sqrt{x^2 + 3x - 2} = x^2 + 3x + 1.$

**19.** $x^2 - 7x + \sqrt{x^2 - 7x + 18} = 24.$

**20.** $6\sqrt{x^2 - 3x - 3} = x^2 - 3x + 2.$

**21.** $\sqrt{\dfrac{3x-3}{2x-5}} + \sqrt{\dfrac{2x-5}{3x-3}} = 2\tfrac{1}{2}.$

# CHAPTER XVIII

## THE FACTOR THEOREM

**281.** If $x^3 - 3x^2 + 4x + 8$ is divided by $x - 2$ and there is a remainder (which does not contain $x$), then

$$x^3 - 3x^2 + 4x + 8 = (x - 2) \times \text{Quotient} + \text{Remainder}.$$

Or, substituting $Q$ and $R$ respectively for "Quotient" and "Remainder," and transposing,

$$R = x^3 - 3x^2 + 4x + 8 - (x - 2)Q.$$

As $R$ does not contain $x$, we could, if $Q$ was known, assign to $x$ any value whatsoever and would always obtain the same answer for $R$.

If, however, we make $x = 2$, then $(x - 2)Q = 0$, no matter what the value of $Q$. Hence, even if $Q$ is unknown, we can find the value of $R$ by making $x = 2$.

$$R = 2^3 - 3 \cdot 2^2 + 4 \cdot 2 + 8 - 0 = 12.$$

**Ex. 1.** Without actual division, find the remainder obtained by dividing $3x^4 + 2x - 5$ by $x - 3$.

$$R = 3x^4 + 2x - 5 - (x - 3)Q.$$

Let $\quad x = 3,$

then $\quad R = 3 \cdot 81 + 2 \cdot 3 - 5 - 0 = 244.$

**Ex. 2.** Without actual division, find the remainder when $ax^4 + bx^3 + cx^2 + dx + e$ is divided by $x - m$.

$$R = ax^4 + bx^3 + cx^2 + dx + e - (x - m)Q.$$

Let $\quad x = m,$

then $\quad R = am^4 + bm^3 + cm^2 + dm + e.$

**282. The Remainder Theorem.** *If an integral rational expression involving x is divided by x — m, the remainder is obtained by substituting in the given expression m in place of x.*

*E.g.* The remainder of the division

$$(4x^5 - 4x + 11) \div (x + 3) \text{ is } 4(-3)^5 - 4(-3) + 11 = -949.$$

The remainder obtained by dividing

$$(x + 4)^4 - (x + 2)(x - 1) + 7 \text{ by } x - 1 \text{ is } 5^4 - 3 \cdot 0 + 7 = 632.$$

### EXERCISE 105

Without actual division find the remainder obtained by dividing:

1. $x^3 - 4x^2 + 2x - 3$ by $x - 2$.

2. $x^4 + 3x^3 - 2x^2 - 32x - 12$ by $x - 3$.

3. $x^4 - x^3 + 4x^2 - 7x + 2$ by $x + 2$.

4. $a^{100} - 50a^{47} + 48a^2 + 2$ by $a - 1$.

5. $x^5 - b^5$ by $x - b$.

6. $x^6 + b^6$ by $x + b$.

7. $a^7 + b^7$ by $a + b$.

8. $y^4 - 14y^3 - 13y^2 - 21y + 25$ by $y - 15$.

**283.** If the remainder is zero, the divisor is a factor of the dividend.

**The Factor Theorem.** *If a rational integral expression involving x becomes zero when m is written in place of x, x — m is a factor of the expression.*

*E.g.* if $x^3 - 3x^2 - 2x - 8$ is divided by $x - 4$, the remainder equals $4^3 - 3 \cdot 4^2 - 2 \cdot 4 - 8' = 0$, hence $(x - 4)$ is a factor of $x^3 - 3x^2 - 2x - 8$.

**284.** Only factors of the absolute term need be substituted for $x$.

**Ex. 1.** Factor $x^3 - 7x^2 + 7x + 15$.

The factors of the absolute term, *i.e.* 15, are $+1, -1, +3, -3, +5,$ $5, +15, -15$.

Let $x = 1$, then $x^3 - 7x^2 + 7x + 15$ does not vanish.

Let $x = -1$, then $x^3 - 7x^2 + 7x + 15 = 0$.

Therefore $x - (-1)$, or $x + 1$, is a factor.

By dividing by $x + 1$, we obtain

$$x^3 - 7x^2 + 7x + 15 = (x + 1)(x^2 - 8x + 15)$$
$$= (x + 1)(x - 3)(x - 5).$$

**EXERCISE 106**

Without actual division, show that

1. $4x^3 + 3x^2 - 2x - 5$ is divisible by $x - 1$.
2. $2x^5 - 2x^4 + 3x^3 - 7x^2 - 5x - 18$ is divisible by $x - 2$.
3. $x^4 - 34x^2 + 225$ is divisible by $x - 5$.

Resolve into factors :

4. $m^3 - 6m^2 + 11m - 6$.       8. $x^3 + 2x^2 - 5x - 6$.

5. $a^3 - 2a + 4$.       9. $2m^3 - 5m^2 - 13m + 30$

6. $p^3 - 5p^2 + 8p - 4$.       10. $a^3 - 8a^2 + 19a - 12$.

7. $p^3 - 9p^2 + 23p - 15$.       11. $b^3 - 8b^2 + 14b + 3$.

        12. $m^4 - 4m^3p - m^2p^2 + 16mp^3 - 12p^4$.

        13. $m^4 + m^3n - 25m^2n^2 + 19mn^3 + 4n^4$.

        14. $a^5 + 32$.

Solve the following equations by factoring :

15. $x^3 + 6x^2 + 11x + 6 = 0$.     21. $x^3 + 9x^2 + 27x + 27$.

16. $x^3 - 5x^2 + x + 15 = 0$.     22. $x^3 - 7x^2 + 16x - 12$.

17. $x^3 - 10x^2 + 29x - 20 = 0$.     23. $y^3 + 7y^2 + 2y - 40 = 0$.

18. $x^3 - 5x^2 + 3x + 9 = 0$.     24. $x^4 - 4x^3 + 2x^2 + 4x - 3 = 0$

19. $x^3 - 8x^2 + 19x - 12 = 0$.     25. $4x^3 - 3x^2 - 25x - 6 = 0$.

20. $x^3 - 7x + 6 = 0$.

**285.** If $n$ is a positive integer, it follows from the Factor Theorem that

**1.** $x^n - y^n$ *is always divisible by* $x - y$.

For substituting $y$ for $x$, $x^n - y^n = y^n - y^n = 0$.

**2.** $x^n + y^n$ *is divisible by* $x + y$, *if* $n$ *is odd*.

For $(-y)^n + y^n = 0$, if $n$ is odd.

By actual division we obtain the other factors, and have for any positive integral value of $n$,

$$x^n - y^n = (x - y)(x^{n-1} + x^{n-2}y + x^{n-3}y^2 \cdots + y^{n-1}).*$$

If $n$ is odd,

$$x^n + y^n = (x + y)(x^{n-1} - x^{n-2}y + x^{n-3}y^2 \cdots + y^{n-1});$$

$$e.g. \ x^5 - y^5 = (x - y)(x^4 + x^3y + x^2y^2 + xy^3 + y^4).$$

$$x^5 + y^5 = (x + y)(x^4 - x^3y + x^2y^2 - xy^3 + y^4).$$

**286.** It can readily be seen that $x^n + y^n$ is not divisible by either $x + y$ or $x - y$, if $n$ is even.

**287.** Two special cases of the preceding propositions are of importance, viz. :

$$x^3 + y^3 = (x + y)(x^2 - xy + y^2),$$

$$x^3 - y^3 = (x - y)(x^2 + xy + y^2).$$

Ex. 1. Factor $27 a^6 + 8$.

$$27 a^6 + 8 = (3 a^2)^3 + 2^3$$
$$= (3 a^2 + 2)(9 a^4 - 6 a^2 + 4).$$

**288.** The difference of two even powers should always be considered as a difference of two squares.

Ex. 2. Factor $m^6 - n^6$.

We may consider $m^6 - n^6$ either a difference of two squares or a dif-

---

* The symbol $\cdots$ means "and so forth to."

ference of two cubes. The first method, however, is preferable, since it leads more directly to the prime factors. Hence

$$m^6 - n^6 = (m^3 + n^3)(m^3 - n^3)$$
$$= (m + n)(m^2 - mn + n^2)(m - n)(m^2 + mn + n^2).$$

**Ex. 3.** Factor $a^{12} + b^{12}$.

$$a^{12} + b^{12} = (a^4)^3 + (b^4)^3$$
$$= (a^4 + b^4)(a^8 - a^4b^4 + b^8).$$

## EXERCISE 107

Resolve into prime factors:

| | | |
|---|---|---|
| 1. $a^3 - 1$. | 9. $216 a^3 + 27 b^3$. | 17. $a^5 + 1$. |
| 2. $x^3 + 1$. | 10. $1000 - x^6$. | 18. $m^{12} + 1$. |
| 3. $1 - a^3b^3$. | 11. $x^4 - x$. | 19. $1 + a^6b^6$. |
| 4. $x^3 - 8$. | 12. $a + a^4$. | 20. $512 - a^3$. |
| 5. $x^3 + 8$. | 13. $a^6 - b^6$. | 21. $x^3y^3 + 125$. |
| 6. $8 a^3 + 1$. | 14. $a^6 + b^6$. | 22. $64 m^3n^3 - y^6$. |
| 7. $8 a^3 - 1$. | 15. $x^6y^6 - 64$. | 23. $27 m^9 - 343$. |
| 8. $a^3 - 64 b^3$. | 16. $a^5 - 1$. | 24. $a^{10} - b^{10}$. |

Solve the following equations:

25. $x^3 - 8 = 0$.　　26. $y^3 + 8 = 0$.　　27. $x^3 - 27 = 0$.　　28. $x = \sqrt[3]{1}$.

# CHAPTER XIX

## SIMULTANEOUS QUADRATIC EQUATIONS

**289.** The **degree of an equation involving several unknown** quantities is equal to the greatest sum of the exponents of the unknown quantities contained in any term.

$xy + y = 4$ is of the second degree.

$x^3y + 5\,x^2y^3 - y^4$ is of the fifth degree.

**290.** Simultaneous quadratic equations involving two unknown quantities lead, in general, to equations of the fourth degree. A few cases, however, can be solved by the methods of quadratics.*

### I. EQUATIONS SOLVED BY FINDING $x + y$ AND $x - y$

**291.** If two of the quantities $x + y$, $x - y$, $xy$ are given, the third one can be found by means of the relation $(x + y)^2 - 4\,xy = (x - y)^2$.

Ex. 1. Solve $\begin{cases} x + y = 5, \\ xy = 4. \end{cases}$  (1)
(2)

Squaring (1), $\qquad x^2 + 2\,xy + y^2 = 25.$  (3)

(2) × 4, $\qquad\qquad\quad 4\,xy = 16.$  (4)

(3) − (4), $\qquad x^2 - 2\,xy + y^2 = 9.$

Hence, $\qquad\qquad\qquad x - y = \pm\,3.$  (5)

Combining (5) with (1), we have

$$\begin{array}{ccc} x + y = 5, & & x + y = 5, \\ x - y = 3. & \text{or} & x - y = -3. \end{array}$$

Hence $\quad \begin{cases} x = 4, \\ y = 1. \end{cases} \quad \text{or} \quad \begin{cases} x = 1, \\ y = 4. \end{cases}$

---

* *The graphic solution of simultaneous quadratic equations has been treated in Chapter XII.

**292.** In many cases two of the quantities $x + y$, $x - y$, and $xy$ are not given, but can be found.

Ex. 2. $\begin{cases} 2x^2 - 3xy + 2y^2 = 8, & (1) \\ x - y = 1. & (2) \end{cases}$

Square (2), $\qquad x^2 - 2xy + y^2 = 1.$ $\qquad\qquad$ (3)

(3) × 2, $\qquad\quad 2x^2 - 4xy + 2y^2 = 2.$ $\qquad\qquad$ (4)

(1) − (4), $\qquad\qquad\qquad xy = 6.$

Hence $\qquad\qquad\qquad 4xy = 24.$ $\qquad\qquad\qquad$ (5)

(3) + (5), $\qquad\quad x^2 + 2xy + y^2 = 25.$

Therefore $\qquad x + y = + 5,$ $\qquad$ or $\; x + y = - 5.$

But $\qquad\qquad x - y = 1,$ $\qquad\qquad x - y = 1.$

Hence $\qquad x = 3, \; y = 2,$ $\qquad\; x = - 2, \; y = - 3.$

Check. $\begin{cases} 2 \cdot 3^2 - 3 \cdot 3 \cdot 2 + 2 \cdot 2^2 = 8, \\ 3 - 2 = 1. \end{cases}$ $\begin{cases} 2 \cdot 2^2 - 3 \cdot 2 \cdot 3 + 2 \cdot 3^2 = 8, \\ -2 + 3 = 1. \end{cases}$

**293.** The roots of simultaneous quadratic equations must be arranged in pairs, *e.g.* the answers of the last example are:

$$\begin{cases} x = 3, \\ y = 2, \end{cases} \text{ or } \begin{cases} x = -2, \\ y = -3. \end{cases}$$

**EXERCISE 108**

Solve:

1. $\begin{cases} x + y = 6, \\ xy = 8. \end{cases}$

5. $\begin{cases} x - y = 19, \\ xy = 66. \end{cases}$

9. $\begin{cases} x^2 + y^2 = 53, \\ xy = 14. \end{cases}$

2. $\begin{cases} x - y = 2, \\ xy = 15. \end{cases}$

6. $\begin{cases} x + y = 29, \\ xy = 100. \end{cases}$

10. $\begin{cases} x^2 - y^2 = 21, \\ xy = 110. \end{cases}$

3. $\begin{cases} x + y = 2, \\ xy = -15. \end{cases}$

7. $\begin{cases} x - y = 61, \\ xy = 876. \end{cases}$

11. $\begin{cases} x - y = 10, \\ x^2 + y^2 = 178. \end{cases}$

4. $\begin{cases} x + y = 40, \\ xy = 300. \end{cases}$

8. $\begin{cases} x^2 + y^2 = 25, \\ x + y = 7. \end{cases}$

12. $\begin{cases} x - y = 14, \\ x^2 + y^2 = 436. \end{cases}$

**13.** $\begin{cases} x+y=8, \\ x^2+y^2=32. \end{cases}$      **16.** $\begin{cases} x^2-xy+y^2=19, \\ x+y=7. \end{cases}$

**14.** $\begin{cases} x^2+y^2=37, \\ x+y=7. \end{cases}$      **17.** $\begin{cases} x^2+3xy+y^2=61, \\ x+y=7, \end{cases}$

**15.** $\begin{cases} x^2+xy+y^2=21, \\ x-y=3. \end{cases}$      **18.** $\begin{cases} x^2+y^2=a^2-2ab+2b^2, \\ x+y=a. \end{cases}$

**19.** $\begin{cases} \dfrac{1}{x}+\dfrac{1}{y}=\dfrac{1}{5}, \\ \dfrac{10}{xy}=\dfrac{1}{18}. \end{cases}$

## I. ONE EQUATION LINEAR, THE OTHER QUADRATIC

**294.** A system of simultaneous equations, one linear and ne quadratic, can be solved by eliminating one of the unknown uantities by means of substitution.

Ex.    Solve $2x+3y=7$,            (1)

$\qquad\qquad x^2+2y^2-y=5.$           (2)

From (1) we have,                  $x=\dfrac{7-3y}{2}.$     (3)

Substituting in (2),    $\left(\dfrac{7-3y}{2}\right)^2+2y^2-y=5.$

Simplifying,     $49-42y+9y^2+8y^2-4y=20.$

Transposing, etc.,       $17y^2-46y+29=0.$

Factoring,          $(y-1)(17y-29)=0.$

Hence               $y=1,$ or $\frac{29}{17}.$

Substituting in (3),       $x=2,$ or $\frac{14}{17}.$

**EXERCISE 109**

Solve :

**1.** $\begin{cases} x^2+y^2=17, \\ x-4y=0. \end{cases}$    **3.** $\begin{cases} x^2+xy=6, \\ x-y=1. \end{cases}$    **5.** $\begin{cases} x^2+4xy=28, \\ x+y=5. \end{cases}$

**2.** $\begin{cases} x^2+2y^2=11, \\ x-3y=0. \end{cases}$    **4.** $\begin{cases} x^2-2y^2=1, \\ 2x-y=4. \end{cases}$    **6.** $\begin{cases} x^2+y^2=5, \\ x+2y=5. \end{cases}$

**7.** $\begin{cases} 7\,x^2 + 5\,xy - 6\,y^2 = 32, \\ x - y = 1. \end{cases}$

**11.** $\begin{cases} x^2 + 3\,xy + y^2 + x + y = 68, \\ x + y = 7. \end{cases}$

**8.** $\begin{cases} 7\,xy - 10\,x - 30\,y = 0, \\ \dfrac{11\,x}{12} - y = \dfrac{y}{10}. \end{cases}$

**12.** $\begin{cases} 2\,x^2 + y^2 = 54, \\ x - 2\,y = 1. \end{cases}$

**9.** $\begin{cases} 6\,y^2 - xy = 2\,x^2, \\ 9\,y - 12 = -4\,x. \end{cases}$

**13.** $\begin{cases} x^2 + y^2 = 29, \\ \dfrac{x + y}{x - y} = \dfrac{7}{3}. \end{cases}$

**10.** $\begin{cases} x^2 + 3\,xy = 10, \\ x + y = 4. \end{cases}$

**14.** $\begin{cases} x : y = 2 : 3, \\ 3\,x - 12 = -3\,xy. \end{cases}$

## III. HOMOGENEOUS EQUATIONS

**295.** A **homogeneous equation** is an equation all of whose terms are of the same degree with respect to the unknown quantities.

$4\,x^3 - 3\,x^2y = 3\,y^3$ and $x^2 - 2\,xy - 5\,y^2 = 0$ are homogeneous equations.

**296.** If one equation of two simultaneous quadratics is homogeneous, the example can always be reduced to an example of the preceding type.

Ex. 1. Solve $\qquad \begin{cases} x^2 - 3\,y^2 + 2\,y = 3, & (1) \\ 2\,x^2 - 7\,xy + 6\,y^2 = 0. & (2) \end{cases}$

Factor (2), $\qquad (x - 2\,y)(2\,x - 3\,y) = 0.$

Hence we have to solve the two systems :

$\begin{cases} x - 2\,y = 0, \\ x^2 - 3\,y^2 + 2\,y = 3. \end{cases}$ or $\begin{cases} 2\,x - 3\,y = 0, & (3) \\ x^2 - 3\,y^2 + 2\,y = 3. & (1) \end{cases}$

From (3), $\qquad x = 2\,y.$

$x = \tfrac{3}{2}\,y.$

Substituting in (1),

$\dfrac{9\,y^2}{4} - 3\,y^2 + 2\,y = 3,$

$4\,y^2 - 3\,y^2 + 2\,y = 3,$

$3\,y^2 - 8\,y + 12 = 0,$

$y^2 + 2\,y - 3 = 0.$

$y = \dfrac{8 \pm \sqrt{-80}}{6}.$

$(y - 1)(y + 3) = 0.$

$y = \dfrac{4 + 2\sqrt{-5}}{3}, \quad \dfrac{4 - 2\sqrt{-5}}{3},$

Hence $\quad \begin{matrix} y = 1, \\ x = 2. \end{matrix} \Big\} \quad \begin{matrix} -3, \\ -6. \end{matrix} \Big\}$

$x = 2 + \sqrt{-5}, \quad 2 - \sqrt{-5}. \Big\}$

**297.** If both equations are homogeneous with exception of the absolute term, the problem can be reduced to the preceding case by eliminating the absolute term.

Ex. 2. Solve $\begin{cases} 3\,x^2 - 4\,xy + 3\,y^2 = 2, & (1) \\ 2\,x^2 - 2\,xy + 5\,y^2 = 5. & (2) \end{cases}$

Eliminate 2 and 5 by subtraction.

(1) × 5,      $15\,x^2 - 20\,xy + 15\,y^2 = 2 \times 5.$      (3)

(2) × 2,      $4\,x^2 - 4\,xy + 10\,y^2 = 5 \times 2.$      (4)

Subtracting,      $11\,x^2 - 16\,xy + 5\,y^2 = 0.$

Factoring,      $(x - y)(11\,x - 5\,y) = 0.$

Hence solve:

$\begin{cases} x - y = 0, \\ 2\,x^2 - 2\,xy + 5\,y^2 = 5. \end{cases}$    or   $\begin{cases} 11\,x - 5\,y = 0, & (3) \\ 2\,x^2 - 2\,xy + 5\,y^2 = 5. & (2) \end{cases}$

From (3),      $x = y.$                           $y = \tfrac{11}{5}\,x,$

Substituting $y$ in (2),         $2\,x^2 - \tfrac{22}{5}\,x^2 + \tfrac{121}{5}\,x^2 = 5,$

     $2\,y^2 - 2\,y^2 + 5\,y^2 = 5,$           $\dfrac{109\,x^2}{5} = 5,$

            $y^2 = 1,$                       $x^2 = \tfrac{25}{109},$

            $y = \pm 1,$

            $x = \pm 1.$       $x = \pm \tfrac{5}{109}\sqrt{109},\; y = \pm \tfrac{11}{109}\sqrt{109}.$

EXERCISE 110

Solve:

1. $\begin{cases} 6\,x^2 - 7\,xy + 2\,y^2 = 0, \\ 4\,y^2 - 5\,xy = 6. \end{cases}$      5. $\begin{cases} 10\,x^2 - 37\,xy + 7\,y^2 = 0, \\ 3\,x^2 - 9\,xy + 3\,y^2 = 33. \end{cases}$

2. $\begin{cases} 12\,x^2 - 17\,xy + 6\,y^2 = 0, \\ 5\,xy - 6\,x^2 = 6. \end{cases}$      6. $\begin{cases} x^2 - 4\,xy + 3\,y^2 = 0, \\ xy - y^2 = 8. \end{cases}$

3. $\begin{cases} 10\,x^2 - 11\,xy + 3\,y^2 = 0, \\ 8\,xy - 13\,x^2 = 3. \end{cases}$      7. $\begin{cases} 10\,x^2 - 11\,xy + 3\,y^2 = 0, \\ 7\,y^2 - 4\,xy - 7\,x^2 = 52. \end{cases}$

4. $\begin{cases} 2\,x^2 + 3\,xy - 2\,y^2 = 0, \\ 4\,x^2 + 3\,xy = 10. \end{cases}$      8. $\begin{cases} 5\,xy - 6\,x^2 = 6, \\ 6\,y^2 - 7\,xy = 12. \end{cases}$

9. $\begin{cases} x^2 + 2\,xy = 24, \\ 11\,xy - 2\,y^2 = 60. \end{cases}$

13. $\begin{cases} 27\,xy - 36\,x^2 = 2, \\ 45\,xy - 36\,y^2 = -6. \end{cases}$

10. $\begin{cases} 9\,x^2 + 4\,xy = 105, \\ 41\,xy - 9\,y^2 = 210. \end{cases}$

14. $\begin{cases} x^2 - xy + y^2 = 39, \\ 2\,x^2 - 3\,xy + 2\,y^2 = 43. \end{cases}$

11. $\begin{cases} 2\,x^2 + 9\,xy = 140, \\ 6\,xy - y^2 = 56. \end{cases}$

15. $\begin{cases} 3\,x^2 - xy - 2\,y^2 = 8, \\ 2\,x^2 - 3\,xy + y^2 = 3. \end{cases}$

12. $\begin{cases} 150\,x^2 - 125\,xy = -6, \\ 150\,y^2 - 175\,xy = 12. \end{cases}$

16. $\begin{cases} x^2 + 2\,y^2 = 17, \\ 3\,x^2 - 5\,xy + 4\,y^2 = 13. \end{cases}$

17. $\begin{cases} (9\,x + y)(x + y) = 273, \\ (9\,x - y)(x - y) = 33. \end{cases}$

## IV. SPECIAL DEVICES

**298.** Many examples belonging to the preceding types, and others not belonging to them, can be solved by special devices, which in most cases must be left to the ingenuity of the student.

Some of the more frequently used devices are the following:

**299.** *A.* **Division of one equation by the other.** Equations of higher degree can sometimes be reduced to equations of the second degree by dividing member by member.

**Ex. 1.** Solve $\qquad \begin{cases} x^3 + y^3 = 28, & \text{(1)} \\ x + y = 4. & \text{(2)} \end{cases}$

Dividing (1) by (2), $\qquad x^2 - xy + y^2 = 7.$     (3)

Squaring (2), $\qquad x^2 + 2\,xy + y^2 = 16.$     (4)

(4) − (3), $\qquad\qquad\qquad 3\,xy = 9,$

$\qquad\qquad\qquad\qquad xy = 3.$     (5)

Solve:

**1.** $\begin{cases} x^3 - y^3 = 152, \\ x - y = 2. \end{cases}$

**5.** $\begin{cases} x^3 + y^3 = 133, \\ x + y = 7. \end{cases}$

**2.** $\begin{cases} x^2 - y^2 = 8, \\ x + y = 4. \end{cases}$

**6.** $\begin{cases} x^2 y + x y^2 = 20, \\ x + y = 5. \end{cases}$

**3.** $\begin{cases} x^3 + y^3 = 189, \\ x + y = 9. \end{cases}$

**7.** $\begin{cases} x^4 - y^4 = 65, \\ x^2 - y^2 = 5. \end{cases}$

**4.** $\begin{cases} x^2 y = 12, \\ x y^2 = 18. \end{cases}$

**8.** $\begin{cases} x^3 - y^3 = 37, \\ x^2 + xy + y^2 = 37. \end{cases}$

**9.** $\begin{cases} x^4 + x^2 y^2 + y^4 = 21, \\ x^2 + xy + y^2 = 7. \end{cases}$

**300.** *B.* Some simultaneous quadratics can be solved by considering not $x$ or $y$, but expressions involving $x$ and $y$, as $\dfrac{1}{x}$, $xy$, $x^2$, $x + y$, etc., at first as the unknown quantities. In more complex examples it is advisable to substitute another letter for such expressions.

**Ex. 1.** Solve $\begin{cases} x + y + \sqrt{x + y} = 20, & (1) \\ x - y - \sqrt{x - y} = 6. & (2) \end{cases}$

Considering $\sqrt{x + y}$ and $\sqrt{x - y}$ as unknown quantities and solving, we have

from (1),          $\sqrt{x + y} = 4 \text{ or } - 5,$

from (2),          $\sqrt{x - y} = 3 \text{ or } - 2.$

But the negative roots being extraneous, we obtain by squaring,

$$x + y = 16,$$
$$x - y = 9.$$

Therefore          $x = 12\frac{1}{2}, \ y = 3\frac{1}{2}.$

Ex. 2. Solve
$$\begin{cases} \sqrt{\dfrac{x+y}{x}} + \sqrt{\dfrac{x}{x+y}} = 4\frac{1}{4}, & (1) \\ 2x + y = 17. & (2) \end{cases}$$

Let
$$z = \sqrt{\frac{x+y}{x}}.$$

Then
$$z + \frac{1}{z} = 4\frac{1}{4},$$

or
$$4z^2 - 17z + 4 = 0.$$

Hence
$$z = \tfrac{1}{4}, \text{ or } z = 4.$$

I.e.
$$\sqrt{\frac{x+y}{x}} = \frac{1}{4}, \text{ or } \sqrt{\frac{x+y}{x}} = 4.$$

Hence we have to solve the two systems:

$$\begin{cases} \dfrac{x+y}{x} = \dfrac{1}{16}, \\ 2x + y = 17. \end{cases} \qquad \begin{cases} \dfrac{x+y}{x} = 16, \\ 2x + y = 17. \end{cases}$$

The solution produces the roots:

$$\begin{cases} x = 16, \\ y = -15. \end{cases} \qquad \begin{cases} x = 1, \\ y = 15. \end{cases}$$

### EXERCISE 112

Solve:

1. $\begin{cases} 3x^2 + 2y^2 = 11, \\ 5x^2 - y^2 = 1. \end{cases}$

2. $\begin{cases} x^2y^2 - 5xy + 6 = 0, \\ 2x + y = 5. \end{cases}$

3. $\begin{cases} x + y = 5, \\ \sqrt{x} + \sqrt{y} = 3. \end{cases}$

4. $\begin{cases} x + y + \sqrt{x+y} = 12, \\ x^2 + y^2 = 45. \end{cases}$

5. $\begin{cases} \dfrac{1}{x} + \dfrac{1}{y} = \dfrac{5}{6}, \\ \dfrac{1}{x^2} + \dfrac{1}{y^2} = \dfrac{13}{36}. \end{cases}$

6. $\begin{cases} \dfrac{1}{x^3} + \dfrac{1}{y^3} = 35, \\ \dfrac{1}{x} + \dfrac{1}{y} = 5. \end{cases}$

7. $\begin{cases} x + y = 520, \\ \sqrt[3]{x} + \sqrt[3]{y} = 10. \end{cases}$

8. $\begin{cases} \sqrt{\dfrac{x+y}{x}} + \sqrt{\dfrac{x}{x+y}} = 2\frac{1}{2}, \\ 2x + y = 5. \end{cases}$

Solve by any method:

**9.** $\begin{cases} x^2 + xy = 126, \\ y^2 + xy = 198. \end{cases}$

**12.** $\begin{cases} 3x^2 + 4xy + 5y^2 = 71, \\ 5x + 7y = 29. \end{cases}$

**10.** $\begin{cases} 4x^2 + 9xy = 190, \\ 4x - 5y = 10. \end{cases}$

**13.** $\begin{cases} 8xy - 13x^2 = 3, \\ 13y^2 - 21xy = 10. \end{cases}$

**11.** $\begin{cases} x^3 + y^3 = 91, \\ x^2 - xy + y^2 = 13. \end{cases}$

**14.** $\begin{cases} 4x^2 + 3xy = 10, \\ 4y^2 - 3xy = 10. \end{cases}$

**15.** $\begin{cases} x^2 + y^2 = 225, \\ xy = 108. \end{cases}$

**16.** $\begin{cases} 4x - 5y = 1, \\ 2x^2 - xy + 3y^2 + 3x - 4y = 47. \end{cases}$

**17.** $\begin{cases} x^3 - y^3 = 189, \\ x - y = 3. \end{cases}$

**24.** $\begin{cases} x + y = 2m, \\ xy = m^2 - n^2. \end{cases}$

**18.** $\begin{cases} \dfrac{x+y}{x-y} + \dfrac{x-y}{x+y} = \dfrac{10}{3}, \\ x^2 - y^2 = 3. \end{cases}$

**25.** $\begin{cases} (7 + x)(6 + y) = 80, \\ x + y = 5. \end{cases}$

**19.** $\begin{cases} \sqrt[3]{x} + \sqrt[3]{y} = 6, \\ x + y = 126. \end{cases}$

**26.** $\begin{cases} \dfrac{1}{x} + \dfrac{1}{y} = \dfrac{9}{20}, \\ \dfrac{1}{x^2} + \dfrac{1}{y^2} = \dfrac{41}{400}. \end{cases}$

**20.** $\begin{cases} 4x^2 + 9y^2 = 34, \\ 6xy = 15. \end{cases}$

**21.** $\begin{cases} x^2 + 3xy = 7, \\ xy + 3y^2 = 14. \end{cases}$

**27.** $\begin{cases} \dfrac{1}{x^3} + \dfrac{1}{y^3} = 91, \\ \dfrac{1}{x} + \dfrac{1}{y} = 7. \end{cases}$

**22.** $\begin{cases} (x - 2)(y - 3) = 1, \\ \dfrac{x-2}{y-3} = 1. \end{cases}$

**28.** $\begin{cases} 3x^2 + 4xy = 20, \\ 5xy + 2y^2 = 12. \end{cases}$

**23.** $\begin{cases} 14x^2 - 122y^2 = 100, \\ x - 3y = 0. \end{cases}$

**29.** $\begin{cases} x^3 + y^3 = 152, \\ x^2 - xy + y^2 = 19. \end{cases}$

**30.** $\begin{cases} \dfrac{x^2}{y} + \dfrac{y^2}{x} = 9, \\ x + y = 6. \end{cases}$

**31.** $\begin{cases} 2\,x^2 + 2\,y^2 = 65, \\ \dfrac{x+y}{x-y} = 8. \end{cases}$

**32.** $\begin{cases} 2\,x^2 + 7\,xy + 2\,y^2 = 41, \\ 3\,x^2 - 10\,xy + 3\,y^2 = 0. \end{cases}$

**33.** $\begin{cases} x^3 - y^3 = 63, \\ x - y = 3. \end{cases}$

**34.** $\begin{cases} \dfrac{x^2 + y^2}{x^2 - y^2} = \dfrac{25}{7}, \\ xy = 48. \end{cases}$

**35.** $\begin{cases} x^2 + y^2 = 34, \\ x^2 - y^2 + \sqrt{x^2 - y^2} = 20. \end{cases}$

**36.** $\begin{cases} x + y + \sqrt{x + y} = 6, \\ x^2 + y^2 = 10. \end{cases}$

**37.** $\begin{cases} x^2 + xy = 10, \\ xy + y^2 = 15. \end{cases}$

**38.** $\begin{cases} \sqrt{\dfrac{x}{y}} + \sqrt{\dfrac{y}{x}} = 2\tfrac{1}{2}, \\ x + y = 10. \end{cases}$

**39.** $\begin{cases} \dfrac{x}{a} + \dfrac{y}{b} = 1, \\ \dfrac{a}{x} + \dfrac{b}{y} = 4. \end{cases}$

Solve graphically (see §§ 201, 203):

**40.** $\begin{cases} y = 2\,x^2 + 1, \\ 2\,y + 3\,x = 9. \end{cases}$

**41.** $\begin{cases} x^2 + y^2 = 25, \\ x + y = 7. \end{cases}$

## INTERPRETATION OF NEGATIVE RESULTS AND THE FORMS OF $\dfrac{0}{0},\ \dfrac{a}{0},\ \dfrac{0}{\infty}$.

**301.** The results of problems and other examples appear sometimes in forms which require a special interpretation, as $\dfrac{a}{0},\ \dfrac{0}{0},\ \dfrac{0}{\infty}$, etc.

**302. Interpretation of** $\dfrac{0}{0}$. According to the definition of division, $\dfrac{0}{0} = x$, if $0 = 0\,x$. But this equation is satisfied by any finite value of $x$, hence $\dfrac{0}{0}$ may be any finite number, or $\dfrac{0}{0}$ is **indeterminate.**

R

**303. Interpretation of $\dfrac{a}{0}$.** The fraction $\dfrac{a}{x}$ increases if $x$ decreases; *e.g.* $\dfrac{a}{\frac{1}{100}} = 100\,a$, $\dfrac{a}{\frac{1}{10000}} = 10,000\,a$. By making $x$ sufficiently small, $\dfrac{a}{x}$ can be made larger than any·assigned number, however great. If $x$ approaches the value zero, $\dfrac{a}{x}$ becomes infinitely large. It is customary to represent this result by the equation $\dfrac{a}{0} = \infty$.

The symbol $\infty$ is called infinity.

**304. Interpretation of $\dfrac{a}{\infty}$.** The fraction $\dfrac{a}{x}$ decreases if $x$ increases, and becomes infinitely small, or *infinitesimal*, if $x$ is infinitely large. This result is usually written:

$$\frac{a}{\infty} = 0.$$

**305.** *In solving a problem the result* $\dfrac{a}{0}$ *or* $\infty$ *indicates that the problem has no solution. If in an equation all terms containing the unknown quantity cancel, while the remaining terms do not cancel, the root is infinity.*

**306.** *The solution* $x = \dfrac{0}{0}$ *indicates that the problem is indeterminate, or that $x$ may equal any finite number. If all terms of an equation, without exception, cancel, the answer is indeterminate. Hence such an equation is satisfied by any number, i.e. it is an identity.*

Ex. **1.** Find three consecutive numbers such that the square of the second exceeds the product of the first and third by 1.

Let $x$, $x + 1$, $x + 2$, be the numbers.

Then $\qquad\qquad (x + 1)^2 - x(x + 2) = 1.$ $\qquad$ (1)

Simplifying, $\qquad x^2 + 2x + 1 - x^2 - 2x = 1.$

Or, $\qquad\qquad\qquad\qquad\qquad 0 = 0.$

Hence any number will satisfy equation (1), *i.e.* (1) is an identity, and the given problem is indeterminate.

**Ex. 2.** Solve the system:

$$\begin{cases} x^2 + 4\,y^2 = 4\,xy, & (1) \\ x - 2\,y = 1. & (2) \end{cases}$$

From (2), $\qquad\qquad x = 1 + 2\,y.$

Substituting, $\qquad 1 + 4\,y + 4\,y^2 + 4\,y^2 = 4\,y + 8\,y^2.$

Or, $\qquad\qquad\qquad 1 = 0.$

Hence $\qquad\qquad y = \infty,$ and $x = \infty.$

*I.e.* no finite numbers can satisfy the given system.

### EXERCISE 113

**1.** One half of a certain number is equal to the sum of its third and sixth parts. Find the number.

**2.** Find three consecutive numbers such that the square of the second exceeds the product of the first and third by 2.

**3.** Solve $\dfrac{x-2}{x-3} = \dfrac{x-5}{x-6}.$

**4.** Solve $\dfrac{x-4}{x-3} - \dfrac{x-6}{x-5} = \dfrac{2}{x^2 - 8\,x + 15}.$

**5.** Solve $\begin{cases} x^2 - 3\,xy + 2\,y^2 = 0, \\ x - 2\,y = 4. \end{cases}$

**6.** Solve $\begin{cases} (x+y)^2 - (x-y)^2 = 2 + 4\,xy, \\ 2\,x - y = 0. \end{cases}$

**7.** Solve $(x+1) : (x+2) = (x+3) : (x+4).$

### EXERCISE 114

#### PROBLEMS

**1.** The sum of two numbers is 76, and the sum of their squares is 2890. Find the numbers.

**2.** The sum of two numbers is 42 and their product is 377. Find the numbers.

**3.** The difference between two numbers is 17 and the sum of their squares is 325. Find the numbers.

**4.** Find two numbers whose product is 255 and the sum of whose squares is 514.

**5.** The sum of the areas of two squares is 208 square feet, and the side of one increased by the side of the other equals 20 feet. Find the side of each square.

**6.** The hypotenuse of a right triangle is 73, and the sum of the other two sides is 103. Find these sides. (§ 228.)

**7.** The area of a right triangle is 210 square feet, and the hypotenuse is 37. Find the other two sides.

**8.** To inclose a rectangular field 1225 square feet in area, 148 feet of fence are required. Find the dimensions of the field.

**9.** The area of a rectangle is 360 square feet, and the diagonal 41 feet. Find the lengths of the sides. (Ex. 12. p. 190.)

**10.** The diagonal of a rectangular field is 53 yards, and its perimeter is 146 yards. Find the sides.

**11.** The mean proportional between two numbers is 6, and the sum of their squares is 328. Find the numbers.

**12.** The area of a rectangle remains unaltered if its length is increased by 20 inches while its breadth is diminished by 10 inches. But if the length is increased by 10 inches and the breadth is diminished by 20 inches, the area becomes $\frac{5}{12}$ of the original area. Find the sides of the rectangle.

**13.** Two cubes together contain $30\frac{3}{8}$ cubic inches, and the edge of one, increased by the edge of the other, equals $4\frac{1}{2}$ inches. Find the edge of each cube.

**14.** The volumes of two cubes differ by 98 cubic centimeters, and the edge of one exceeds the edge of the other by 2 centimeters. Find the edges.

**15.** The sum of the radii of two circles is equal to 47 inches, and their areas are together equal to the area of a circle whose radius is 37 inches. Find the radii. (Area of circle $= \pi R^2$.)

**16.** The radii of two spheres differ by 8 inches, and the difference of their surfaces is equal to the surface of a sphere whose radius is 20 inches. Find the radii. (Surface of sphere $= 4 \pi R^2$.)

**17.** If a number of two digits be divided by the product of its digits, the quotient is 2, and if 27 be added to the number, the digits will be interchanged. Find the number.

# CHAPTER XX

## PROGRESSIONS

**307.** A **series** is a succession of numbers formed according to some fixed law.

The **terms** of a series are its successive numbers.

### ARITHMETIC PROGRESSION

**308.** An **arithmetic progression** (A. P.) is a series, each term of which, except the first, is derived from the preceding by the addition of a constant number.

The **common difference** is the number which added to each term produces the next term.

Thus each of the following series is an A. P.:

$$3, 7, 11, 15, 19, \cdots.$$
$$17, 10, 3, -4, -11, \cdots.$$
$$a, a+d, a+2d, a+3d, \cdots.$$

The common differences are respectively 4, $-7$, and $d$.

The first is an ascending, the second a descending, progression.

**309.** To find the $n$th term $l$ of an A. P., the first term $a$ and the common difference $d$ being given.

The progression is $a, a+d, a+2d, a+3d$.

Since $d$ is added to each term to obtain the next one,

$2d$ must be added to $a$, to produce the 3d term,

$3d$ must be added to $a$, to produce the 4th term,

$(n-1)d$ must be added to $a$, to produce the $n$th term.

Hence $\qquad l = a + (n-1)d.$ $\qquad$ (I)

Thus the 12th term of the series 9, 12, 15 is $9+11 \cdot 3$ or 42.

**310. To find the sum $s$ of the first $n$ terms of an A. P.**, the first term $a$, the last term $l$, and the common difference $d$ being given.

$$s = a + (a + d) + (a + 2d) \cdots (l - d) + l.$$

Reversing the order,

$$s = l + (l - d) + (l - 2d) \cdots (a + d) + a.$$

Adding,  $2s = (a + l) + (a + l) + (a + l) \cdots (a + l) + (a + l).$

Or  $2s = n(a + l).$

Hence  $s = \dfrac{n}{2}(a + l).$  (II)

Thus to find the sum of the first 50 odd numbers, 1, 3, 5 $\cdots$ we have from (I)  $l = 1 + 49 \cdot 2 = 99.$

Hence  $s = \tfrac{50}{2}(1 + 99) = 2500.$

### EXERCISE 115.

1. Which of the following series are in A. P.?
   (a) 1, 3, 5, 7, $\cdots$ ;
   (b) 2, 4, 8, 16, $\cdots$ ;
   (c) $-3$, 1, 5, 9, $\cdots$ ;
   (d) $1\frac{1}{2}$, $-\frac{1}{2}$, $-2\frac{1}{2}$, $-4\frac{1}{2}$ $\cdots$.

2. Write down the first 6 terms of an A. P., if
   (a) $a = 5$, $d = 3$ ;
   (b) $a = 2$, $d = -3$ ;
   (c) $a = -1$, $d = -2$.

3. Find the 5th term of the series 2, 5, 8, $\cdots$.

4. Find the 10th term of the series 17, 19, 21, $\cdots$.

5. Find the 7th term of the series $1\frac{1}{2}$, 2, $2\frac{1}{2}$, $\cdots$.

6. Find the 21st term of the series 10, 8, 6, $\cdots$.

7. Find the 12th term of the series $-4$, $-7$, $-10$, $\cdots$.

8. Find the 101th term of the series 1, 3, 5, $\cdots$.

9. Find the $n$th term of the series 2, 4, 6, $\cdots$.

Find the last term and the sum of the following series:

**10.** 3, 7, 11, ···, to 8 terms.

**11.** −2, −4, −6, ···, to 7 terms.

**12.** 8, 12, 16, ···, to 20 terms.

**13.** 3, $2\frac{1}{3}$, $1\frac{2}{3}$, ···, to 10 terms.

Sum the following series:

**14.** 7, 11, 15, ···, to 20 terms.

**15.** 33, 31, 29, ···, to 16 terms.

**16.** 15, 11, 7, ···, to 20 terms.

**17.** 1, $1\frac{1}{3}$, $1\frac{2}{3}$, ···, to 15 terms.

**18.** $2 + 1\frac{1}{3} + \frac{2}{3} + \cdots$, to 10 terms.

**19.** $2.5 + 3.1 + 3.7 + \cdots$, to 12 terms.

**20.** $(x+1) + (x+2) + (x+3) + \cdots$, to $a$ terms.

**21.** $1 + 2 + 3 + 4 + \cdots + 100$.

**22.** $1 + 2 + 3 + 4 + \cdots + n$.

**23.** Find the sum of the first $n$ odd numbers.

**24.** How many times does a clock, striking hours only, strike in 12 hours?

**25.** For boring a well 60 yards deep a contractor receives $1 for the first yard, and for each yard thereafter 10 ¢ more than for the preceding one. How much does he receive all together?

**26.** A bookkeeper accepts a position at a yearly salary of $1000, and a yearly increase of $120. How much does he receive (a) in the 21st year; (b) during the first 21 years?

**311.** *In most problems relating to A. P., five quantities are involved; hence if any three of them are given, the other two may be found by the solution of the simultaneous equations:*

$$\begin{cases} l = a + (n-1)\,d. & \text{(I)} \\ s = \dfrac{n}{2}(a + l). & \text{(II)} \end{cases}$$

**Ex. 1.** The first term of an A. P. is 12, the last term 144, and the sum of all terms 1014.  Find the series.

$$s = 1014, \; a = 12, \; l = 144.$$

Substituting in (I) and (II),

$$144 = 12 + (n-1)d. \tag{1}$$

$$1014 = \frac{n}{2}(12 + 144). \tag{2}$$

From (2),  $\qquad 78\,n = 1014, \text{ or } n = 13.$

Substituting in (1),  $\qquad 144 = 12 + 12 \cdot d.$

Hence  $\qquad d = 11.$

The series is, 12, 23, 34, 45, 56, 67, 78, 89, 100, 111, 122, 133, 144.


**Ex. 2.** Find $n$, if $s = 204, \; d = 6, \; l = 49.$

Substituting,  $\qquad 49 = a + (n-1) \cdot 6. \tag{1}$

$$204 = \frac{n}{2}(a + 49). \tag{2}$$

From (1),  $\qquad a = 49 - 6(n - 1).$

Substituting in (2),  $\qquad 204 = \frac{n}{2}(98 - \overline{n-1} \cdot 6).$

$$408 = n(104 - 6\,n).$$

$$6\,n^2 - 104\,n + 408 = 0.$$

$$3\,n^2 - 52\,n + 204 = 0.$$

Solving,  $\qquad n = 6, \text{ or } 11\frac{1}{3}.$

But evidently $n$ cannot be fractional, hence $n = 6$.


**312.** When three numbers are in A. P., the second one is called the **arithmetic mean** between the other two.

Thus $x$ is the arithmetic mean between $a$ and $b$, if $a, x,$ and $b$ form an A. P., or if

$$x - a = b - x.$$

Solving,  $\qquad x = \frac{a + b}{2}.$

I.e. *the arithmetical mean between two numbers is equal to half their sum.*

## EXERCISE 116

Find the arithmetic means between:

**1.** $a + b$ and $a - b$.  **3.** $\dfrac{1}{m}$ and $\dfrac{1}{n}$.

**2.** $x - y$ and $x + 5y$.  **4.** $\dfrac{1}{a+b}$ and $\dfrac{1}{a-b}$.

**5.** Between 4 and 8 insert 3 terms (arithmetic means) so that an A. P. of 5 terms is produced.

**6.** Between 10 and 6 insert 7 arithmetic means

**7.** How many terms has the series $\frac{1}{24}, \frac{1}{12}, \frac{1}{8}, \cdots, \frac{2}{3}$?

**8.** How many terms has the series 82, 78, 74, $\cdots$, 6?

**9.** Given $d = 3$, $n = 16$, $s = 440$. Find $a$ and $l$.

**10.** Given $s = 44$, $n = 4$, $l = 17$. Find $a$

**11.** Given $a = 7$, $l = 83$, $n = 20$. Find $d$.

**12.** Given $a = -3$, $n = 13$, $l = 45$. Find $d$.

**13.** Given $a = 4$, $n = 17$, $l = 52$. Find $d$ and $s$.

**14.** Given $a = 1700$, $d = 5$, $l = 1870$. Find $n$.

**15.** Given $a = \frac{1}{3}$, $l = \frac{1}{15}$, $s = 1$. Find $n$.

**16.** Given $a = 1$, $n = 16$, $s = 70$. Find $l$.

**17.** Find $l$ in terms of $a$, $n$, and $s$.

**18.** A man saved each month $2 more than in the preceding one, and all his savings in 5 years amounted to $6540. How much did he save the first month?

**19.** $300 is divided among 6 persons in such a way that each person receives $10 more than the preceding one. How much did each receive?

## GEOMETRIC PROGRESSION

**313.** A **geometric progression** (G. P.) is a series each term of which, except the first, is derived from the preceding one by multiplying it by a constant number, called the *ratio*.

$$E.g. \ 4, \ 12, \ 36, \ 108, \ \cdots.$$
$$4, \ -2, \ +1, \ -\tfrac{1}{2}, \ \cdots.$$
$$a, \ ar, \ ar^2, \ ar^3, \ \cdots.$$

The ratios are respectively 3, $-\tfrac{1}{2}$, and $r$.

**314.** To find the *n*th term *l* of a G. P., the first term $a$ and the ratios $r$ being given.

The progression is $a, \ ar, \ ar^2, \ \cdots.$

To obtain the $n$th term $a$ must evidently be multiplied by $r^{n-1}$.

Hence $\qquad\qquad\qquad l = ar^{n-1}.$ \hfill (I)

Thus the 5th term of the series 16, 24, 36, $\cdots$, is $16(\tfrac{3}{2})^4$, or 81

**315.** To find the sum *s* of the first *n* terms of a G. P., the first term $a$ and the ratio $r$ being given.

$$s = a + ar + ar^2 \cdots ar^{n-1}. \tag{1}$$

Multiplying by $r$, $rs = \qquad ar + ar^2 \cdots + ar^n.$ \hfill (2)

Subtracting (1) from (2),

$$s(r-1) = ar^n - a.$$

Therefore $\qquad\qquad s = \dfrac{a(r^n - 1)}{r - 1}.$ \hfill (II)

Thus the sum of the first 6 terms of the series 16, 24, 36, $\cdots$.

$$s = \frac{16\left[(\tfrac{3}{2})^6 - 1\right]}{\tfrac{3}{2} - 1} = 32(\tfrac{729}{64} - 1) = 332\tfrac{1}{2}.$$

**NOTE.** If $n$ is less than unity, it is convenient to write formula (II) in the following form :

$$s = \frac{a(1 - r^n)}{1 - r} \tag{III}$$

**316.** *In most problems relating to G. P. five quantities are involved ; hence, if any three of them are given, the other two may be found by the solution of the simultaneous equations :*

$$\begin{cases} l = ar^{n-1}, & \text{(I)} \\ s = \dfrac{a(r^n - 1)}{r - 1}. & \text{(II)} \end{cases}$$

**Ex. 1.** To insert 5 geometric means between 9 and 576.

Evidently the total number of terms is $5 + 2$, or 7.

Hence $n = 7$, $a = 9$, $l = 576$.

Substituting in I,      $576 = 9\, r^6.$

$$r^6 = 64.$$
$$r = \pm 2.$$

Hence the series is      9, 18, 36, 72, 144, 288, 576,

or             9, $-$ 18, 36, $-$ 72, 144, $-$ 288, 576.

And the required means are $\pm$ 18, 36, $\pm$ 72, 144, $\pm$ 288.

**EXERCISE 117**

**1.** Which of the following series are in G. P. ?

     (a) 2, 6, 18, 54, $\cdots$;       (c) $\frac{3}{2}$, 1, $\frac{2}{3}$, $\frac{4}{9}$, $\cdots$;

     (b) 1, 4, 9, 25, $\cdots$;       (d) 5, $-5$, $+5$, $-5$, $\cdots$.

**2** Write down the first 5 terms of a G. P. whose first term is 3, and whose common ratio is 4.

**3.** Write down the first 6 terms of a G. P. whose first term is 16, and whose second term is 8.

**4.** Find the 6th term of the series $\frac{9}{4}$, $\frac{3}{2}$, 1, $\cdots$.

**5.** Find the 7th term of the series $\frac{1}{64}$, $-\frac{1}{32}$, $+\frac{1}{16}$, $\cdots$.

**6.** Find the 6th term of the series 6, $4\frac{1}{2}$, $3\frac{3}{8}$, $\cdots$.

**7.** Find the 9th term of the series 5, 20, 80, $\cdots$.

**8.** Find the 11th term of the series $\frac{1}{64}$, $\frac{1}{16}$, $\frac{1}{4}$, $\cdots$.

**9.** Find the 7th term of the series $-\frac{2}{3}$, $\frac{1}{3}$, $-\frac{1}{6}$, $\cdots$.

**10.** Find the 5th term of a G. P. whose first term is 125 and whose common ratio is $\frac{2}{5}$.

Find the sum of the following series:

**11.** 32, 48, 72, ···, to 6 terms.

**12.** 243, 81, 27, ···, to 6 terms.

**13.** 14, 42, 126, ···, to 8 terms.

**14.** 1, $-2$, 4, ···, to 7 terms.

**15.** 81, 54, 36, ···, to 6 terms.

**16.** $\frac{1}{36}$, $\frac{1}{18}$, $\frac{1}{9}$, ···, to 6 terms.

**17.** $\frac{2}{9}$, $\frac{4}{9}$, $\frac{8}{9}$, ···, to 12 terms.

**18.** $x^9$, $x^8$, $x^7$, ···, to 5 terms.

**19.** Given $r = 4$, $n = 3$, $l = 160$. Find $a$ and $s$.

**20.** Given $r = \frac{1}{3}$, $n = 4$, $l = 3$. Find $a$ and $s$.

**21.** Given $r = 2$, $n = 5$, $s = 310$. Find $a$ and $l$.

**22.** Given $r = 3$, $n = 5$, $s = 605$. Find $a$ and $l$.

**23.** Find the geometric mean between $7\frac{1}{2}$ and 270.

**24.** Prove that the geometric mean between $a$ and $b$ equals $\sqrt{ab}$.

## INFINITE GEOMETRIC PROGRESSION

**317.** If the value of $r$ of a G. P. is less than unity, the value of $r^n$ decreases, if $n$ increases. The formula for the sum may be written

$$s = \frac{a - ar^n}{1 - r} = \frac{a}{1 - r} - \frac{ar^n}{1 - r}.$$

By taking $n$ sufficiently large, $r^n$, and hence $\frac{ar^n}{1 - r}$, may be made less than any assignable number.

Consequently the sum of an infinite decreasing series is

$$s = \frac{a}{1 - r}.$$

**Ex. 1.** Find the sum to infinity of the series 1, $-\frac{1}{3}$, $\frac{1}{9}$, ···.

$$a = 1, \ r = -\frac{1}{3}.$$

Therefore 
$$s_\infty = \frac{1}{1 + \frac{1}{3}} = \frac{3}{4}.$$

**Ex. 2.** Find the value of .3727272 ···.

$$.3727272 \cdots = .3 + .072 + .00072 + \cdots.$$

The terms after the first form an infinite G. P.

$$a = .072, \quad r = .01.$$

Hence $$s = \frac{.072}{1 - .01} = \frac{.072}{.99} = \frac{72}{990} = \frac{4}{55}.$$

Therefore $$.37272 \cdots = \frac{3}{10} + \frac{4}{55} = \frac{41}{110}.$$

### EXERCISE 118

Find the sum to infinity of the following series:

1. $1, \frac{1}{2}, \frac{1}{4}, \cdots$.    3. $16, 12, 9, \cdots$.    5. $5, 1, \frac{1}{5}, \cdots$.

2. $1, \frac{1}{3}, \frac{1}{9}, \cdots$.    4. $3, -1, \frac{1}{3}, \cdots$.    6. $250, 100, 40, \cdots$

7. $9, 6, 4, \cdots$.

8. If $a = 40$, $r = \frac{3}{7}$. Find the sum to infinity.

Find the value of:

9. $.555 \cdots$.    11. $.191919 \cdots$.    13. $.27777 \cdots$.

10. $.717171 \cdots$.    12. $.272727 \cdots$.    14. $.3121212 \cdots$.

15. The sum of an infinite G. P. is 9, and the common ratio is $\frac{1}{3}$. Find the first term.

16. The sum of an infinite G. P. is 16, and the first term is 8. Find $r$.

17. Given an infinite series of squares, the diagonal of each equal to the side of the preceding one. If the side of the first square is 2 inches, what is (a) the sum of the areas, (b) the sum of the perimeters, of all squares?

### EXERCISE 119

Expand the following:

1. $(a+b)^5$.  3. $(1+x)^8$.  5. $(x+\frac{1}{2})^5$.  7. $\left(1-\frac{m}{3}\right)^6$.

2. $(x-y)^6$.  4. $(x-2)^7$.  6. $\left(2a+\frac{b}{3}\right)^4$.  8. $\left(x^2-\frac{1}{x^2}\right)^5$.

Simplify:

9. $(1+\sqrt{x})^4+(1-\sqrt{x})^4$.  10. $(\sqrt{a}+\sqrt{b})^5-(\sqrt{a}-\sqrt{b})^5$.

11. Find the 5th term of $(a+b)^9$.

12. Find the 3d term of $(a-b)^{10}$.

13. Find the 4th term of $(m+n)^{12}$.

14. Find the 5th term of $(1+x)^{11}$.

15. Find the 4th term of $(a+2b)^7$.

16. Find the 6th term of $(x-x^2)^{25}$.

17. Find the 5th term of $\left(\sqrt{x}+\dfrac{1}{\sqrt{x}}\right)^8$.

18. Find the 3d term of $\left(a+\dfrac{3}{\sqrt{a}}\right)^7$.

19. Find the coefficient of $a^2b^{13}$ in $(a+b)^{15}$.

20. Find the coefficient of $a^4b^{12}$ in $(a+b)^{16}$.

21. Find the coefficient of $a^5b^{15}$ in $(a-b)^{20}$.

22. Find the coefficient of $a^3b^{97}$ in $(a-b)^{100}$.

23. Find the coefficient of $a^8b^{16}$ in $(a^2-b^2)^{12}$.

24. Find the middle term of $(x+y)^4$.

25. Find the middle term of $(a-b)^6$.

26. Find the middle term of $\left(x-\dfrac{1}{x}\right)^8$.

27. Find the middle term of $(m-n)^{16}$.

28. Find the 99th term of $(a+b)^{100}$.

29. Find the 1000th term of $(a+b)^{1000}$.

# REVIEW EXERCISE

Find the numerical values of:

**1.** $27\,x^8 - 27\,x^2y + 9\,xy^2 - y^3$, if

$$\left.\begin{array}{l} x=1 \\ y=2 \end{array}\right\} \text{ or } \left.\begin{array}{l} 1 \\ 3 \end{array}\right\} \left.\begin{array}{l} 2 \\ 1 \end{array}\right\} \left.\begin{array}{l} 2 \\ 2 \end{array}\right\} \left.\begin{array}{l} 2 \\ 3 \end{array}\right\} \left.\begin{array}{l} 2 \\ 4 \end{array}\right\} \left.\begin{array}{l} 3 \\ 2 \end{array}\right\} \left.\begin{array}{l} 3 \\ 4 \end{array}\right\} \left.\begin{array}{l} 4 \\ 5 \end{array}\right\}.$$

**2.** $16\,x^4 - 32\,x^3y + 24\,x^2y^2 - 8\,xy^3 + y^3$, if

$$x = 1,\ 1,\ 2,\ 2,\ 2,\ 3,\ 3,\ 4.$$
$$y = 2,\ 3,\ 2,\ 3,\ 4,\ 3,\ 4,\ 5.$$

**3.** $4\,x^2 - 4\,xy - 4\,xz + y^2 + 2\,yz + z^2$, if

$$x = 2,\ 2,\ 3,\ 3,\ 3,\ 4,\ 4,\ 5.$$
$$y = 1,\ 2,\ 4,\ 2,\ 3,\ 4,\ 3,\ 2.$$
$$z = 2,\ 1,\ 1,\ 2,\ 2,\ 1,\ 3,\ 6.$$

**4.** $(2\,a^4 - 13\,a^3b + 31\,a^2b^2 - 38\,ab^3 + 24\,b^4) \div (2\,a^2 - 3\,ab + 4\,b^2)$, if

$$a = 2,\ 2,\ 3,\ 3,\ 4,\ 4,\ 5,\ 6.$$
$$b = 1,\ 2,\ 1,\ 2,\ 2,\ 4,\ 2,\ 3.$$

**5.** $\dfrac{a^3 + b^3 + c^3 + 3\,a^2b + 3\,ab^2}{a^2 + b^2 + c^2 + 2\,ab - ac - bc}$, if

$$a = 1,\ 2,\ 4,\ 4,\ 5,\ 5,\ 5,\ 5.$$
$$b = 2,\ 2,\ 1,\ 5,\ 1,\ 3,\ 5,\ 6.$$
$$c = 3,\ 3,\ 2,\ 1,\ 2,\ 4,\ 2,\ 7.$$

**6.** $(b - c)(c + a) - (c - a)(a + b) - a(a + b - c) + c(a + c)$, if

$$a = 3,\ 3,\ 4,\ 1,\ 2,\ \ \ 4,\ \ \ 2,\ \ \ 2.$$
$$b = 2,\ 1,\ 2,\ 2,\ 1,\ \ \ 1,\ \ \ 3,\ \ \ 3.$$
$$c = 1,\ 2,\ 1,\ 3,\ 3,\ -3,\ -5,\ -6.$$

**7.** $\dfrac{a^3}{(a - b)(a - c)} + \dfrac{b^3}{(b - c)(b - a)} + \dfrac{c^3}{(c - a)(c - b)}$, if

$$a = 2,\ -2,\ \ \ 1,\ -1,\ \ \ 3,\ \ \ 4,\ -4\ \ -1.$$
$$b = 1,\ \ \ 1,\ -2,\ -2,\ \ \ 2,\ \ \ 2,\ \ \ 2,\ +1.$$
$$c = 3,\ -3,\ \ \ 4,\ -3,\ -1,\ -1,\ -3,\ \ \ 2.$$

**8.** $\dfrac{a(x-b)(x-c)}{(a-b)(a-c)} + \dfrac{b(x-c)(x-a)}{(b-c)(b-a)} + \dfrac{c(x-a)(x-b)}{(c-a)(c-b)}$, if

$$a = 1, \ 1, \ -1, \ -2, \ 3, \ -3, \ 2, \ 4.$$
$$b = 2, \ 3, \ +1, \ +3, \ 4, \ -4, \ 6, \ -2.$$
$$c = 3, \ 2, \ -2, \ -4, \ 5, \ -5, \ 1, \ 2.$$
$$x = 4, \ 5, \ +2, \ +4, \ 6, \ +2, \ 3, \ 1.$$

**9.** The radius $r$ of a circle inscribed in a triangle whose sides are $a$, $b$, and $c$ is represented by the formula

$$r = \tfrac{1}{2}\sqrt{\frac{(a+b-c)(a-b+c)(-a+b+c)}{a+b+c}}.$$

Find $r$, if
$$a = 3, \ 10, \ 8, \ 25, \ 29, \ 41.$$
$$b = 4, \ 21, \ 17, \ 24, \ 21, \ 9.$$
$$c = 5, \ 26, \ 15, \ 7, \ 20, \ 40.$$

Add the following expressions and check the answers:

**10.** $x^3 - 2\,ax^2 + a^2x + a^3, \ x^3 + 3\,ax^2, \ 2\,a^3 - ax^2 - 2\,x^3.$

**11.** $x^2 + y^4 + z^3, \ -4\,x^2 - 5\,z^3, \ 8\,x^2 - 7\,y^4 + 10\,z^3, \ 6\,y^4 - 6\,z^3.$

**12.** $x^4 - 4\,x^3y + 6\,x^2y^2 - 4\,xy^3 + y^4, \ 4\,x^3y - 12\,x^2y^2 + 12\,xy^3 - 4\,y^4,$
$6\,x^2y^2 - 12\,xy^3 + 6\,y^4, \ 4\,xy^3 - 4\,y^4, \ y^4.$

**13.** $x^3 + xy^2 + xz^2 - x^2y - xyz - x^2z, \ x^2y + y^3 + yz^2 - xy^2 - y^2z - xyz,$
$x^2z + y^2z + z^3 - xyz - yz^2 - xz^2.$

**14.** $3\,x^2 - 10\,y^2 + 5\,z^2 - 7\,yz, \ -x^2 + 4\,y^2 - 10\,z^2 + 3\,xy,$
$z^2 + 11\,yz + 8\,xz - 2\,xy, \ 4\,z^2 - 4\,yz + xz,$
and $-2\,x^2 + 6\,y^2 - 9\,xz - xy.$

**15.** $1 + 3\,x + 2\,x^3 - x^5, \ 4 - 2\,x^2 - 8\,x^3 + 7\,x^4,$
$12\,x - 4\,x^2 + 12\,x^5, \text{ and } 5\,x^2 + 7\,x^3 - 11\,x^5.$

**16.** $11\,x^4 - 2\,xy^3, \ 7\,xy^3 - 2\,x^2y^2 + 3\,x^3y - 8\,y^4, \ y^4 + 8\,x^4 - 7\,x^3y,$
$-12\,x^4 + 5\,x^2y^2 + 4\,x^3y, \ 7\,y^4 - 3\,x^2y^2 - 5\,xy^3.$

**17.** $4\,a^5 - a + 12\,a^3 - 10, \ 6\,a^4 - a^3 - 7 + 2\,a^2, \ 4\,a + 9\,a^2 - 3\,a^5,$
$9 - 2\,a^4 + 11\,a - a^5, \ 4\,a^3 - a^4 - 5.$

**18.** $11\,x^3 + 14\,x^2y - 7\,xy^2 + z^3, \ x^3 - 2\,x^2y + 3\,xy^2 - 7\,y^3$
$+ 3\,y^2z - 2\,z^3, \ 4\,x^3 + 2\,y^3 - 11\,z^3 + 4\,yz^2 - 3\,xyz,$
and $3\,y^3 + 12\,z^3 - 7\,y^2z + 4\,xyz + 4\,xy^2 - 4\,yz^2.$

**19.** $6\sqrt{1+x} - 5\sqrt{1+y} + 8$, $-4\sqrt{1+x} - 2\sqrt{1+y} - 7$,
$-3\sqrt{1+x} + 4\sqrt{1+y} + 3$, and $2\sqrt{1+x} + 4\sqrt{1+y} - 3$.

**20.** Take the sum of $2x^3 + 4x^2 + 9$ and $4x - 2 - x^2 + 4x^3$ from $6x^3 + 4x + 7$.

**21.** Take the sum of $6a^3 + 4a^2x + 3ax^2$, $x^3 - ax^2 + 2a^2x$, and $a^2x - 2ax^2 + 3x^3$ from $6a^3 + 2a^2x - 4x^3$.

**22.** Take the sum of $3x + x^3 - 2$, $2x^3 - x + x^2 + 5$, and $4 - 7x + 2x^2 + 4x^3$ from the sum of $9x + 2x^3$ and $5x^3 + 3x^2$.

**23.** From the sum of $12x^5 + 4xy^4 + y^5$, $2x^5 - 4x^4y - xy^4 + 3y^5$, and $6x^4y + 2x^2y^3 - 3xy^4$ take the sum of $6x^5 + 2x^2y^3 - y^5$, $x^5 - 2x^4y + 2y^5 + x^3y^2$, $4x^4y + 6x^5 - 2x^3y^2 + 3y^5$.

**24.** From the sum of $4 - x - x^2$, $6x - 4 - 5x^2$, $5 + 2x^2$, and $7x - x^2$ take $4 - 4x^2 + 11x$.

**25.** From the sum of $b + c + 3a$, $c + a - 3b$, and $a + b - 3c$ take the sum of $2b - 2c - a$, $2c + 2a - b$, and $2a - 2b - c$.

**26.** Find what expression added to $3x^2 - 5x - 2x^3 + 3$ will give $3 - x^3 - 4x + 4x^2$.

**27.** From the sum of $b + c - 3a$, $c + a - 3b$, and $a + b - 3c$ take the sum of $2b - 2c - a$, $2c - 2a - b$, and $2a - 2b - c$.

**28.** Subtract the difference of $x^3 + 3x^2 - 3x - 1$ and $x^3 - 3x^2 + 3x - 1$ from $6x^2 + 6x$.

**29.** Add $2 \cdot 5^{10} + 6^{11} + 3 \cdot 7^{12}$, $3 \cdot 5^{10} - 5 \cdot 6^{11} + 4 \cdot 7^{12}$, $-4 \cdot 5^{10} + 7 \cdot 6^{11} - 9 \cdot 7^{12}$, and $-5^{10} - 3 \cdot 6^{11} + 2 \cdot 7^{12}$.

**30.** If $a = x + y + z$, $b = x + y - z$, $c = x - y + z$, and $d = -x + y + z$, find

    (a) $a + b$,　　　(c) $a + b + c$,　　　(e) $a + b + c + d$,

    (b) $b - c$,　　　(d) $a - b + c$,　　　(f) $a - b + c - d$.

Simplify:

**31.** $2x - [3y - \{3x - (5y - \overline{6x - 7y})\}]$.

**32.** $a - [5b - \{a - (5c - \overline{2c - b} - 4b) + 2a - (a - \overline{2b + c})\}]$.

**33.** $x^4 - [4x^3 - \{6x^2 - (4x - 1)\}] - (x^4 + [4x^3 + 6x^2] + 4x + 1)$.

**34.** $13 - 3b - \left[17a - 5b - [7a - 3b - \{4a - 4b - (2a - 3b)\}]\right]$.

**35.** $3x^2 - (x^2 - 4) - [4x - 5 - \{2x^2 - (7x + 2) - (4x^2 - \overline{2x - 7})\}]$.

**36.** $2a + 7c - (7b + 4c) - [6a - \overline{3b + 2c} + 4c - \{2a - (b - \overline{2a - 2})\}]$.

**37.** $7\,a^2 - \{5\,a^2 - 2\,a + (2\,a^2 - \overline{7\,a - 5})\} + (3\,a^2 - \overline{4\,a - 12})$.

**38.** $\{5\,a^2 - \overline{2\,a - 7}\} - \{3\,a^2 - (2\,a + \overline{5\,a - 6})\} - (2\,a^2 - 7)$.

**39.** $2\,x - [3\,y - 2\,z + \{4\,x - (5\,y - \overline{3\,x - 2\,z})\} + 5\,z]$.

**40.** $a - [2\,b + \{3\,c - (3\,a - \overline{a - b + c})\} + (2\,b - 3\,c)]$.

**41.** $3\,x^2 - [4\,x - 5 - \{2\,x^2 - (7\,x + 2) - (4\,x^2 - \overline{2\,x - 7})\}]$.

**42.** $5\,a - (7\,b + 4\,c) + [6\,a - \overline{3\,b + 2\,c} + 4\,c - \{2\,a - (b - \overline{2\,a - c})\}]$.

**43.** $a - \{-b - (c - d)\} + a - [-b + \{-2\,c - (d - \overline{e - f})\}]$
$$- (2\,a + 2\,b - 3\,c).$$

**44.** $5\,a - (7\,b + 4\,c) - [6\,a - \overline{3\,b + 2\,c} + 4\,c - \{2\,a - (b - \overline{2\,a - c})\}]$.

**45.** $13\,a - 9\,b - \left[17\,a - 5\,b - [7\,a - 3\,b - \{4\,a - 4\,b - (2\,a - 3\,b)\}]\right]$.

**46.** $a - [2\,b - \{3\,c - (4\,d - 5\,e)\}] + \{4\,c - (2\,d - 2\,e)\}$
$$- [b - \{2\,c - (3\,d + 7\,e) - a\}].$$

Perform the operations indicated:

**47.** $(x^2 + x + 1)(x + 2)$.

**48.** $(x^2 - 2\,x + 1)(x - 1)$.

**49.** $(x^2 + 4\,x + 5)(x - 3)$.

**50.** $(1 - 6\,x + 5\,x^2)(2 - 3\,x)$.

**51.** $(1 - x + x^2)(1 - x^2)$.

**52.** $(x^2 + 2\,x + 3)(x^2 - 2\,x + 3)$.

**53.** $(2\,x^2 - 3\,x + 1)(3\,x^2 - 2\,x + 1)$.

**54.** $(b^4 - 2\,b^2 + 1)(b^4 + 2\,b^2 + 1)$.

**55.** $(a^4 + 3\,a^2 - 2)(1 - 4\,a^2 + a^4)$.

**56.** $(a^2 + b^2 + c^2 - ab - ac - bc)(a + b + c)$.

**57.** $(a^2 + b^2 + c^2 + ab + ac - bc)(a - b - c)$.

**58.** $(x^2 + 4\,y^2 + 3\,z^2)(x^2 - 2\,y^2 - 3\,z^2)$.

**59.** $(a^2 + b^2 + 9 - 3\,a + 3\,b + ab)(a - b + 3)$.

**60.** $(4\,x^2 + 9\,y^2 + z^2 - 6\,xy - 2\,xz - 3\,yz)(2\,x + 3\,y + z)$.

**61.** $(x + 7)(x + 5)(x + 3)$.

**62.** $(x - 3)(x - 5)(x - 7)$.

**63.** $(x - 2)(x - 4)(x - 9)$.

**64.** $(x^2 + x + 1)(x^2 - x + 1)(x^4 - x^2 + 1)$.

**65.** $(x - a)(x + a)(x^2 + a^2)(x^4 + a^4)$.

**66.** $(x^2 + xy + y^2)(x^2 - xy + y^2)(x^4 - x^2y^2 + y^4)$.

**67.** $(1 - x)(1 + x)(1 + x^2)(1 + x^4)$.

**68.** $(x - y)(x + y)(x^2 - xy + y^2)(x^2 + xy + y^2)$.

**69.** $(a^8 - x^8)(a^8 + x^8)(a^6 + x^6)(a^{12} + x^{12})$.

**70.** $(a^2 - 2a + 1)(a^2 + 2a + 1)(a^2 + 1)$.

**71.** $(a - b)^2(a + b)^2$.　　　　**73.** $(a - 2b)^3(a + 2b)$.

**72.** $(2x - y)^2(x + y)$.　　　　**74.** $(x - 3y)^2(x^2 + 6xy - 9y^2)$.

**75.** $(a^2 - 2a + 1)(a^2 + 2a + 1)(a^2 + 1)$.

**76.** $(x^{2n} + x^n y^m + y^{2m})(x^n - y^m)$.

**77.** $(a^{p+1} + b^q)(a^{2p+2} - a^{p+1}b^q + b^{2q})$.

**78.** $(a^{2m} + b^{2n} + c^2 + 2a^m b^n - a^m c - b^n c)(a^m + b^n + c)$.

**79.** $(m^a + n^b + p^c)(m^{2a} + n^{2b} + p^{2c} - m^a n^b - m^a p^c - n^b p^c)$.

Simplify:

**80.** $4(a + b)(a + 4b)(a - b) + 4(2b - a)(2b^2 + a^2)$.

**81.** $p(p + q)^2 + q(p - q)^2 + q^2(p - q)$.

**82.** $a(2a + 3b)^2 - (2a + b)^3$.

**83.** $(x + 2y)(2x + y)(x - y) + (x + y)^2(x - y)$.

**84.** $(p + 3q)^2(p - 3q) - (p - 3q)^2(p + 3q)$.

**85.** $(x^2 + y^2)^2 - (x - y)(x + y)(x^2 + y^2)$.

**86.** $(x + y)^4 - (x^2 + y^2)^2 - 4xy(x^2 + xy + y^2)$.

**87.** $(a + b + c)^2 - (a + b - c)^2$.

**88.** $(x + y)(y + z)(z + x) - x^2(y + z) - z^2(x + y)$.

**89.** $(x + y + z)(x + y - z)(x - y + z)(- x + y + z)$.

**90.** $(a - b)(x + a)(x + b) + (b - c)(x + b)(x + c)$
$$+ (c - a)(x + c)(x + a).$$

**91.** $(a + b + c)^2 - (b + c)^2 - (c + a)^2 - (a + b)^2 + a^2 + b^2 + c^2$.

**92.** $(a + b + c)^3 - (b + c)^3 - (c + a)^3 - (a + b)^3 + a^3 + b^3 + c^3$.

**93.** $3a - (4b - \{3a - 2b\}) + \{3a - 5b - \overline{b - a}\}$.

**94.** $3[a\{2a - 3(b - c)\} - 2b(a - c)]$.

**95.** Prove the following identities, by multiplying out each side of the equality.

(a) $(a + b + c)^3 = a^3 + b^3 + c^3 + 3(b + c)(c + a)(a + b)$.

(b) $(x - y)(x - 2y)(x - 3y) + 9y(x - y)(x - 2y) + 18y^2(x - y) + 6y^3$
$$= x(x + y)(x + 2y).$$

Simplify:

**96.** $[10(a + b)^3 - 5(a + b)^2] \div 5(a + b).$

**97.** $\dfrac{6(a + b)^{2n} - 12(a + b)^n}{6(a + b)^n}.$

**98.** $(10\,a^{3x} - 5\,a^{2x} + 5\,a^x) \div 5\,a^x.$

**99.** $2^{3a} - 2^{2a} + 2^{4a} - 2^{a+1}) \div 2^a.$

**100.** $(3^m + 3^{m+n} - 3^{3n} - 3^n) \div 3^n.$

**101.** $(5^{m+4} + 5^{m+3} - 5^{m+2} + 5^{m+1}) \div 5^m.$

**102.** $(6\,x^4 + 23\,x^3 + 42\,x^2 + 41\,x + 20) \div (3\,x^2 + 4\,x + 5).$

**103.** $(20\,x^4 - 33\,x^3 + 72\,x^2 - 35\,x + 30) \div (4\,x^2 - 5\,x + 10).$

**104.** $(2\,y^4 - 16\,y^3 + 2\,y^2 + 92\,y + 48) \div (y^2 - 5\,y - 12).$

**105.** $(5\,x^4 - 14\,x^3y + 31\,x^2y^2 - 22\,xy^3 + 12\,y^4) \div (5\,x^2 - 4\,xy + 3\,y^2).$

**106.** $(2\,a^4 - 2\,a^3b - 5\,a^2b^2 + 4\,ab^3 + 5\,b^4) \div (2\,a^2 - 6\,ab + 5\,b^2).$

**107.** $(6\,x^4 - 7\,ax^3 - 36\,a^2x^2 + 17\,a^3x + 10\,a^4) \div (3\,x^2 + 4\,ax - 5\,a^2).$

**108.** $(x^4 - 8\,x^3y + 21\,x^2y^2 - 16\,xy^3 - 7\,y^4) \div (x^2 - 5\,xy + 7\,y^2).$

**109.** $(x^4 - 9\,ax^3 + 12\,a^2x^2 + 35\,a^3x + 15\,a^4) \div (x^2 - 4\,ax - 3\,a^2).$

**110.** $(2\,y^4 + 2\,y^2 + 92\,y - 16\,y^3 + 48) \div (y^2 - 5\,y - 12).$

**111.** $(80\,a + 3\,a^4 - 23\,a^3 + 50 - 5\,a^2) \div (a^2 - 6\,a - 10).$

**112.** $(4\,a^4 - 4\,a^2b^2 + 25\,b^4 - 16\,a^3b + 40\,ab^3) \div (2\,a^2 - 4\,ab - 5\,b^2).$

**113.** $(25\,xy^3 - 3\,x^3y - 6\,y^4 + 27\,x^4 - 35\,x^2y^2) \div (7\,xy - 9\,x^2 - 2\,y^2)$

**114.** $(8\,x^4 - 2\,x^3y + y^4 - 2\,xy^3 - 21\,x^2y^2) \div (4\,x^2 - y^2 + 5\,xy).$

**115.** $(y^3 - 27) \div (y^2 + 3\,y + 9).$

**116.** $(x^4 + x^2y^2 + y^4) \div (x^2 - xy + y^2).$

**117.** $(a^4 + 16\,a^2 + 256) \div (a^2 + 4\,a + 16).$

**118.** $(a^8 + a^4b^4 + b^8) \div (a^4 - a^2b^2 + b^4).$

**119.** $(a^3 - 8\,b^3 + c^3 + 6\,abc) \div (a - 2\,b + c).$

**120.** $(x^3 - 27\,y^3 - 1 - 9\,xy) \div (x - 3\,y - 1).$

**121.** $(x^6 - y^6) \div (x^2 + xy + y^2).$

**122.** $(x^{10} - x) \div (x^3 - 1).$

**123.** $(a^{n+3} - 3\,a^{n+1} + a^{n-1}) \div (a^2 - a - 1).$

**124.** $(a^{x+1} - a^{x-2}b^3) \div (a - b).$

**125.** $(1 - a^3 - 6\,ax - 8\,x^3) \div (1 - a - 2\,x)$.

**126.** $(x^3 + y^3 + z^3 - 3\,xyz) \div (x + y + z)$.

**127.** $(x^{3a+1} + x^{2a} - x^{a+1} + x^a) \div (x^{a+1} + x)$.

**128.** $(1 - 27\,a^{3m} - 8\,b^{3n} - 18\,a^m b^n) \div (1 - 3\,a^m - 2\,b^n)$.

**129.** What is the remainder when $a^4 - 3\,a^3 b + 2\,a^2 b^2 - b^4$ is divided by $a^2 - ab + 2\,b^2$?

**130.** By what expression must $x + 3$ be multiplied to give $x^7 + 2187$?

**131.** By what expression must $3\,a^4 - 8\,a^3 b + 4\,a^2 b^2 - 8\,ab^3 - 12\,b^4$ be divided to give the quotient $3\,a^2 - 2\,ab + 6\,b^2$?

**132.** By what expression must $x^3 + 6\,x^2 - 4\,x - 1$ be divided to give $x^2 + 5\,x - 9$ as quotient, with 8 as remainder?

Solve the following equations and check the answers:

**133.** $3(2\,x - 1) - 4(6\,x - 5) = 12(4\,x - 5) - 22$.

**134.** $4(x - 3) - 5(2\,x - 3) = 12 - (x + 9)$.

**135.** $7(2\,x - 3) - 2(x - 2) = 3\,x - 2(5\,x - 9) + 3$.

**136.** $10(2\,x - 9) + 7(4\,x - 19) + 5 = 4\,x - 3(2\,x - 3)$.

**137.** $5\,x + 3(7\,x - 4) - 2(10\,x - 7) = 4 - (x - 5)$.

**138.** $\dfrac{x}{2} + \dfrac{x}{3} + \dfrac{x}{4} = 13$.        **139.** $\dfrac{x}{8} + \dfrac{x}{3} + \dfrac{x}{6} = 15$.

**140.** $10(2\,x - 9) - 7(6\,x - 32) + 5 = 4\,x - 3(2\,x - 3)$.

**141.** $4(x - 6) - 2\{3\,x - (x - 8)\} = 5(13 - 3\,x)$.

**142.** $2(3\,x + 4) = 5\{2\,x - 3(x + 4) + 9\} - (1 - 3\,x)$.

**143.** $8[2(x - 1) - (x + 3)] = 5\{x + 7[x - 2(4 - x)]\}$.

**144.** $4 - 2(3\,x - 2\{x + 1\} + 8) = 6(x + 2)$.

**145.** $5\,x - \{3\,x - 2[2\,x - \overline{2\,x - 5}] + 4\} = 2(3\,x - 1)$.

**146.** $x^2 + (x + 1)(x - 1) = 2(x - 2)(x + 3)$.

**147.** $(x - 1)(x + 2)(x - 3) = x^2(x - 2) + 2(x + 4)$.

**148.** $(2\,x - 3)(6\,x - 7) = (4\,x - 5)(3\,x - 4)$.

**149.** $(5\,x + 2)(x + 7) - (3\,x - 1)(x + 10) = (2\,x - 1)(x + 14)$.

**150.** $(4\,x - 3)(3\,x + 7) = (7\,x - 11)(3\,x - 4) - (9\,x + 10)(x - 3)$.

**151.** $(x+4)(2x+5) - (x+2)(7x+1) = (x-3)(3-5x) + 47.$

**152.** $(x-2)(x+1) + (x-1)(x+4) = (2x-1)(x+3).$

**153.** $(x-3)(x-4)(x-5) = (x-1)(x-14)(x+3) - 24.$

**154.** $(x-2)(7-x) + (x-5)(x+3) - 2(x-1) + 12 = 0.$

**155.** $(2x-7)(x+5) = (9-2x)(4-x) + 229.$

**156.** $(7-6x)(3-2x) = (4x-3)(3x-2).$

**157.** $14 - x - 5(x-3)(x+2) + (5-x)(4-5x) = 45x - 76.$

**158.** $(x+5)^2 - (4-x)^2 = 21x.$

**159.** $5(x-2)^2 + 7(x-3)^2 = (3x-7)(4x-19) + 42.$

**160.** $(3x-17)^2 + (4x-25)^2 - (5x-29)^2 = 1.$

**161.** $(x+5)(x-9) + (x+10)(x-8) = (2x+3)(x-7) - 113.$

**162.** $\dfrac{x}{10} + \dfrac{x}{5} = 13 + \dfrac{x}{25}.$      **163.** $\dfrac{x}{2} - \dfrac{x}{5} + \dfrac{x}{4} = x - 9.$

**164.** Write down four consecutive numbers of which $y$ is the greatest.

**165.** By how much does 15 exceed $a$?

**166.** How much must be added to $k$ to make 23?

**167.** A man is 30 years old; how old will he be in $x$ years?

**168.** Find five consecutive numbers whose sum equals 100.

**169.** There are 63 sheep in three flocks. The second contains 3 sheep more than the first, and the third twice as many as the first. How many sheep are there in each flock?

**170.** The sum of the three angles of a triangle is 180°. The second angle of a triangle is twice as large as the first, and if 15° were taken from the third and added to the first, these two angles would be equal. What are the three angles?

**171.** A picture which is 3 inches longer than wide is surrounded by a frame 2 inches wide. If the area of the frame is 108 square inches, how wide is the picture?

**172.** The formula which transforms Fahrenheit (F.) readings of a thermometer into Centigrade readings is C. $= \frac{5}{9}$(F. $- 32$).

(a) If C. $= 15°$, find the value of F.

(b) At what temperature do the Centigrade scale and the Fahrenheit scale indicate equal numbers?

(c) How many degrees C. transformed into F. will produce F. $= 2$ C.?

**173.** A number increased by 3 gives the same result as the number multiplied by 3. Find the number.

**174.** A number divided by 3 gives the same result as the number diminished by 3. Find the number.

**175.** An express train runs 7 miles an hour faster than an ordinary train. The two trains run a certain distance in 4 h. 12 m. and 5 h. 15 m. respectively. What is the distance?

**176.** A square grass plot would contain 73 square feet more if each side were one foot longer. Find the side of the plot.

**177.** A boy is $\frac{1}{4}$ as old as his father and 3 years younger than his sister; the sum of the ages of the three is 57 years. Find the age of the father.

**178.** A house has 3 rows of windows, 6 in each row; the lowest row has 2 panes of glass in each window more than the middle row, and the middle row has 4 panes in each window more than the upper row; there are in all 168 panes of glass. How many are there in each window?

**179.** Four years ago a father was three times as old as his son is now, and the father's present age is twice what the son will be 8 years hence. What are their ages?

**180.** Two engines are together of 80 horse power; one of the two is 16 horse power more than the other. Find the power of each.

**181.** The length of a floor exceeds its width by 2 feet; if each dimension is increased 2 feet, the area of the floor will be increased 48 square feet. Find the dimensions of the floor.

**182.** The age of the elder of two boys is twice that of the younger; three years ago it was three times that of the younger. Find the age of each.

**183.** A boy is 5 years older than his sister and $\frac{1}{6}$ as old as his father; the sum of the ages of all three is 51. Find the age of the father.

Resolve into prime factors:

**184.** $x^2 + x - 2$.          **187.** $a^2 + a - 110$.          **190.** $4 a^2 + 13 a + 3$.

**185.** $y^2 - y - 42$.          **188.** $y^2 - 11 y - 102$.          **191.** $10 x^2 + 11 x - 6$.

**186.** $x^2 - 9 x - 36$.          **189.** $a^2 b^2 + 11 ab - 242$.          **192.** $x^2 + x - 56$.

**193.** $y^2 - 77\,y + 150.$     **196.** $6\,x^2 - 5\,xy - 6\,y^2.$     **199.** $2\,x^2 + xy - 10\,y^2.$

**194.** $2\,a^2 - 19\,a - 10.$     **197.** $3\,y^2 - 13\,y + 4.$     **200.** $x^2 - 12\,x - 64.$

**195.** $a^2 + 3\,a - 28.$     **198.** $x^4 + 8\,x^2 + 15.$     **201.** $y^2 - 29\,y + 120.$

**202.** $x^2 - 2\,xy + y^2 - 9.$     **213.** $60\,a^2 + 11\,a^2b - a^2b^2.$

**203.** $x^5 - 19\,x^4y + 84\,x^3y^2.$     **214.** $12\,x^3y - 14\,x^2y - 10\,xy.$

**204.** $5\,x^2 - 8\,xy + 3\,y^2.$     **215.** $x^3 + 5\,x^2 - 6\,x.$

**205.** $14\,x^2 - 25\,xy + 6\,y^2.$     **216.** $2\,x^2 - 22\,x + 48.$

**206.** $3\,x^3 - x^2 - 12\,x + 4.$     **217.** $2\,x^2y - 28\,xy + 66\,y.$

**207.** $16\,x^4 - 81.$     **218.** $x^3 + 13\,x^2 + 30\,x.$

**208.** $2\,a^3 - 8\,ab^2.$     **219.** $(a^2 + x^2)^2 - (a + x^2)^2.$

**209.** $y^2 - b^2 - y + b.$     **220.** $(x + y + xy)^2 - (x - y)^2.$

**210.** $x^3 - 11\,x^2 + 10\,x.$     **221.** $6\,a^2 + 5\,a - 6.$

**211.** $20\,x^4 - 20\,x^3 - 5\,x^2.$     **222.** $x^3y + 6\,x^2y^2 + 9\,x^3y^3.$

**212.** $3\,x^2 - 21\,x - 54.$     **223.** $6\,x^2 + 5\,xy - 6\,y^2.$

**224.** $7\,x^2 + 48\,xy - 7\,y^2.$

**225.** $a^2(a + b - c)^2 - c^2(b + c - a)^2.$

**226.** $x^4 + yx^3 + z^3x + z^3y.$     **230.** $15\,x^2 + 26\,x + 8.$

**227.** $7\,x^2 - 22\,xy + 3\,y^2.$     **231.** $9\,a^3 - 4\,ab^2.$

**228.** $x^2(x + y)^2 - (x^2 + y^2)^2.$     **232.** $(a + b - c)^2 - (a - c)^2.$

**229.** $x^2 + 2\,xy + y^2 - 4.$     **233.** $y^3 + y^2 - y - 1.$

**234.** $(a + 2\,b + 3\,c)^2 - 4\,(a + b - c)^2.$

**235.** $x^6 - x^2 + 2\,x - 1.$     **237.** $x^4 - x^2 - x + 1.$

**236.** $24\,x^2 - 23\,x - 12.$     **238.** $x^2y^2 - x^2 - y^2 + 1.$

**239.** $(13\,x^2 - 5\,y^2)^2 - (12\,x^2 + 4\,y^2)^2.$

**240.** $4\,a^2b^2 - (a^2b^2 - c^2)^2.$

**241.** $(a + b)^2 + (a + c)^2 - (c + d)^2 - (b + d)^2.$

**242.** $x^{m+1} - x^my + xy^m - y^{m+1}.$     **245.** $3\,ap - 6\,aq - 3\,cp + 6\,cq.$

**243.** $4\,x^3 - 12\,x^2y - xy + 3\,y^2.$     **246.** $3\,x^2y^2 - 3\,xy^2 + 3\,x^2y - 3\,xy.$

**244.** $2\,x^3 - 3\,x^2y - 2\,b^2x + 3\,b^2y.$     **247.** $a^6 + a^4 + a^2 + 1.$

**248.** $lx - ly + la - mx + my - am.$

**249.** $7\,ax + 14\,bx - 3\,ay + a + 2\,b - 6\,by.$

**250.** $a^3b^2c^3 - 3\,abc - a^2bc^2 + 3.$

**251.** $2\,x^3 - 2\,ax^2 + 2\,bx^2 - 2\,abx.$

**252.** $2\,ax + ay + az + 2\,bx + by + bz.$

Find the H. C. F. of :

**253.** $6(x+1)^3,\ 9(x^2-1).$

**254.** $3\,x^2 + 10\,x + 8,\ 6\,x^2 + 23\,x + 20.$

**255.** $5\,x^2 + 7\,x + 2,\ 15\,x^2 + 26\,x + 8.$

**256.** $9\,x^2y^2 + 18\,xy + 5,\ 18\,x^2y^2 + 39\,xy + 15.$

**257.** $a^2 - 11\,a + 10,\ a^2 - 10\,a + 9.$

**258.** $3\,a^2b^2 - 5\,ab + 2,\ 3\,a^2b^2 - 4\,ab + 1.$

**259.** $10\,x^2 - 23\,x + 12,\ 30\,x^2 - 67\,x + 33.$

**260.** $7\,x^2 + 16\,x - 15,\ 28\,x^2 + 71\,x - 65.$

**261.** $12\,x^2y^2 - 19\,xy - 21,\ 48\,x^2y^2 - 73\,xy - 91.$

**262.** $x^2 + 8\,x + 15,\ x^2 + 9\,x + 20.$

**263.** $x^2 - 9\,x + 14,\ x^2 - 11\,x + 28.$

**264.** $x^2 + 2\,x - 120,\ x^2 - 2\,x - 80.$

**265.** $x^2 - 15\,x + 36,\ x^2 - 9\,x - 36.$

Find the L. C. M. of :

**266.** $x^2 + 3\,x + 2,\ x^2 + 4\,x + 3,\ x^2 + 5\,x + 6.$

**267.** $x^2 - 3\,x - 4,\ x^2 - x - 12.$

**268.** $2\,x^2 - 7\,x + 5,\ 6\,x^2 - 23\,x + 20.$

**269.** $2\,x^2 + 13\,x + 15,\ 8\,x^2 + 10\,x - 3.$

**270.** $x^2 - 18\,xy + 32\,y^2,\ x^2 - 9\,xy + 14\,y^2.$

Reduce to lowest terms :

**271.** $\dfrac{6\,a^2 - 7\,ab - 3\,b^2}{6\,a^2 + 11\,ab + 3\,b^2}.$

**272.** $\dfrac{x^2 + 4\,x - 77}{x^2 + 18\,x + 77}.$

**273.** $\dfrac{p^2 - 5\,p + 4}{p^2 - 7\,p + 12}.$

**274.** $\dfrac{x^2 - 8\,xy + 15\,y^2}{x^2 + 3\,xy - 40\,y^2}.$

**275.** $\dfrac{5\,x^2 - 7\,x - 6}{5\,x^2 - 17\,x + 14}.$

**276.** $\dfrac{12\,x^2 + 7\,xy + y^2}{28\,x^2 + 3\,xy - y^2}.$

**277.** $\dfrac{8\,x^2 - 6\,x - 9}{6\,x^3 - 17\,x^2 + 12\,x}.$

**284.** $\dfrac{4\,x^2 - 8\,x + 3}{4\,x^2 + 4\,x - 3}.$

**278.** $\dfrac{m^4 - n^4}{m^4 + 2\,m^2 n^2 + n^4}.$

**285.** $\dfrac{x^4 - x^3 - x + 1}{x^4 + x^3 - x - 1}.$

**279.** $\dfrac{x^3 - 3\,x^2 + 2\,x - 6}{x^2 - 9}.$

**286.** $\dfrac{1 - 2\,x + x^2 - 2\,x^3}{1 + 3\,x + x^2 + 3\,x^3}.$

**280.** $\dfrac{m^2 + n^2 + 2\,mn - p^2}{m^2 + p^2 + 2\,mp - n^2}.$

**287.** $\dfrac{x^3 + (x + y)xz + yz^2}{x^4 - y^2 z^2}.$

**281.** $\dfrac{2\,ax(a^2 - x^2)}{6\,(a^2 x + ax^2)}.$

**288.** $\dfrac{6\,ac + 10\,bc + 9\,ax + 15\,bx}{6\,c^2 + 9\,cx - 2\,c - 3\,x}.$

**282.** $\dfrac{x^2 - (a + b)x + ab}{x^2 - (a + c)x + ac}$

**289.** $\dfrac{x^2 - (y - z)^2}{(x - y)^2 - z^2}.$

**283.** $\dfrac{6\,x^2 + xy - y^2}{8\,x^2 + 2\,xy - y^2}.$

**290.** $\dfrac{x^2 + y^2 - z^2 + 2\,xy}{x^2 - y^2 + z^2 + 2\,xz}.$

**291.** $\dfrac{x^2 + y^2 + z^2 + 2\,yz + 2\,zx + 2\,xy}{x^2 - y^2 - z^2 - 2\,yz}.$

**292.** $\dfrac{2\,x^2 + xy - y^2}{x^3 + x^2 y - x - y}.$

**293.** $\dfrac{acx^2 + (ad + bc)x + bd}{a^2 x^2 - b^2}.$

**294.** $\dfrac{2\,b^2 c^2 + 2\,c^2 a^2 + 2\,a^2 b^2 - a^4 - b^4 - c^4}{a^2 + b^2 - c^2 + 2\,ab}.$

**295.** $\dfrac{x^3 + x^2 y + x^2 z - xyz - y^2 z - yz^2}{(x^2 - yz)(x + y + z)}.$

**296.** $\dfrac{(ac + bd)^2 - (ad + bc)^2}{(a - b)(c - d)}.$

**297.** $\dfrac{(1 + x + 2\,x^2)^2 - (1 - x - 2\,x^2)^2}{(1 + x - 2\,x^2)^2 - (1 - x + 2\,x^2)^2}.$

Find the value of :

**298.** $\dfrac{23^3 \cdot 19 - 23 \cdot 19^3}{23 \cdot 19(23 + 19)}$.

**299.** $\dfrac{x+3}{x+4} + \dfrac{x-4}{x-3} + \dfrac{x+5}{x+7}$.

**300.** $\dfrac{x-1}{x-2} + \dfrac{x-2}{x-3} + \dfrac{x-3}{x-4}$.

**301.** $\dfrac{1}{(x+1)(x+2)} - \dfrac{3}{(x+1)(x+2)(x+3)}$.

**302.** $\dfrac{x^2}{x^2-1} + \dfrac{x}{x-1} + \dfrac{x}{x+1}$.

**303.** $\dfrac{1}{(a+c)(a+d)} - \dfrac{1}{(a+c)(a+e)}$.

**304.** $\dfrac{x-a}{x-b} + \dfrac{x-b}{x-a} - \dfrac{(a-b)^2}{(x-a)(x-b)}$.

**305.** $\dfrac{a+b}{(b-c)(c-a)} + \dfrac{b+c}{(c-a)(a-b)} + \dfrac{c+a}{(a-b)(b-c)}$.

**306.** $\dfrac{1}{x^2+9x+20} + \dfrac{1}{x^2+12x+35}$.

**307.** $\dfrac{1}{x^2+7x-44} + \dfrac{1}{x^2-2x-143}$.

**308.** $\dfrac{m}{n} + \dfrac{2m}{m+n} - \dfrac{2mn}{(m+n)^2}$.

**309.** $\dfrac{1+x}{1+x+x^2} + \dfrac{1-x}{1-x+x^2} - \dfrac{2}{1+x^2+x^4}$.

**310.** $\dfrac{5x+4}{x-2} - \dfrac{3x-2}{x-3} - \dfrac{x^2-2x-17}{x^2-5x+6}$.

**311.** $\dfrac{a-b}{2(a+b)} + \dfrac{a+b}{2(a-b)} - \dfrac{a^2+b^2}{a^2-b^2}.$

**312.** $\dfrac{3\,x^2-8}{x^3-1} - \dfrac{5\,x+7}{x^2+x+1} + \dfrac{2}{x-1}.$

**313.** $\dfrac{1}{a-b} - \dfrac{b}{a^2+ab+b^2} - \dfrac{2\,b^2}{a^3-b^3}.$

**314.** $\dfrac{1}{x^2-7\,x+12} + \dfrac{2}{x^2-9\,x+20} - \dfrac{3}{x^2-8\,x+15}.$

**315.** $\dfrac{10\,x^2}{(1+x^2)(1-4\,x^2)} + \dfrac{2}{1+x^2} - \dfrac{1}{1-2\,x}.$

**316.** $\dfrac{1}{(x-2)(x-1)x(x+1)} + \dfrac{1}{(x-1)x(x+1)(x+2)}.$

**317.** $\dfrac{x-1}{(x+2)(x+5)} - \dfrac{2(x+2)}{(x+5)(x-1)} + \dfrac{x+5}{(x-1)(x+2)}.$

**318.** $1 - \dfrac{x-y}{x+y} + \dfrac{2\,y^2}{x^2-y^2} + \dfrac{2\,xy}{x^2+y^2}.$

**319.** $\dfrac{1-x}{x} + \dfrac{x}{8(1-x)} - \dfrac{x^2}{4(1+x)^2} + \dfrac{9\,x}{8(1+x)} - \dfrac{x(1-x)}{4(1+x^2)}.$

**320.** $\dfrac{x^2-(y-z)^2}{(x+z)^2-y^2} + \dfrac{y^2-(z-x)^2}{(y+x)^2-z^2} + \dfrac{z^2-(x-y)^2}{(z+y)^2-x^2}.$

**321.** $\dfrac{(x-a)^2-(b-c)^2}{(x+c)^2-(a+b)^2} + \dfrac{(x-b)^2-(c-a)^2}{(x+a)^2-(b+c)^2} + \dfrac{(x-c)^2-(a-b)^2}{(x+b)^2-(c+a)^2}.$

**322.** $\dfrac{b-c}{(b-c)^2-(x-a)^2} + \dfrac{c-a}{(c-a)^2-(x-b)^2} + \dfrac{a-b}{(a-b)^2-(x-c)^2}.$

**323.** $\dfrac{2}{y-z} + \dfrac{2}{z-x} + \dfrac{2}{x-y} + \dfrac{(y-z)^2+(z-x)^2+(x-y)^2}{(y-z)(z-x)(x-y)}.$

Simplify:

**324.** $\dfrac{2\,x^3 - 3\,x^2}{x^2 - 4\,x - 5} \div \dfrac{4\,x^3 + 2\,x^2 - 12\,x}{10\,x^2 - 250}$.

**325.** $\dfrac{x^2 + 12\,x + 35}{x^3 + 3\,x^2 - 10\,x} \times \dfrac{3\,x^2 - 12}{5\,x^2 + 10\,x} \div \dfrac{x^2 - 49}{2\,x^2}$.

**326.** $\dfrac{a^2 - 5\,a}{a^2 - 2\,a - 63} \times \dfrac{a^4 - 4\,a^2}{a^2 + a - 42} \div \dfrac{a^3 + 2\,a^2 + a}{a^2 - 15\,a + 54}$.

**327.** $\dfrac{8\,a^2 - 28\,a + 12}{2\,a^3 - 11\,a^2 + 12\,a} \times \dfrac{a^2 - 8\,a + 16}{4\,a^2 + 6\,a - 4} \div \dfrac{27\,a^3 - 12\,a}{6\,a^2 - 5\,a - 6}$.

**328.** $\dfrac{2\,x^2 - 18\,x + 40}{3\,x^2 - 75} \times \dfrac{x^2 + 5\,x}{(x+3)^2 - 9} \div \dfrac{x^3 - 16\,x}{x^2 + x - 12}$.

**329.** $\dfrac{4\,x^2 - 9\,y^2}{4\,x^2 - 10\,xy - 24\,y^2} \times \dfrac{6\,x^2 - 9\,xy + 27\,y^2}{x^2 - 7\,xy + 12\,y^2} \times \dfrac{(x - 4\,y)^2}{3(2\,x - 3\,y)^2}$.

**330.** $\dfrac{y^3 - y^2 + y - 1}{3\,y^2 + y - 4} \times \dfrac{3\,y^2 - 11\,y - 20}{2\,y^4 - y^2 - 3} \times \dfrac{2\,y^3 + 4\,y^2 - 3\,y - 6}{6\,y^2 + 7\,y - 10}$.

**331.** $\dfrac{2\,ay + 3\,y - 6\,a - 9}{6\,y^2 + 11\,y - 10} \times \dfrac{6\,y^2 + 5\,y - 6}{6\,a^2 + 5\,a - 6} \times \dfrac{6\,ay - 4\,y + 15\,a - 10}{5\,y^2 - 12\,y - 9}$.

**332.** $\dfrac{3\,x^2 - 4\,xy - 4\,y^2}{2\,x^5 - 8\,x^3 y^2} \times \dfrac{5\,x^3 + 8\,x^2 y - 4\,xy^2}{10\,x^2 - 19\,xy + 6\,y^2} \div \dfrac{6\,x^2 + 13\,xy + 6\,y^2}{8\,x^4 - 18\,x^2 y^2}$

**333.** $\dfrac{x^2 - 7\,xy + 12\,y^2}{x^2 + 5\,xy + 6\,y^2} \div \dfrac{x^2 - 5\,xy + 4\,y^2}{x^2 + xy - 2\,y^2}$.

**334.** $\dfrac{x^2 + 2\,x - 15}{x^2 + 8\,x - 33} \div \dfrac{x^2 + 9\,x + 20}{x^2 + 7\,x - 44}$.

**335.** $\left(x + \dfrac{xy}{x - y}\right) \times \left(x - \dfrac{xy}{x + y}\right) \div \dfrac{x^2 + y^2}{x^2 - y^2}$.

**336.** $\left\{ \dfrac{b^2 - c^2}{a} + \dfrac{c^2 - a^2}{b} + \dfrac{a^2 - b^2}{c} \right\} \div \left\{ \dfrac{b - c}{a} + \dfrac{c - a}{b} + \dfrac{a - b}{c} \right\}.$

**337.** $(a + 2b)^2.$  **338.** $\left( a + \dfrac{1}{a} \right)^2.$  **339.** $\left( \dfrac{1}{x} - \dfrac{1}{y} \right)^2.$  **340.** $\left( \dfrac{3x}{5y} \right)^3.$

**341.** $(m + n)\left( \dfrac{1}{m} + \dfrac{1}{n} \right).$    **345.** $\left( \dfrac{2x}{3y} - \dfrac{3y}{4x} \right)^2.$

**342.** $(a - b)\left( \dfrac{1}{a} + \dfrac{1}{b} \right).$    **346.** $\left( \dfrac{1}{x^2} - \dfrac{1}{x} + 1 \right)(x^2 + x).$

**343.** $\left( x + 1 + \dfrac{1}{x} \right)^2.$    **347.** $\left( 1 + \dfrac{2}{x} \right)(x^2 - 2x).$

**344.** $\left( \dfrac{m}{n} + \dfrac{n}{m} \right)^2.$    **348.** $1 \div \dfrac{m + n}{m - n}.$

**349.** $\dfrac{x^2 - 11x + 24}{x^2 - 1} \div \dfrac{x^2 - x - 6}{x^2 + 3x + 2}.$

**350.** $\dfrac{2^3 \cdot 7^4 \cdot 5^2}{11^3 \cdot 13^2 \cdot 17^4} \times \dfrac{2^4 \cdot 11^4 \cdot 17^4}{7^5 \cdot 5^2 \cdot 13} \div \dfrac{2^7 \cdot 7^2}{13^3 \cdot 11}.$

Find the numerical values of:

**351.** $1 + \dfrac{1}{1 + \dfrac{1}{a}}$, if $a = 3.$

**352.** $1 + \dfrac{1}{a + \dfrac{1}{b + \dfrac{1}{c}}}$, if $a = 2, b = 3, c = \dfrac{1}{2}.$

Simplify:

**353.** $\dfrac{x - 3}{2} - \left[ \dfrac{x + 1}{3} - \left( \dfrac{2x - 1}{4} - \dfrac{5x - 2}{10} \right) \right].$

**354.** $\dfrac{\dfrac{x}{y} + \dfrac{y}{x}}{\dfrac{x}{y} - \dfrac{y}{x}} + \dfrac{\dfrac{x}{y} - \dfrac{y}{x}}{\dfrac{x}{y} + \dfrac{y}{x}}.$

T

**355.** $\left[\dfrac{5}{2(x-3)} - \dfrac{3}{2(x-1)}\right] \times \left[\dfrac{5}{x+2} - \dfrac{4}{x+1}\right]$.

**356.** $\left(1 - \dfrac{1-x}{1+x} - \dfrac{1-10\,x^2}{1-x^2}\right)\left(\dfrac{x-1}{4\,x-1}\right)$.

**357.** $\left(\dfrac{x^2+xy+y^2}{x^2-xy+y^2} - \dfrac{x^2-xy+y^2}{x^2+xy+y^2}\right)\dfrac{(x^2-y^2)(x^4+x^2y^2+y^4)}{x^4-y^4}$.

**358.** $\left\{\dfrac{a}{a+1} - \dfrac{a-1}{a}\right\} \div \left\{\dfrac{a}{a+1} + \dfrac{a-1}{a}\right\}$.

**359.** $\left(y - \dfrac{a^2-xy}{y-x}\right)\left(x + \dfrac{a^2-xy}{y-x}\right) + \left(\dfrac{a^2-xy}{y-x}\right)^2$.

**360.** $(a+b+c)\left(\dfrac{1}{a} + \dfrac{1}{b} + \dfrac{1}{c}\right) - \dfrac{(b+c)(c+a)(a+b)}{abc}$.

**361.** $\dfrac{a}{1+\dfrac{a}{b}} + \dfrac{b}{1+\dfrac{b}{a}} - \dfrac{2}{\dfrac{1}{a}+\dfrac{1}{b}}$.

**362.** $\left(\dfrac{y}{3\,x-y} + \dfrac{3\,x}{3\,x+y}\right) \times \dfrac{3\,x-y}{9\,x^2+y^2} \div \left(\dfrac{1}{3\,x-y} - \dfrac{1}{3\,x+y}\right)$.

**363.** $\left(\dfrac{a^2}{b^2} + 1 + \dfrac{b^2}{a^2}\right)\left(\dfrac{a^2}{b^2} - 1 + \dfrac{b^2}{a^2}\right) \div \left(\dfrac{a^4}{b^4} + 1 + \dfrac{b^4}{a^4}\right)$.

**364.** $\left(\dfrac{2\,x}{3\,a} + 1 + \dfrac{3\,a}{2\,x}\right)\left(\dfrac{2\,x}{3\,a} - 1 + \dfrac{3\,a}{2\,x}\right) \div \dfrac{16\,x^4 + 36\,a^2x^2 + 81\,a^4}{a^2x^2}$.

**365.** $\left(\dfrac{x}{1+x} + \dfrac{1-x}{x}\right) \div \left(\dfrac{x}{1+x} - \dfrac{1-x}{x}\right)$.

**366.** $\dfrac{a + \dfrac{ab}{a-b}}{a^2 - \dfrac{2\,a^2b^2}{a^2+b^2}} \times \dfrac{\dfrac{1}{a^2} - \dfrac{1}{b^2}}{\dfrac{1}{a} - \dfrac{1}{b}}$.

**367.**
$$\frac{\left(\dfrac{b^2}{b^2 + c^2} + \dfrac{c^2}{b^2 - c^2}\right)(b^2 + c^2)^2}{\dfrac{b}{b - c} - \dfrac{c}{b + c}}.$$

**368.**
$$\frac{\dfrac{a - b}{1 + ab} + \dfrac{b - c}{1 + bc}}{1 - \dfrac{(a - b)(b - c)}{(1 + ab)(1 + bc)}}.$$

**369.**
$$\frac{\left(\dfrac{a^2}{a^2 + b^2} + \dfrac{b^2}{a^2 - b^2}\right)(a^2 + b^2)^2}{\dfrac{a}{a + b} + \dfrac{b}{a - b}}.$$

**370.**
$$\frac{\dfrac{a - b}{1 + ab} + \dfrac{b - c}{1 + bc}}{1 - \dfrac{(a - b)(b - c)}{(1 + ab)(1 + bc)}} \div \frac{1 - \dfrac{c}{a}}{\dfrac{1}{a} + c}.$$

**371.**
$$\frac{\dfrac{2\,yz}{y + z} - y}{\dfrac{1}{z} + \dfrac{1}{y - 2\,z}} - \frac{z - \dfrac{2\,yz}{y + z}}{\dfrac{1}{y} - \dfrac{1}{2\,y - z}}.$$

**372.**
$$\left(2 - \frac{3\,n}{m} + \frac{9\,n^2 - 2\,m^2}{m^2 + 2\,mn}\right) \div \left\{\frac{1}{m} - \frac{1}{m - 2\,n - \dfrac{4\,n^2}{m + n}}\right\}.$$

**373.**
$$\frac{1 + x^3}{1 - x + \dfrac{3\,x - x^3}{3 - x + \dfrac{3}{x + \dfrac{x^2}{1 - x}}}}.$$

**374.**
$$\frac{2(a - x)}{2(a - x) + \dfrac{2\,b^2}{b + \dfrac{b^2 x}{x^2 - a^2}}}.$$

**375.** $\dfrac{\left(\dfrac{3x+x^3}{1+3x^2}\right)^2-1}{\dfrac{3x^2-1}{x^3-3x}+1} \div \dfrac{\dfrac{9}{x^2}-\dfrac{33-x^2}{3x^2+1}}{\dfrac{3}{x^2}-\dfrac{2(x^2+3)}{(x^3-x)^2}}.$

Solve the equations :

**376.** $\dfrac{5x-8}{4}+\dfrac{x-2}{5}=15.$      **377.** $\dfrac{3x-5}{6}+\dfrac{4x-7}{8}=2\tfrac{1}{3}.$

**378.** $2(3x+1)-\tfrac{2}{5}(x+4)+20+\tfrac{2}{3}(x+7)=0.$

**379.** $4(x+6)-\dfrac{x}{3}-\dfrac{2}{5}(x+10)=\dfrac{13}{10}(x+5)-1\tfrac{1}{4}.$

**380.** $5\{2x+1-3(x+1)\}-\dfrac{x+2}{3}+\dfrac{10}{7}(3x+5\tfrac{1}{2})+2\tfrac{2}{3}=0.$

**381.** $\dfrac{x^2+7x-6}{x^2+5x-10}=\dfrac{x+1}{x-1}.$     **382.** $\dfrac{1}{x+2}+\dfrac{2}{x-2}=\dfrac{3}{x-3}.$

**383.** $\dfrac{4}{2x-5}-\dfrac{3}{3x-7}=\dfrac{1}{x}.$     **384.** $\dfrac{2x+7}{x+1}+\dfrac{3x-5}{x+2}=\dfrac{5x+9}{x+3}.$

**385.** $\dfrac{x^2-6x+10}{x^2+8x+17}-\dfrac{(x-3)^2}{(x+4)^2}=0.$

**386.** $\dfrac{1}{15}(3x+13)-\dfrac{3x+10}{10x-50}=\dfrac{x}{5}.$

**387.** $\dfrac{8x+5}{14}-\dfrac{3-7x}{6x+2}=\dfrac{16x+15}{28}+\dfrac{2\tfrac{1}{4}}{7}.$

**388.** $\dfrac{6x-7\tfrac{1}{4}}{13-2x}+\dfrac{1+16x}{24}=\dfrac{53-24x}{12}-\dfrac{12\tfrac{1}{2}-8x}{3}.$

**389.** $\dfrac{2-\tfrac{5}{3}x}{5}=\dfrac{7-2x^2}{14(x-1)}+\dfrac{x+\tfrac{1}{3}}{7}-\dfrac{6x-6\tfrac{3}{5}}{18}+\dfrac{51}{105}$

**390.** $\dfrac{\dfrac{3}{x}-1}{2}-\dfrac{9\left(\dfrac{1}{2x}-1\right)-\dfrac{2}{5}\left(\dfrac{9}{2x}-4\right)}{\dfrac{3}{x}-4}=\dfrac{\dfrac{9}{x}-16}{6}.$

**391.** $\dfrac{x-4}{x-5} - \dfrac{x-5}{x-6} = \dfrac{x-7}{x-8} - \dfrac{x-8}{x-9}.$

**392.** $\dfrac{x}{x-2} + \dfrac{x-9}{x-7} = \dfrac{x+1}{x-1} + \dfrac{x-8}{x-6}.$

**393.** $\dfrac{3-2x}{1-2x} - \dfrac{2x-5}{2x-7} = 1 - \dfrac{4x^2-1}{7-16x+4x^2}.$

**394.** $\dfrac{x-5}{7} + \dfrac{x^2+6}{3} = \dfrac{x^2-2}{2} - \dfrac{x^2-x+1}{6} + 3.$

**395.** $(x+1)(x+2)(x+3)$
$$= (x-1)(x-2)(x-3) + 3(4x-2)(x+1).$$

**396.** $(8x-3)^2(x-1) = (4x-1)^2(4x-5).$

**397.** $\dfrac{x^2-x+1}{x-1} + \dfrac{x^2+x+1}{x+1} = 2x.$

**398.** $.5x - 2 = .25x + .2x - 1.$

**399.** $.5x + .6x - .8 = .75x + .25.$

**400.** $3^x = 177{,}147.$

**401.** $\dfrac{x-a}{a} = \dfrac{b-c}{c}.$

**402.** $\dfrac{x-1}{x+1} = \dfrac{1-a}{1+a}.$

**403.** $\dfrac{mx+n}{mx-n} = \dfrac{b+c-a}{c+a-b}.$

**404.** $a\dfrac{a-x}{b} - b\dfrac{b+x}{a} = x.$

**405.** $\dfrac{x^2-a^2}{bx} - \dfrac{a-x}{b} = \dfrac{2x}{b} - \dfrac{a}{x}.$

**406.** $\gamma(x-a) + x(x-b) = 2(x-a)(x-b).$

**407.** $x-a)(x-b)(x+2a+2b) = (x+2a) \; x + 2b)(x-a-b).$

**408.** $(x-a)(x+b) + c = (x+a)(x-b).$

**409.** $\dfrac{x-a}{2} + \dfrac{x-b}{3} = \dfrac{a+3x}{3} - \dfrac{2x-b}{2}.$

**410.** $(x + a)(x - b) - 2 a^2b = (x + b)(x - a) - 2 b^2a.$

**411.** $(x - a)(x - b) = (x - a - b)^2.$

**412.** $\dfrac{a}{x - a} - \dfrac{b}{x - b} = \dfrac{a - b}{x - c}.$

**413.** $\dfrac{a}{x + a} + \dfrac{b}{x + b} = \dfrac{a + b}{x + c}.$

**414.** $\dfrac{1}{x - a} - \dfrac{1}{x - b} = \dfrac{a - b}{x^2 - ab}.$

**415.** $\dfrac{1}{x - a} - \dfrac{1}{x - a + c} = \dfrac{1}{x - b - c} - \dfrac{1}{x - b}.$

**416.** $\dfrac{mx - a - b}{nx - c - d} = \dfrac{mx - a - c}{nx - b - d}.$

**417.** $(a - b)(x - c) - (b - c)(x - a) - (c - a)(x - b) = 0.$

**418.** $\dfrac{a}{a - 2 x} - \dfrac{12 x}{a - 3 x} = \dfrac{16 x}{4 x - a} + \dfrac{a}{a - 6 x}.$

**419.** $\dfrac{x - 2 a}{b + c - a} + \dfrac{x - 2 b}{c + a - b} + \dfrac{x - 2 c}{a + b - c} = \dfrac{3 x}{a + b + c}.$

**420.** In 6 hours A walks 2 miles more than B walks in 7 hours; in 9 hours B walks 11 miles more than A walks in 5 hours. Find the number of miles an hour that A and B each walk.

**421.** In a number of two digits the first digit is twice the second, and if 18 be subtracted from the number, the order of the digits will be inverted. Find the number.

**422.** A man drives to a certain place at the rate of 8 miles an hour. Returning by a road 3 miles longer at the rate of 9 miles an hour, he takes $7\frac{1}{2}$ minutes longer than in going. How long is each road?

**423.** A person walks up a hill at the rate of $2\frac{1}{4}$ miles an hour, and down again at the rate of $3\frac{1}{2}$ miles an hour, and was out 5 hours. How far did he walk all together?

**424.** A steamer which goes at the rate of 264 miles a day is followed in 2 days by another which goes 286 miles a day. When will the second steamer overtake the first?

**425.** Find two consecutive numbers such that the sum of the fifth and eleventh parts of the greater may exceed by 1 the sum of the sixth and ninth parts of the less.

Find the fourth proportional:

**426.** $x - y$, $x^2 - y^2$, $x^2 - xy + y^2$.

**427.** $\frac{1}{3}$, $\frac{1}{4}$, $\frac{5}{8}$.　　　　**428.** $a + b$, $a - b$, $a^2 + 2ab + b^2$.

Find the mean proportional to

**429.** $3\frac{3}{8}$ and $1\frac{1}{2}$.　　　　**430.** $x^2 - \dfrac{1}{z^2}$ and $z^2 - \dfrac{1}{x^2}$.

**431.** Find the ratio $x : y$, if

$$5x = 7y; \quad mx + y = ny; \quad ax + by = cx + dy.$$

**432.** A line 10 inches long is divided in the ratio $m : n$. Find the length of the parts.

**433.** The sum of the three angles of any triangle is 180°. If one angle of a triangle is to another as $4 : 5$ and the third angle is equal to the sum of the first two, find the angles of the triangle.

**434.** If $a : b = 5 : 7$, and $b : c = 14 : 15$, find $a : c$.

**435.** Solve: $n - m : n(n - x) = p - m : n(p - x)$.

**436.** Which ratio is greater, $5 : 7$ or $151 : 208$?

**437.** Prove that the number of miles one can see from an elevation of $h$ feet is very nearly equal to $\sqrt{\dfrac{3h}{2}}$ miles.

**438.** Which of the following proportions are true?

a.  $(3a + 4b) : (9a + 8b) = (a - 2b) : (3a - 4b)$.

b.  $(9a^2 - 4b^2) : (15a^2 - 31ab + 14b^2)$
$$= (15a^2 + 31ab + 14b^2) : (25a^2 - 49b^2).$$

c.  $(a^3 + b^3) : (a^2 + b^2) = (a^2 - b^2) : (a - b)$.

d.  $(a^8 + b^8) : (a + b) = (a^6 - a^4b + a^3b^2 - a^2b^3 + ab^4 - b^5) : (a^3 - b^3)$.

**439.** Find the value of $x$, if

*a.* $29(a + b) : x = 551(a^2 - b^2) : 19(a - b)$.

*b.* $[a - b] : \left[ \dfrac{(a + b)^2}{2\,ab} - 1 \right] = x : \left( a + b + \dfrac{2\,b^2}{a - b} \right)$.

*c.* $(3\,a^2 + 2\,ab - 8\,b^2) : (5\,a^2 + 4\,ab - 12\,b^2) = x : (5\,a - 6\,b)$.

**440.** The volumes of two spheres are to each other as the cubes of their diameters. If a sphere 2 inches in diameter weighs 12 ounces, what is the weight of a sphere of the same material having a diameter of 3 inches?

Solve the following systems:

**441.** $7\,x - 2\,y = 1$; $3\,x + 5\,y = 59$.

**442.** $x + 17\,y = 53$; $8\,x + y = 19$.

**443.** $33\,x + 35\,y = 4$; $55\,x - 55\,y = -16$.

**444.** $7\,x - 9\,y = 17$; $9\,x - 7\,y = 71$.

**445.** $7\,x - y = 3$; $5\,x + 4\,y = 10$.

**446.** $7\,x - 3\,y = 3$; $5\,x + 7\,y = 25$.

**447.** $x + 5\,y = 49$; $3\,x - 11\,y = 95$.

**448.** $ax + by = 2$; $a^2x + b^2y = a + b$.

**449.** $5\,x - 4\,y = 3\,x + 2\,y = 1$.

**450.** $ax + by = 2$; $ab(x + y) = a + b$.

**451.** $28 = 5\,a - 4\,b$; $8\,a + 21 + 3\,b = 0$.

**452.** $5\,p + 7\,q = 0$; $12\,p - 89 = q$.

**453.** $20\,y + 21\,z = 28$; $4\,z = 15\,y + 137$.

**454.** $18\,a = 50 + 25\,y$; $5\,y + 10 = -27\,a$.

**455.** $56 + 10\,y = 7\,x$; $15\,x = 26 + 8\,y$.

**456.** $9\,p = 2 - 11\,q$; $21\,q = 27 + 6\,p$.

**457.** $3\,x - 7\,y = 25$; $4\,x = 5\,y + 29$.

**458.** $8\,x - 59 = 3\,z$; $5\,z = 7\,x - 35$.

**459.** $\dfrac{x}{5} - \dfrac{1}{7}(x - 2\,y) = 0$; $\dfrac{1}{10}(3\,x + 5\,y) - \dfrac{1}{8}(x + y) = 1$.

**460.** $\dfrac{3x+7y}{4} - \dfrac{x+y}{8} = 5$; $7x = 56 - 3y$.

**461.** $\tfrac{1}{12}(x - 2y) - \tfrac{1}{8}(2x - 7) = 1$; $y = x + 17$.

**462.** $\dfrac{3x-y}{12} - \dfrac{7x-3y}{15} = \dfrac{1}{10}$; $\dfrac{x}{4} - \dfrac{y}{10} = \dfrac{3}{10}$.

**463.** $4x + \dfrac{x-y}{8} = 21$; $2y + \dfrac{3x-5y}{3} = 4$.

**464.** $\dfrac{x}{4} + \dfrac{2x-y}{5} = 2\tfrac{1}{2}$; $\dfrac{y}{7} + \dfrac{x-2y}{8} = 0$.

**465.** $\tfrac{1}{5}(x - 4y) - \tfrac{1}{4}(x + 9y) = 2\tfrac{1}{2}$; $\tfrac{1}{3}(2x + 7y) = \tfrac{1}{2}(x + y)$.

**466.** $\dfrac{4x-y}{2} = \dfrac{7x+3y}{11} + \dfrac{1}{2}$; $\dfrac{x}{4} + \dfrac{2x+3y}{8} = x$.

**467.** $\dfrac{5x}{7} - \dfrac{1}{10}(7x + 4y) = 2$; $\dfrac{1}{10}(x - 2y) - \dfrac{1}{4}(2x + y) = 2\tfrac{1}{4}$.

**468.** $\dfrac{5}{x} + \dfrac{12}{y} = 7$; $\dfrac{1}{x} + \dfrac{6}{y} = 2$.     **469.** $\dfrac{2}{x} - \dfrac{7}{y} = 5$; $\dfrac{7}{x} - \dfrac{12}{y} = 5$.

**470.** $\dfrac{x+3y}{3} - \dfrac{2x-7y}{8} = \dfrac{1}{3} = \dfrac{x}{12} - \dfrac{y}{4}$.

**471.** $ax + by = c$;
$dx - ey = f$.

**472.** $ax - by = m$;
$cx + ey = n$.

**473.** $cx = dy$;
$x + y = e$.

**474.** $ax + by = c$;
$dx + fy = c^2$.

**475.** $\dfrac{a}{b+y} = \dfrac{b}{3a+x}$;
$ax + 2by = d$.

**476.** In a certain proper fraction the difference between the nu merator and the denominator is 12, and if each be increased by 5 the fraction becomes equal to ⅔. Find the fraction.

**477.** What is that fraction which becomes ⅔ when its numerator is doubled and its denominator is increased by 1, and becomes ⅜ when its denominator is doubled and its numerator increased by 4?

**478.** If 1 be added to the numerator of a fraction it becomes equal to ⅓, if 1 be added to the denominator it becomes equal to ¼. Find the fraction.

**479.** The sum of three numbers is 21. The greatest exceeds the least by 4, and the other number is half the sum of the greatest and least. Find the numbers.

**480.** There are two numbers the half of the greater of which exceeds the less by 2, also a third of the greater exceeds half the less by 2. Find the numbers.

**481.** Of the ages of two brothers one exceeds half the other by 4 years, and a fifth part of one brother's age is equal to an eighth of that of the other. Find their ages.

**482.** If 31 years were added to the age of a father it would be thrice that of his son; also if one year were taken from the son's age and added to the father's, the latter would then be twice the son's age. Find their ages.

**483.** A and B together have \$6000. A spends ⅓ of his money and B spends ⅕ of his. B then has ⅔ as much as A. How much money had each at first?

**484.** Find two numbers such that twice the greater exceeds the less by 30, and 5 times the less exceeds the greater by 3.

**485.** A sum of money at simple interest amounted in 10 months to \$2100, and in 18 months to \$2180. Find the sum and the rate of interest.

**486.** A sum of money at simple interest amounts in 8 months to \$260, and in 20 months to \$275. Find the principal and the rate of interest.

**487.** A number consists of two digits whose difference is 4; if the sum of the digits be multiplied by 4, the digits will be inverted. Find the number.

**488.** There is a number of two digits which is equal to seven times the sum of the digits; also if the digits be transposed the new number will exceed 10 times the difference of the digits by 6. Find the number.

**489.** Find two numbers whose sum equals $s$ and whose difference equals $d$.

**490.** The sum of two numbers is $a$, and the difference of their squares is $b$. Find the numbers.

Solve the following systems:

**491.** $x + y + z = 29\frac{1}{4}$; $x + y - z = 18\frac{1}{4}$; $x - y + z = 13\frac{3}{4}$.

**492.** $y + \dfrac{x}{2} = 41$; $x + \dfrac{z}{4} = \dfrac{41}{2}$; $y + \dfrac{z}{5} = 34$.

**493.** $5x + 3y - 6z = 4$; $3x - y + 2z = 8$; $x - 2y + 2z = 2$.

**494.** $4x - 5y + 2z = 6$; $2x + 3y - z = 20$; $7x - 4y + 3z = 35$.

**495.** $8x + 4y - 3z = 6$; $x + 3y - z = 7$; $4x - 5y + 4z = 8$.

**496.** $4x - 3y + z = 9$; $9x + y - 5z = 16$; $x - 4y + 3z = 2$.

**497.** $y - x + z = -5$; $z - y - x = -25$; $x + y + z = 35$.

Solve:

**498.** $x + y = 11$; $y + z = -1$; $x - z = 12$.

**499.** $3x + 5y = 161$; $7x + 2z = 209$; $2x + z = 79$.

**500.** $\dfrac{1}{x} + \dfrac{1}{y} = a$; $\dfrac{1}{y} + \dfrac{1}{z} = b$; $\dfrac{1}{z} + \dfrac{1}{x} = c$.

**501.** $4x - 3y + 2z = 40$; $5x + 9y - 7z = 47$; $9x + 8y - 3z = 97$

**502.** $3x + 2y = 8$; $4x + 3z = 20$; $3y + 6z = 27$.

**503.** $2x + 7y = 15$; $2y + 3z = 11$; $4x + 9z = 23$.

**504.** $7p - 2q = 20$; $2p - 3r = 4$; $3q - 4r = -9$.

**505.** $3x - 5y = 17$; $2y + 5z = -12$; $x - 2z = 8$.

**506.** $2x - 3y = 8$; $5y - 9z = 10$; $x + 4y - 2z = 15$.

**507.** $\dfrac{x}{3} + \dfrac{y}{4} + \dfrac{z}{2} = 2$; $\dfrac{x}{6} + y + \dfrac{z}{10} = 2$; $x + y + \dfrac{z}{4} = \dfrac{1}{4}$.

**508.** $\dfrac{x}{2} + y + \dfrac{z}{5} = 1$; $\dfrac{x}{4} - \dfrac{y}{8} - \dfrac{z}{10} = 2$; $3x - 1z = 4$.

**509.** $\frac{x}{5} - \frac{y}{3} = 0$; $\frac{x}{2} + \frac{z}{5} = 5$; $x - y + z = 4$.

**510.** $\frac{p}{5} + \frac{q}{3} = \frac{r}{4}$; $\frac{p}{10} - \frac{q}{12} = \frac{1}{4}$; $p + q = r$.

**511.** $x + 2z = 4x + 2y + 1 = 5y + 3z - 2 = 13$.

**512.** $\frac{x}{2} + 2y + z = 0$; $3y - 2z = \frac{x}{8} = \frac{x-y}{3} + \frac{z}{2} - 1$.

**513.** $x + \frac{z}{2} = 0$; $\frac{x+z}{7} = \frac{y}{6}$; $\frac{x}{2} = \frac{y-z}{3} - \frac{5}{6}$.

**514.** $\frac{1}{x} + \frac{1}{y} = 1$; $\frac{1}{y} + \frac{1}{z} = \frac{3}{2}$; $\frac{1}{z} + \frac{1}{x} = 2$.

**515.** $\frac{1}{x} - \frac{1}{y} = \frac{1}{6}$; $\frac{1}{y} + \frac{1}{z} = 3\frac{5}{6}$; $\frac{4}{x} + \frac{3}{y} = \frac{4}{z}$.

**516.** $\begin{cases} \dfrac{a}{x} + \dfrac{b}{y} + \dfrac{c}{z} = a, \\ \dfrac{b}{y} - \dfrac{c}{z} + \dfrac{a}{x} = b, \\ \dfrac{a}{x} - \dfrac{b}{y} - \dfrac{c}{z} = c. \end{cases}$

**517.** $\begin{cases} \dfrac{1}{x} - \dfrac{1}{y} = a, \\ \dfrac{1}{y} + \dfrac{1}{z} = b, \\ \dfrac{1}{z} - \dfrac{1}{x} = c. \end{cases}$

**518.** $\begin{cases} \dfrac{1}{x} - \dfrac{1}{y} = \dfrac{1}{6}, \\ \dfrac{1}{y} + \dfrac{1}{z} = \dfrac{7}{12}, \\ \dfrac{4}{x} - \dfrac{3}{y} = \dfrac{4}{z}. \end{cases}$

**519.** $\begin{cases} \dfrac{1}{x} - \dfrac{5}{y} + \dfrac{1}{z} = -3, \\ \dfrac{3}{x} + \dfrac{4}{y} - \dfrac{2}{z} = 5, \\ \dfrac{5}{x} + \dfrac{7}{y} + \dfrac{6}{z} = 18. \end{cases}$

**520.** $\begin{cases} y + z - 3x = 2a, \\ z + x - 3y = 2b, \\ x + y - 3z = 2c. \end{cases}$

**521.** $\begin{cases} bz + cy = a, \\ cx + az = b, \\ ay + bx = c. \end{cases}$

**522.** $\begin{cases} \dfrac{yz}{y+z} = \dfrac{bc}{b+c}, \\ \dfrac{zx}{z+x} = \dfrac{ca}{c+a}, \\ \dfrac{xy}{x+y} = \dfrac{ab}{a+b}. \end{cases}$

**523.** $\begin{cases} \dfrac{x}{3} + \dfrac{y}{5} + \dfrac{z}{7} + \dfrac{u}{9} = 2800, \\ \dfrac{x}{5} + \dfrac{y}{7} + \dfrac{z}{9} + \dfrac{u}{11} = 2144, \\ \dfrac{x}{7} + \dfrac{y}{9} + \dfrac{z}{11} + \dfrac{u}{13} \quad 1744, \\ \dfrac{x}{9} + \dfrac{y}{11} + \dfrac{z}{13} + \dfrac{u}{15} = 1472. \end{cases}$

**524.** $\begin{cases} x + y + z = 3a + b + c, \\ x + y + t = a + 3b + c, \\ x - z - t = a + b - c, \\ y + z - t = 3a - b - c. \end{cases}$

**525.** When weighed in water, 37 pounds of tin lose 5 pounds, and 23 pounds of lead lose 2 pounds.

(a) How many pounds of tin and lead are in a mixture weighing 120 pounds in air, and losing 14 pounds when weighed in water?

(b) How many pounds of tin and lead are in an alloy weighing 226 pounds in air and 201 pounds in water?

**526.** A and B together can do a piece of work in 2 days, B and C in 3 days, and C and A in 4 days. In how many days can each alone do the same work?

**527.** Three numbers are such that the sum of the reciprocals of the first and second equals $\frac{1}{2}$; the sum of the reciprocals of the first and third equals $\frac{1}{3}$; the sum of the reciprocals of the second and third equals $\frac{1}{4}$. Find the numbers.

**528.** A vessel can be filled by three pipes, $L$, $M$, $N$. If $M$ and $N$ run together, it is filled in 35 minutes; if $N$ and $L$, in 28 minutes; if $L$ and $M$, in 20 minutes. In what time will it be filled if all run together?

**529.** A boy is $a$ years old; his mother was $b$ years old when he was born; his father is half as old again as his mother was $c$ years ago. Find the present ages of his father and mother.

**530.** A can do a piece of work in 12 days; B and C together can do the same piece of work in 4 days; A and C can do it in half the time in which B alone can do it. How long will B and C take to do it separately?

**531.** A number of three digits whose first and last digits are the same has 7 for the sum of its digits; if the number be increased by 90, the first and second digits will change places. Find the number.

**532.** In $\triangle ABC$, $AB = 6$, $BC = 5$, and $CA = 7$. An (escribed) circle touches $AC$ in $D$, and the prolongations of $BA$ and $BC$ in $E$ and $F$ respectively. Find $AD$, $CD$, and $BE$.

**533.** Two persons start to travel from two stations 24 miles apart, and one overtakes the other in 6 hours. If they had walked toward each other, they would have met in 2 hours. What are their rates of travel?

**534.** Represent the following table graphically:

TABLE OF POPULATION (IN MILLIONS) OF UNITED STATES, FRANCE, GERMANY, AND BRITISH ISLES

| Year | 1800 | 1810 | 1820 | 1830 | 1840 | 1850 | 1860 | 1870 | 1880 | 1890 | 1900 |
|---|---|---|---|---|---|---|---|---|---|---|---|
| U.S. | 5.3 | 7.2 | 9.6 | 12.9 | 17.0 | 23.2 | 31.4 | 38.6 | 50.2 | 62.6 | 76.3 |
| France | 27.2 | 28.8 | 30.5 | 32.4 | 34.0 | 35.6 | 37.3 | 36.1 | 37.6 | 38.6 | 38.9 |
| Germany | 22.0 | 23.4 | 26.2 | 29.7 | 32.4 | 35.2 | 38.1 | 40.5 | 45.2 | 49.4 | 56.4 |
| British Isles | 16.0 | 17.6 | 20.5 | 24.0 | 26.4 | 27.2 | 28.7 | 31.2 | 34.5 | 37.5 | 41.2 |

**535.** One dollar equals 4.16 marks. Draw a graph for the transformation of dollars into marks.

**536.** The number of workmen required to finish a certain piece of work in $D$ days is $\dfrac{24}{D}$. Draw the graph from $D = 1$ to $D = 12$. How long will it take 11 men to do the work?

**537.** If $l$ feet is the length of a pendulum, the time of whose swing is $t$ seconds, then $l = 3.3\, t^2$. Draw a graph for the formula from $t = 0$ to $t = 3$ and write down the time of swing for a pendulum of length 8 feet.

Draw the graphs of the following functions:

**538.** $3x + 5$.　　　　**542.** $x^2 + x$.　　　　**546.** $2 - x - x^2$.

**539.** $2x - 7$.　　　　**543.** $x^2 - x - 5$.　　　　**547.** $x^3$.

**540.** $2 - 3x$.　　　　**544.** $x^2 + x - 3$.　　　　**548.** $x^3 - 2x$.

**541.** $-3x$.　　　　**545.** $x^2 - x + 2$.　　　　**549.** $x^3 - x + 1$.

**550.** Draw the graph of $y = 2 + 2x - x^2$, from $x = -2$ to $x = 4$, and from the diagram determine:

*a.* The values of $y$, i.e. the function, if $x = \frac{1}{2}, -1\frac{1}{2}, 2\frac{1}{4}$.

*b.* The values of $x$ if $y = -2$.

*c.* The greatest value of the function.

*d.* The value of $x$ that produces the greatest value of $y$.

*e.* The roots of the equation $2 + 2x - x^2 = 1$.

**551.** The formula for the distance traveled by a falling body is $S = \frac{1}{2} g t^2$.

*a.* Represent $\frac{1}{2} g t^2$ graphically from $t = 0$ to $t = 5$. (Assume $g = 10$ meters, and make the scale unit of the $t$ equal to 10 times the scale unit of the $\frac{1}{2} g t^2$.)

*b.* How far does a body fall in $2\frac{1}{2}$ seconds?

*c.* In how many seconds does a body fall 25 meters?

Solve graphically the following equations:

**552.** $x^2 - 2x - 7 = 0$.

**553.** $x^2 - 6x + 9 = 0$.

**554.** $x^2 + 5x - 4 = 0$.

**555.** $x^2 - 5x - 3 = 0$.

**556.** $x^2 - 3x - 6 = 0$.

**557.** $x^2 - 2x - 9 = 0$.

**558.** $3x^2 - 3x - 17 = 0$.

**559.** $2x^2 - 4x - 15 = 0$.

**560.** $2x^2 + 10x - 7 = 0$

**561.** $3x^2 - 6x - 13 = 0$.

**562.** $x^3 - 3x - 1 = 0$.

**563.** $x^3 + 3x - 11 = 0$.

**564.** $2x^3 - 6x + 3 = 0$.

**565.** $x^4 - 10x^2 + 8 = 0$.

**566.** $x^4 - 4x^2 + 4x - 4 = 0$.

**567.** $x^5 - x^4 - 11x^3 + 9x^2 + 18x - 4 = 0$.  **568.** $2^x + x - 4 = 0$.

**569.** If $y = x^3 + 5x^2 - 10$,

*a.* Solve $y = 0$.

*b.* Solve $y = 5$.

*c.* Solve $y = -5$.

*d.* Solve $y = -15$.

*e.* Determine the number of real roots of the equation $y = -2$.

*f.* Determine the limits between which $m$ must lie, if $y = m$ has three real roots.

*g.* Find the value of $m$ that will make two roots equal if $y = m$.

*h.* Find the greatest value which $y$ may assume for a negative $x$.

*i.* Which negative value of $x$ produces the greatest value of $y$?

Solve graphically:

**570.** $\begin{cases} 7x - 3y = 3, \\ 5x + 7y = 25. \end{cases}$

**571.** $\begin{cases} 7x - y = 3, \\ 5x + 4y = 10. \end{cases}$

**572.** $\begin{cases} x^2 + y^2 = 16, \\ x + y = 2. \end{cases}$

**573.** $\begin{cases} x + y = 5, \\ xy = 6. \end{cases}$

**574.** $\begin{cases} x - y = 1, \\ x^2 + y^2 = 25. \end{cases}$

**575.** $\begin{cases} x - y = 2, \\ xy = 8. \end{cases}$

**576.** $\begin{cases} 4\,x - 5\,y = 10, \\ xy = 6. \end{cases}$

**580.** $\begin{cases} x^2 + y^2 = 4, \\ x + y = 3. \end{cases}$

**577.** $\begin{cases} x^2 - y^2 = 4, \\ x = 2\,y. \end{cases}$

**581.** $\begin{cases} x^2 + y^2 = 50, \\ x - y = -6. \end{cases}$

**578.** $\begin{cases} xy = 6, \\ x^2 + y^2 = 25. \end{cases}$

**582.** $\begin{cases} x^2 - 2\,x + y^2 - 4\,y = 0, \\ y = 2\,x. \end{cases}$

**579.** $\begin{cases} x^2 + xy = 12, \\ x^2 - y^2 = 8. \end{cases}$

**583.** $\begin{cases} x^2 - 4\,x + y^2 + 2\,y + 3 = 0, \\ x - y = 3. \end{cases}$

Perform the operations indicated:

**584.** $\left( -\dfrac{4\,x}{3\,y^2} \right)^3$

**586.** $(a + b)^7$.

**590.** $(2 + x)^3$.

**587.** $(a - b)^7$.

**591.** $(3 - 2\,x)^3$.

**585.** $\left( -\dfrac{x^3}{y^2 z^2} \right)^4$.

**588.** $(a + b)^3 (a - b)^3$.

**592.** $(1 + x)^4$.

**589.** $(1 - x)^3$.

**593.** $(x - 2)^4$.

**594.** $(ax + by)^3 + (ax - by)^3$.

**597.** $(1 + x)^4 (1 - x)^4$.

**595.** $(ax + by)^4 + (ax - by)^4$.

**598.** $(1 + x + x^2)^2$.

**596.** $(1 + x)^5 - (1 - x)^5$.

**599.** $(1 - x + x^2)^2$.

**600.** $(2 + 3\,x + 4\,x^2)^2 + (2 - 3\,x + 4\,x^2)^2$.      **601.** $(1 + x + x^2)^3$.

Extract the square roots of the following expressions:

**602.** $64\,a^{12} - 128\,a^{10}b + 160\,a^8b^2 - 160\,a^6b^3 + 100\,a^4b^4 - 48\,a^2b^5 + 16\,b^6$.

**603.** $4 - 8\,xy^2 + 36\,x^2y^4 - 64\,x^3y^6 + 96\,x^4y^8 - 128\,x^5y^{10} + 64\,x^6y^{12}$.

**604.** $x^{16} - 2\,x^{14}y + 3\,x^{12}y^2 - 4\,x^{10}y^3 + 5\,x^8y^4 - 4\,x^6y^5 + 3\,x^4y^6 - 2\,x^2y^7 + y^8$.

**605.** $a^8 - 4\,a^7b + 4\,a^6b^2 + 6\,a^5b^3 - 14\,a^4b^4 + 4\,a^3b^5 + 9\,a^2b^6 - 6\,ab^7 + b^8$.

**606.** $9 - 6\,x + 13\,x^2 - 4\,x^3 + 4\,x^4$.

**607.** $(2\,a^2b + b^2)^2 + (a^2 - 2\,ab^2)(4\,ab + 1)$.

**608.** $x^6 + \dfrac{1}{x^6} + 10\left( \dfrac{1}{x^2} - 1 \right) + 6\,x^4 - \dfrac{4}{x^4} + 5\,x^2$.

**609.** $4\,x^4 + y^6 + 6\,xy^4 + 16\,x^2y^4 + 16\,x^3y^2 + 12\,x^3y + 28\,x^2y^3 + 9\,x^2y^2 + 8\,xy^5$.

**610.** $a^2x^2 + 2\,abx + b^2 - 2\,bx + x^2 - 2\,ax^2$.

**611.** $a^2 + 9\,b^2 + 25\,c^2 - 30\,bc + 10\,ac - 6\,ab$.

**612.** $x^2 + \dfrac{a^2}{9} + \dfrac{b^2}{4} + \dfrac{2\,ax}{3} - \dfrac{ab}{3} - bx$.

Find the fourth root of:

**613.** $a^4 + 4 a^3b + 6 a^2b^2 + 4 ab^3 + b^4$.

**614.** $16 x^4 + 96 x^3 + 216 x^2 + 216 x + 81$.

**615.** $81 x^8 - 108 x^6y + 54 x^4y^2 - 12 x^2y^3 + y^4$.

**616.** $16 a^4 - 32 a^3b^2 + 24 a^2b^4 - 8 ab^6 + b^8$.

Find the eighth root of:

**617.** $a^8 - 8 a^7b + 28 a^6b^2 - 56 a^5b^3 + 70 a^4b^4 - 56 a^3b^5 + 28 a^2b^6$
$$- 8 ab^7 + b^8.$$

**618.** $x^8 - 16 x^7 + 112 x^6 - 448 x^5 + 1120 x^4 - 1792 x^3 + 1792 x^2$
$$- 1024 x + 256.$$

Find the square root of:

**619.** 942841.   **621.** 6090.2416.   **623.** 49042009.

**620.** 25623844.   **622.** 4376464.   **624.** 44352.36.

**625.** $\sqrt{61009} + \sqrt{582169}$.   **626.** $\sqrt{956484} - \sqrt{256} - \sqrt{4096}$.

Find to three decimal places the square roots of the following numbers:

**627.** 49.871844.   **629.** 635.191209.   **631.** 494210406001.

**628.** 371246.49.   **630.** 216.15174441.   **632.** $2\frac{1}{4}$.

**633.** $2\frac{7}{9}$.   **634.** $3\frac{6}{25}$.   **635.** $4\frac{1}{5}$.   **636.** $9\frac{3}{8}$.   **637.** $40\frac{1}{4}$.   **638.** $66\frac{11}{18}$.

**639.** According to Kepler's law, the cubes of the distances of the planets from the sun have the same ratio as the squares of their periods of revolution about the sun. If the distances of Earth and Jupiter from the sun are at $1 : 5.2$, and the Earth's period equals $365\frac{1}{4}$ days, find Jupiter's period.

Solve the following equations:

**640.** $x^2 + 9 x = 70$.

**641.** $x^2 - 21 x = 100$.

**642.** $x^2 + x = 156$.

**643.** $x^2 - 53 x = -150$.

**644.** $8 x^2 + 24 x = 32$.

**645.** $9 x^2 + 189 x = 900$.

**646.** $x^2 + 6 x - 16 = 0$.

**647.** $x^2 + 9 x - 22 = 0$.

**648.** $x^2 - 5 x - 66 = 0$.

**649.** $x^2 + \frac{2}{3} x = 87$.

**650.** $3 x^2 + x = 14$.

**651.** $(x - 2)^2 + (x + 5)^2 = (x + 7)^2$.

U

**652.** $12\,x^2 + 7\,x = 62.$

**653.** $-x^2 + 5\,x = 6.$

**654.** $7\,x + 3\,x^2 = 6.$

**655.** $6\,x^2 + 7\,x = 20.$

**656.** $3\,x^2 - 28 = 17\,x.$

**657.** $4\,x^2 = x + 5.$

**658.** $\dfrac{1}{x-2} - \dfrac{2}{x+2} = \dfrac{3}{5}.$

**659.** $\dfrac{10}{x} - \dfrac{14 - 2\,x}{x^2} = \dfrac{22}{9}.$

**660.** $\dfrac{12}{5-x} + \dfrac{8}{4-x} = \dfrac{32}{x+2}.$

**661.** $x^2 + (a+b)x + ab = 0.$

**662.** $x^2 - 2\,ax + a^2 - b^2 = 0.$

**663.** $x^2 + \dfrac{a}{b}\,x - \dfrac{2\,a^2}{b^2} = 0.$

**664.** $\dfrac{x+1}{x+2} + \dfrac{x-1}{x-2} = \dfrac{2\,x-1}{x-1}.$

**665.** $\dfrac{x-2}{x+2} + \dfrac{x+2}{x-2} = 2\,\dfrac{x+3}{x-3}.$

**666.** $\dfrac{x-1}{x+1} - \dfrac{5}{6} = \dfrac{2}{7(x-1)}.$

**667.** $\dfrac{4}{x+2} + \dfrac{5}{x+4} = \dfrac{12}{x+6}.$

**668.** $\dfrac{x-1}{x+1} + \dfrac{x-2}{x+2} = \dfrac{2\,x+13}{x+16}.$

**669.** $\dfrac{x+1}{x-1} + \dfrac{x+2}{x-2} = \dfrac{2\,x+13}{x+1}.$

**670.** $\dfrac{2\,x-1}{x+1} + \dfrac{3\,x-1}{x+2} = \dfrac{5\,x-11}{x-1}.$

**671.** $x - \dfrac{14\,x-9}{8\,x-3} = \dfrac{x^2-3}{x+1}.$

**672.** $a^2x^2 - 2\,a^3x + a^4 - 1 = 0.$

**673.** $4\,a^2x = (a^2 - b^2 + x)^2.$

**674.** $\dfrac{x}{a} + \dfrac{a}{x} = \dfrac{x}{b} + \dfrac{b}{x}.$

**675.** $\dfrac{1}{x} + \dfrac{1}{x+b} = \dfrac{1}{a} + \dfrac{1}{a+b}.$

**676.** $\dfrac{1}{x+a} + \dfrac{1}{x+b} = \dfrac{1}{m+a} + \dfrac{1}{m+b}.$

**677.** $\dfrac{1}{a} + \dfrac{1}{a+x} + \dfrac{1}{a+2\,x} = 0.$

**678.** $\dfrac{x+a+2\,b}{x+a-2\,b} = \dfrac{b-2\,a+2\,x}{b+2\,a-2\,x}.$

**679.** $\dfrac{x+a-b}{x-a+b} = \dfrac{a(x+a+5b)}{b(x+5\,a+b)}.$

**680.** $\dfrac{1}{x-1} + \dfrac{2}{x-2} = \dfrac{3}{x-3}.$

**681.** $\dfrac{x}{x - \dfrac{1}{2 + \dfrac{x}{x-3}}} = 1.$

**682.** $\dfrac{a+1}{x} - \dfrac{x+1}{a} + a + 1 = 0.$

**683.** $\dfrac{a-x}{x-b} + \dfrac{x-b}{a-x} = \dfrac{13}{6}.$

**684.** $\dfrac{a^3 + b^3}{x^2 + ab} - \dfrac{ab}{x} = 0.$

**685.** $\dfrac{1}{a+x} + \dfrac{1}{b+x} = \dfrac{a+b}{ab}.$

**686.** $\dfrac{1}{2\,x-5\,a} + \dfrac{5}{2\,x-a} = \dfrac{2}{a}.$

**687.** $\dfrac{1}{x-a} + \dfrac{1}{x-b} = \dfrac{a+b}{ab}.$

**688.** $\dfrac{x}{x+a} + \dfrac{x+b}{x} = \dfrac{c}{a+c} + \dfrac{b+c}{c}.$

**689.** $\dfrac{1}{x+a+b} = \dfrac{1}{x} + \dfrac{1}{a} + \dfrac{1}{b}.$

**690.** $\dfrac{1+6\,x+20\,x^2}{1+4\,x+12\,x^2} = \dfrac{9(3+4\,x+4\,x^2)}{4(5+6\,x+4\,x^2)}.$

**691.** $\dfrac{(a-x)^2+(b-x)^2}{(a-x)(b-x)} = \dfrac{5}{2}.$

**692.** $\dfrac{(x+a)^2+(x-b)^2}{(x+a)^2-(x-b)^2} = \dfrac{a^2+b^2}{2\,ab}.$

**693.** $\dfrac{(x+1)^3}{x^3-x^2-x+1} = \dfrac{a^2}{b^2}.$

**694.** $\dfrac{2\,a+(1+a^2)x}{1+a^2+2\,ax} = \dfrac{2\,b-(1+b^2)x}{1+b^2-2\,bx}.$

**695.** $(1-a^2b^2)x^2 - 2\,a(1+b^2)x + a^2 - b^2 = 0.$

**696.** $ax^2 + bx^2 + cx^2 - ax - bx - c = 0.$

**697.** $ax^2 + bx + cx - a - b - c = 0.$

**698.** $x^2 - 2\sqrt{3}\,x + 1 = 0.$

**699.** $x^2 - 2\sqrt{5}\,x + 4 = 0.$

**700.** $\left(\dfrac{x+3}{x+2}\right)^2 + \left(\dfrac{x+2}{x+3}\right)^2 = 4\tfrac{1}{4}.$

**701.** $(x^2+3\,x)^2 - 2\,x^2 - 6\,x - 8 = 0.$

**702.** $2(4\,x^2 - 3\,x)^2 - 28\,x^2 + 21\,x + 5 = 0.$

**703.** $(x^2 - 5\,x + 7)^2 - (x-2)(x-3) = 1.$

**704.** $(x^2+x)(x^2+x+1) = 42.$

**705.** $\left(\dfrac{x}{x+1}\right)^2 + \left(\dfrac{x+1}{x}\right)^2 = 2\tfrac{1}{4}.$

**706.** $\left(\dfrac{x-1}{x+1}\right)^2 + \left(\dfrac{x+1}{x-1}\right)^2 = 2\dfrac{a^2+b^2}{a^2-b^2}.$

**707.** $x(x+2)(x+3)(x+5) = 40.$

**708.** $x(x-1)^2(x-2) = 42.$

**709.** $x^4 - 13x^2 + 36 = 0.$

**710.** $16x^4 - 40a^2x^2 + 9a^4 = 0.$

**711.** $x^{2n} + 2ax^n + a^2 - b^2 = 0.$

**712.** $\dfrac{x^2 + 16}{25} + \dfrac{25}{x^2 - 16} = 2.$

**713.** $3x^4 - 44x^2 + 121 = 0.$

**714.** $\dfrac{4}{2x^2 - 5} - \dfrac{6}{3x^2 - 7} = 1.$

**715.** $\left(\dfrac{x^2 + x + 1}{x^2 - x + 1}\right)^2 - 10\left(\dfrac{x^2 + x + 1}{x^2 - x + 1}\right) + 21 = 0.$

**716.** Find two consecutive numbers whose product equals 600.

**717.** What number exceeds its reciprocal by $1\frac{9}{10}$.

**718.** Find two numbers whose sum is $a$ and whose product equals $b$.

**719.** A needs 15 days longer to build a wall than B, and working together they can build it in 18 days. In how many days can A build the wall?

**720.** The area of a rectangle is 221 square feet and its perimeter equals 60 feet. Find the dimensions of the rectangle.

**721.** Find the price of an apple, if 1 more for 30¢ would diminish the price of 100 apples by \$1.

**722.** The difference of the cubes of two consecutive numbers is 217 ; find them.

**723.** Find four consecutive integers whose product is 7920.

**724.** Find the altitude of an equilateral triangle whose side equals $a$.

**725.** A man bought a certain number of shares in a company for \$375; if he had waited a few days until each share had fallen \$6.25 in value, he might have bought five more for the same money. How many shares did he buy ?

**726.** What two numbers are those whose sum is 47 and product 312?

**727.** A man bought a certain number of pounds of tea and 10 pounds more of coffee, paying \$12 for the tea and \$9 for the coffee. If a pound of tea cost 30¢ more than a pound of coffee, what is the price of the coffee per pound?

Find the numerical value of:

**728.** $12^0 - 4^{\frac{1}{2}} + 3^{-1} + 0^5 - 8^{-\frac{2}{3}} + 1^{\frac{3}{4}}.$

**729.** $\left(\frac{1}{4}\right)^{-\frac{1}{2}} - \left(3\frac{3}{8}\right)^{\frac{1}{3}} + (a + b)^0 + 64^{.5} + \dfrac{1}{2^{-2}}.$

implify:

**30.** $(y^{\frac{3}{4}} + y^{\frac{1}{2}} + y^{\frac{1}{4}} + 1)(y^{\frac{1}{4}} - 1)$.

**31.** $(a^{\frac{2}{3}} - x^{\frac{2}{3}})(a^{\frac{4}{3}} + a^{\frac{2}{3}}x^{\frac{2}{3}} + x^{\frac{4}{3}})$.

**32.** $(a^{\frac{2}{3}} + b^{\frac{2}{3}} + c^{\frac{2}{3}} - a^{\frac{1}{3}}b^{\frac{1}{3}} - a^{\frac{1}{3}}c^{\frac{1}{3}} - b^{\frac{1}{3}}c^{\frac{1}{3}})(a^{\frac{1}{3}} + b^{\frac{1}{3}} + c^{\frac{1}{3}})$.

**33.** $(m^{\frac{4}{5}} + m^{\frac{3}{5}}n^{\frac{1}{5}} + m^{\frac{2}{5}}n^{\frac{2}{5}} + m^{\frac{1}{5}}n^{\frac{3}{5}} + n^{\frac{4}{5}})(m^{\frac{1}{5}} - n^{\frac{1}{5}})$.

**34.** $(m^{\frac{2}{3}} - 2\,d^{\frac{1}{4}}m^{\frac{1}{3}} + 4\,d^{\frac{1}{2}})(m^{\frac{2}{3}} + 2\,d^{\frac{1}{4}}m^{\frac{1}{3}} + 4\,d^{\frac{1}{2}})$.

**35.** $(x^2 - 1 + x^{-2})(x^2 + 1 + x^{-2})$.

**36.** $(a^{-2} + b^{-2})(a^{-2} - b^{-2})$.

**37.** $(a^{-1} - b^{-1} + c^{-1})(a^{-1} + b^{-1} + c^{-1})$.

**38.** $(1 + ab^{-1} + a^2b^{-2})(1 - ab^{-1} + a^2b^{-2})$.

**39.** $(4\,x^{-3} + 3\,x^{-2} + 2\,x^{-1} + 1)(x^{-2} - x^{-1} + 1)$.

**40.** $(64\,x^{-1} + 27\,y^{-2}) \div (4\,x^{-\frac{1}{3}} + 3\,y^{-\frac{2}{3}})$.

**41.** $(x^{\frac{3}{2}} - xy^{\frac{1}{2}} + x^{\frac{1}{2}}y - y^{\frac{3}{2}}) \div (x^{\frac{1}{2}} - y^{\frac{1}{2}})$.

**42.** $(a^{\frac{2}{3}} + a^{\frac{1}{3}}b^{\frac{1}{3}} + b^{\frac{2}{3}}) \div (a^{\frac{1}{3}} + a^{\frac{1}{6}}b^{\frac{1}{6}} + b^{\frac{1}{3}})$.

**43.** $(a^{\frac{2}{3}} + b^{\frac{2}{3}} - c^{\frac{2}{3}} + 2\,a^{\frac{1}{3}}b^{\frac{1}{3}}) \div (a^{\frac{1}{3}} + b^{\frac{1}{3}} + c^{\frac{1}{3}})$.

**44.** $(x^{\frac{3}{4}} - 2\,a^{\frac{3}{2}}x^{\frac{3}{8}} + a^3) \div (x^{\frac{1}{4}} - 2\,a^{\frac{1}{2}}x^{\frac{1}{8}} + a)$.

**45.** $\left(x^2 + \dfrac{9}{x^2} - 4\,x - \dfrac{12}{x} + 10\right)^{\frac{1}{2}}$.

**46.** $(4\,x^2 - 12\,x^{\frac{4}{3}} + 28\,x + 9\,x^{\frac{2}{3}} - 42\,x^{\frac{1}{3}} + 49)^{\frac{1}{2}}$.

**47.** $(a^6 + b^{-6} + 4\,a^5b^{-1} + 2\,b^{-5}a + 2\,a^4b^{-2} - 3\,b^{-4}a^2 - 6\,a^3b^{-3})^{\frac{1}{2}}$.

**48.** $(x^{\frac{1}{2}} \times x^{\frac{1}{3}})^6$, $(x^{\frac{2}{3}} \times x^{-\frac{4}{3}} \times x^{\frac{4}{3}})^n$, $x^{b-c} \times x^{c-a} \times x^{a-b}$.

**49.** $[(x^{-\frac{1}{3}})^{\frac{3}{4}}]^4$, $[\sqrt{x^2y^{-4}\sqrt[3]{x^6y^{12}}}]^{\frac{1}{2}}$, $[a^{\frac{1}{2}}b^{-\frac{1}{4}}\sqrt{a^{\frac{1}{3}}b^{\frac{1}{2}}\sqrt{b^{\frac{2}{3}}}}]^8$.

**50.** $(\sqrt{\sqrt[4]{8\,a^3}})^4$, $(\sqrt{4\,a^2\sqrt[3]{9\,a^2}})^3$, $\dfrac{y}{\sqrt{x}} \cdot \dfrac{\sqrt[4]{z^8}}{\sqrt{y}} \cdot \sqrt[3]{xz}$.

**51.** $(\sqrt[3]{\sqrt[5]{a^2}} \cdot \sqrt[4]{\sqrt[6]{a^9}})^{60}$, $\left(\sqrt{\dfrac{ay}{x^3}} \cdot \sqrt[3]{\dfrac{bx^2}{y^4}} \cdot \sqrt[6]{\dfrac{y^8}{a^{\frac{5}{2}}b^{\frac{5}{2}}}}\right)^{\frac{4}{7}}$.

**52.** $(\sqrt[bc]{x})^{b-c} \times (\sqrt[ca]{x})^{c-a} \times (\sqrt[ab]{x})^{a-b}$.

**753.** $\sqrt[a-b]{(\sqrt[a-c]{x^a})^a} \times \sqrt[b-c]{(\sqrt[b-a]{x^b})^b} \times \sqrt[c-a]{(\sqrt[c-b]{x^c})^c}$.

**754.** $\dfrac{x + (xy^2)^{\frac{1}{3}} - (x^2y)^{\frac{1}{3}}}{x+y}$.

**755.** $\left[1 - \dfrac{1-x^{\frac{1}{2}}}{1+x^{\frac{1}{2}}} + \dfrac{1+2x}{1-x}\right]\left[\dfrac{x^{\frac{1}{2}}+1}{2x^{\frac{1}{2}}+1}\right]$.

**756.** $(3\sqrt{45} - 7\sqrt{5})(\sqrt{1\tfrac{4}{5}} + 2\sqrt{9\tfrac{1}{5}})$.

**757.** $2\sqrt[3]{3}(\sqrt[3]{9} - 2\sqrt[3]{2\tfrac{2}{3}} + 4\sqrt[3]{\tfrac{1}{3}} - 3\sqrt[3]{2})$.

**758.** $\sqrt[3]{\sqrt{12}-2} \cdot \sqrt[3]{\sqrt{12}+2} + \sqrt[3]{7+\sqrt{22}} \cdot \sqrt[3]{7-\sqrt{22}}$.

**759.** $2\sqrt{48} + 3\sqrt{108} + \sqrt{27}$.

**760.** $4\sqrt{50} + 12\sqrt{288} + 3\sqrt{1000}$.

**761.** $\sqrt[3]{686} + \sqrt[3]{16} - \sqrt[3]{128}$.

**762.** $b\sqrt{b^2x} + c\sqrt{c^2x} + a\sqrt{a^2x}$.

**763.** $3\sqrt[3]{16} + \sqrt[3]{432} + 7\sqrt[3]{54}$.

**764.** $\sqrt[3]{7} - \sqrt[3]{6\tfrac{1}{4}} + {}^{0.3}\!\sqrt{27} - {}^{0.75}\!\sqrt{27} + \sqrt[3]{64}$.

**765.** $2\sqrt[12]{\sqrt[5]{7}} + 3\sqrt[6]{\sqrt[10]{7}} - 3\sqrt[5]{\sqrt[12]{7}} - \sqrt[10]{\sqrt[8]{7}}$.

**766.** $\sqrt[2x]{\sqrt[3y]{a^5}} \cdot \sqrt[6x]{\sqrt[y]{a^8}} \cdot \sqrt[x]{\sqrt[6y]{a^9}} \cdot \sqrt[6y]{\sqrt[x]{a}}$.

**767.** $\sqrt[6]{\sqrt[8]{a^5b^7c^{-11}}} \cdot \sqrt[3]{\sqrt[16]{a^{-48}b^7c^{87}}}$.

**768.** $\sqrt[3]{\tfrac{27}{64}} + \sqrt[3]{\tfrac{64}{125}} - 4\sqrt[3]{3\tfrac{3}{8}} - 2\sqrt[3]{2\tfrac{10}{27}} + 3\sqrt[3]{1\tfrac{61}{64}}$.

**769.** $\sqrt[2]{\sqrt[3]{\tfrac{25^3}{64^3}}} + \sqrt[3]{\sqrt[2]{\tfrac{8^2}{27^2}}} - \sqrt[3]{\sqrt[x]{\tfrac{27^x}{125^x}}}$.

**770.** $\sqrt{\dfrac{m}{a^2} - \dfrac{n}{a^2}} + \sqrt{\dfrac{m}{n^2} - \dfrac{1}{n}}$.

**772.** $\sqrt[x]{\dfrac{a^{x+1}}{b^{x-2}c^{x-3}d^{x-4}}}$.

**771.** $\sqrt{2\dfrac{(a^2+b^2)^2}{c^2} - 2\dfrac{(a^2-b^2)^2}{c^2}}$.

**773.** $\sqrt[4x+6y]{\dfrac{a^{28x^2}a^{10xy}}{a^{48y^2}}}$.

**774.** ${}^{3a-2}\!\sqrt{(x^{5am})^3 \cdot (x^6)^2} : {}^{3a-2}\!\sqrt{x^{10m} \cdot x^{18a}}$.

**775.** ${}^{12x-14y}\!\sqrt{(a^{7x}a^{11x})^{8x}} : {}^{12x-14y}\!\sqrt{(a^{5y} \cdot a^{23y})^{7y}}$.

**776.** $\dfrac{2}{2-\sqrt{2}}$.    **777.** $\dfrac{\sqrt{3}}{2-\sqrt{3}}$.    **778.** $\dfrac{2-\sqrt{2}}{2+\sqrt{2}}$.    **779.** $\dfrac{\sqrt{a}+\sqrt{x}}{\sqrt{a}-\sqrt{x}}$.

**780.** $\dfrac{\sqrt{m^2+1}-\sqrt{m^2-1}}{\sqrt{m^2+1}+\sqrt{m^2-1}}.$

**782.** $\dfrac{\sqrt{a+x}+\sqrt{a-x}}{\sqrt{a+x}-\sqrt{a-x}}.$

**781.** $\dfrac{1+\sqrt{x}}{1-\sqrt{x}}.$

**783.** $\dfrac{a+\sqrt{a^2-x^2}}{a-\sqrt{a^2-x^2}}.$

Find the square roots of the following binomial surds:

**784.** $10+2\sqrt{21}.$  **789.** $38-12\sqrt{10}.$  **794.** $3\frac{1}{2}-\sqrt{10}.$

**785.** $16+2\sqrt{55}.$  **790.** $14-4\sqrt{6}.$  **795.** $57-12\sqrt{15}.$

**786.** $9-2\sqrt{14}.$  **791.** $103-12\sqrt{11}.$  **796.** $16-6\sqrt{7}.$

**787.** $94-42\sqrt{5}.$  **792.** $75-12\sqrt{21}.$  **797.** $43+12\sqrt{7}.$

**788.** $13-2\sqrt{30}.$  **793.** $87-12\sqrt{42}.$  **798.** $2a+2\sqrt{a^2-b^2}.$

**799.** $a-c+2\sqrt{ab-ac+bc-b^2}.$  **800.** $\dfrac{a}{b}+\dfrac{b}{a}+\sqrt{\dfrac{2a}{b}-\dfrac{2b}{a}+3}.$

Simplify:

**801.** $\dfrac{\sqrt{3-\sqrt{5}}}{\sqrt{2}+\sqrt{7-3\sqrt{5}}}.$

**802.** $\dfrac{\sqrt{5+2\sqrt{6}}+\sqrt{5-2\sqrt{6}}}{\sqrt{5+2\sqrt{6}}-\sqrt{5-2\sqrt{6}}}.$

**803.** $\dfrac{1}{\sqrt{3}-\sqrt{2}}+\dfrac{\sqrt{3}}{\sqrt{2}+1}-\dfrac{2\sqrt{2}}{\sqrt{3}-1}.$

**804.** $\dfrac{2+\sqrt{3}}{\sqrt{2}+\sqrt{2+\sqrt{3}}}+\dfrac{2-\sqrt{3}}{\sqrt{2}-\sqrt{2-\sqrt{3}}}.$

**805.** $\left(\dfrac{x+y}{x}+\dfrac{x+y}{y}+\dfrac{\sqrt{x}+\sqrt{y}}{\sqrt{x}}+\dfrac{\sqrt{x}-\sqrt{y}}{\sqrt{y}}+\dfrac{1}{4}\right)^{\frac{1}{2}}.$

**806.** $\dfrac{7+3\sqrt{5}}{7-3\sqrt{5}}+\dfrac{7-3\sqrt{5}}{7+3\sqrt{5}}+\dfrac{2+\sqrt{3}}{4+\sqrt{3}}+\dfrac{4-2\sqrt{3}}{5-2\sqrt{3}}.$

**807.** $\dfrac{1}{x+\sqrt{x^2-a^2}}+\dfrac{1'}{x-\sqrt{x^2-a^2}}.$

**808.** $\dfrac{\sqrt{a+x}+\sqrt{a-x}}{\sqrt{a+x}-\sqrt{a-x}}-\dfrac{\sqrt{a+x}-\sqrt{a-x}}{\sqrt{a+x}+\sqrt{a-x}}.$

**809.** $\left(\sqrt{\dfrac{1+x}{1-x}}-\sqrt{\dfrac{1-x}{1+x}}\right)\div\left(\sqrt{\dfrac{1+x}{1-x}}+\sqrt{\dfrac{1-x}{1+x}}\right).$

**810.** Find the sum and difference of

$$(x + \sqrt{2\,y - x^2})^4 \text{ and } (x - \sqrt{2y - x^2})^4.$$

**811.** $\sqrt{x+9} - \sqrt{x} = 1.$      **813.** $\sqrt{x+6} + \sqrt{x-3} = 9.$

**812.** $\sqrt{4x-3} + 2\sqrt{x} = 3.$      **814.** $\sqrt{x+a^2} - \sqrt{x} = b.$

**815.** $\sqrt{2(x+1)} + \sqrt{2x+15} = 13.$

**816.** $\sqrt{3x-5} + \sqrt{3x+12} = 17.$

**817.** $\sqrt{9x+10} - 3\sqrt{x-1} = 1.$

**818.** $\sqrt{x+60} = 2\sqrt{x+5} + \sqrt{x}.$

**819.** $2\sqrt{x+5} + 3\sqrt{x-7} = \sqrt{25x-79}.$

**820.** $3\sqrt{x+3} - 2\sqrt{x-12} = 5\sqrt{x-9}.$

**821.** $x - 5 = \sqrt{20 + x - x^2}.$

**822.** $\sqrt{x+1} + \sqrt{x-2} = \sqrt{2x+3}.$

**823.** $\sqrt{x+2} + \sqrt{x-3} = \sqrt{2x+11}.$

**824.** $\sqrt{2x+11} - \sqrt{x-3} - \sqrt{x+2} = 0.$

**825.** $\sqrt{x+2} - \sqrt{x-6} = \sqrt{2x-10}.$

**826.** $\sqrt{2x-2} + \sqrt{x} = \sqrt{6x-5}.$

**827.** $\sqrt{x+a} + \sqrt{x} + \sqrt{x-a} = 0.$

**828.** $\sqrt{x+5} + 2\sqrt{x+1} = \sqrt{3x+7}.$

**829.** $\sqrt{x+28} + \sqrt{9x-28} = 4\sqrt{2x-14}.$

**830.** $\sqrt{14x+9} + 2\sqrt{x+1} + \sqrt{3x+1} = 0.$

**831.** $\sqrt{12x-3} + \sqrt{x+2} + \sqrt{7x-13} = 0.$

**832.** $\sqrt{2x+7} + \sqrt{3x-18} = \sqrt{7x+1}.$

**833.** $\sqrt{x+3} - 2\sqrt{x+1} = \sqrt{5x+4}.$

**834.** $\sqrt{3x+1} - \sqrt{4x+5} + \sqrt{x-4} = 0.$

**835.** $\sqrt{8x+1} - \sqrt{x+1} = \sqrt{3x}.$

**836.** $\dfrac{c}{\sqrt{x}-d} + \dfrac{d}{\sqrt{x}-c} = 2.$      **837.** $\sqrt{\dfrac{x+7}{2}} + \sqrt{\dfrac{x-7}{2}} = 7.$

**838.** $5x^2 + 11x - 12\sqrt{(x+4)(5x-9)} = 36.$

**839.** $4x + 4\sqrt{3x^2 - 7x + 3} = 3x(x-1) + 6.$

**840.** $x^2 + \sqrt{x^2 + 3x + 5} = 7 - 3x.$

**841.** $4x^2 + \sqrt{4x^2 - 10x + 1} = 10x + 1.$

**842.** $x^2 - 3x - 3\sqrt{x^2 - 3x - 10} = 118.$

Resolve into prime factors:

**843.** $x^4 + 2x^3 + 5x^2 + 18x + 16.$

**844.** $x^5 + x^4 + x^3 + x^2 + x + 1.$

**845.** $5a^4 + 7a^3 - 28a - 80.$

**846.** $4x^4 - x^3y + 4xy^3 - 64y^4.$

**847.** $x^3 - 3x + 2.$

**848.** $x^3 + 3x^2 - 4.$

**849.** $a^3 - 4ab^2 + 3b^3.$

**850.** $x^3 + 4x^2 + 2x - 3.$

**851.** $x^4 + 2x^3y - 2xy^3 - y^4.$

**852.** $x^3 - 2x^2 - 8x + 15.$

**853.** $x^3 + 11x^2 + 13x - 45.$

**854.** $x^3 - 13x + 12.$

**855.** $x^3 - 8x^2 + 19x - 12.$

**856.** $x^3 - 4x^2 - 19x - 14.$

**857.** $4x^3 + 8x^2 - 3x - 9.$

**858.** $16x^3 - 40x^2 - 7x + 49.$

**859.** $8x^3 + 27y^3.$

**860.** $8 + a^3x^3.$

**861.** $27x^3 - 64.$

**862.** $1 - 64a^3.$

**863.** $x^3y^3 + z^3.$

**864.** $27b^3 - 1.$

**865.** $64a^3 - 1000b^3.$

**866.** $729x^3 + 512y^3.$

**867.** $8x^3 - 27y^3z^3.$

**868.** $a^3 + 8b^3c^3.$

**869.** $a^{3m} - b^{3m}.$

**870.** $a^{6m} + b^{6n}.$

**871.** $a^{12} + 1.$

**872.** $a^3 + 216n^9.$

**873.** $a^3b - b^4.$

**874.** $a^{10} - ab^9.$

**875.** $a^{13} + a.$

**876.** $a^{10m} - 1.$

**877.** Show that $99^{99} + 1$ is divisible by 100.

**878.** Show that $1001^{79} - 1$ is divisible by 1000.

**879.** For what value of $m$ is $2\,x^3 - mx^2 - 5\,x - 3$ exactly divisible by $x - 3$?

**880.** What must be the value of $m$ and $n$ to make

$x^3 + mx^2 + nx + 42$ exactly divisible by $x - 2$ and by $x - 3$?

Solve the following systems:

**881.** $x + 2\,y = 12,\ xy + y^2 = 32.$

**882.** $x^3 + y^3 = 28,\ x^2 - xy + y^2 = 7.$

**883.** $x + y = 7,\ x^3 + y^3 = 133.$

**884.** $x^2 + xy = 10,\ y^2 + xy = 15.$

**885.** $x^2 + xy + y^2 = 37,\ x^4 + x^2y^2 + y^4 = 481.$

**886.** $\dfrac{x}{x - y} - \dfrac{x - y}{x + y} = 1,\ 2 + 3\,xy = 3\,x.$

**887.** $x^2 + y^2 = 34,\ x^2 - y^2 + \sqrt{(x^2 - y^2)} = 20.$

**888.** $x^2 + y^2 - 1 = 2\,xy,\ xy(xy + 1) = 6.$

**889.** $x^2 + xy = 8\,x + 3,\ y^2 + xy = 8\,y + 6.$

**890.** $x^2 - xy = 2\,x + 5,\ xy - y^2 = 2\,y + 2.$

**891.** $\dfrac{1}{x} + \dfrac{1}{y} = \dfrac{3}{4};\ \dfrac{1}{x^2} + \dfrac{1}{y^2} = \dfrac{5}{16}.$

**892.** $\dfrac{1}{x} + \dfrac{1}{y} = 5;\ \dfrac{1}{x^2} + \dfrac{1}{y^2} = 13.$

**893.** $\dfrac{1}{x} - \dfrac{1}{y} = 2\tfrac{1}{2};\ \dfrac{1}{x^2} - \dfrac{1}{y^2} = 8\tfrac{3}{4}.$

**894.** $x^3 - y^3 = 37;\ x^2 + xy + y^2 = 37.$

**895.** $x^3 + y^3 = 152,\ x^2 - xy + y^2 = 19.$

**896.** $x^2 - xy = 35,\ xy + y^2 = 18.$

**897.** $x^2 + xy = 126,\ y^2 + xy = 198.$

**898.** $x^2 + 3\,y^2 = 43,\ x^2 + xy = 28.$

**899.** $x^2 + 2\,y^2 = 17,\ 3\,x^2 - 5\,xy + 4\,y^2 = 13.$

**900.** $x(x^2 + y^2) = 16\,y,\ y(x^2 + y^2) = 25\,x.$

**901.** $x^2 - xy = 2\,ax + b^2,\ xy - y^2 = 2\,ay + a^2.$

**902.** $x^2 + y^2 - x - 2y = 1$, $x^2y^2 - 2x^2y - xy^2 + 2xy + 2 = 0.$

**903.** $x^4 + x^2y^2 + y^4 = 243$, $x^2 + xy + y^2 = 9.$

**904.** $x^2 + xy + y^2 = 84$, $x - \sqrt{xy} + y = 6.$

**905.** $(9x + y)(x + y) = 273$, $(9x - y)(x - y) = 33.$

**906.** $\sqrt[3]{x + 10} + \sqrt[3]{y + 14} = 12$, $x + y = 444.$

**907.** $x + y^2 = ax$, $y + x^2 = by.$

**908.** $x + y = 9$, $x^2 - xy + y^2 = 27.$

**909.** $23 x^2 - y^2 = 22$, $7y - 23x = 200.$

**910.** $\dfrac{m}{x} + \dfrac{n}{y} = 2$, $ny - mx = n^2 - m^2.$

**911.** $bx + ay = a^2 + b^2$, $\dfrac{x^2}{a^2} + \dfrac{y^2}{b^2} = \dfrac{b^2}{a^2} + \dfrac{a^2}{b^2}.$

**912.** $x^2 + 3xy = 2$, $3y^2 + xy = 1.$

**913.** $x^2 + 2xy = 39$, $xy + 2y^2 = 65.$

**914.** $x^2 - 5xy = 11$, $y^2 + 3xy = -2.$

**915.** $x^2 + 2xy = 32$, $2y^2 + xy = 16.$

**916.** $x^2 - xy + y^2 = 3$, $x^2 + xy + y^2 = 7.$

**917.** $(x - 3)^2 + (y - 3)^2 = 34$, $xy - 3(x + y) = 6.$

**918.** $(3x - y)(3y - x) = 21$, $3x(3x - 2y) = 49 - y^2.$

**919.** $(x + 2y)(2x + y) = 20$, $4x(x + y) = 16 - y^2.$

**920.** $(x + y)(x^3 + y^3) = 1216$, $x^3 - y^3 = 49(x - y).$

**921.** $xy = a(x + y)$, $x^2y^2 = b^2(x^2 + y^2).$

**922.** $x^2y + xy^2 = 180$, $x^3 + y^3 = 189.$

**923.** $9x + 8y + 7xy = 0$, $7x + 4y + 6xy = 0.$

**924.** $x^2 + xy = a^2$, $y^2 + xy = b^2.$

**925.** $xy + x = 15$, $xy - y = 8.$

**926.** $\dfrac{x^2 - xy + y^2}{x + y} = 28$, $\dfrac{x^2 + xy + y^2}{x + y} = \dfrac{(x - y)^2}{12}.$

**927.** $2x + y = 2a + b$, $x^2 + 2xy = a^2 + 2ab.$

**928.** $x + \dfrac{1}{x} = y + \dfrac{1}{y} = xy + \dfrac{1}{xy}.$

**929.** $yz = 24$, $zx = 12$, $xy = 8.$

**930.** $(z + x)(x + y) = 10,$  $(x + y)(y + z) = 50,$  $(y + z)(z + x) = 5$

**931.** $x(y + z) = 18,$  $y(z + x) = -102,$  $z(x + y) = 24.$

**932.** $x(x + y + z) = 152,$  $y(x + y + z) = 133,$  $z(x + y + z) = 76.$

**933.** $(y + z)(x + y + z) = 108,$  $(z + x)(x + y + z) = 96,$
$$(x + y)(x + y + z) = 84.$$

**934.** The difference of two numbers is 3; the difference of their cubes is 513. Find the numbers.

**935.** The difference of two numbers is 3, and the difference of their cubes is 279. Find the numbers.

**936.** The sum of two numbers is 20, and the sum of their cubes is 2240. Find the numbers.

**937.** A certain rectangle contains 300 square feet; a second rectangle is 8 feet shorter, and 10 feet broader, and also contains 300 square feet. Find the length and breadth of the first rectangle.

**938.** The sum of the perimeters of two squares is 23 feet, and the sum of the areas of the squares is $16\frac{1}{8}$ feet. Find the sides of the squares.

**939.** The perimeter of a rectangle is 92 feet, and its diagonal is 34 feet. Find the area of the rectangle.

**940.** A plantation in rows consists of 10,000 trees. If there had been 20 less rows, there would have been 25 more trees in a row. How many rows are there?

**941.** The sum of the perimeters of two squares equals 140 feet; the sum of their areas equals 617 square feet. Find the side of each square.

**942.** The sum of the circumferences of two circles is 44 inches, and the sum of their areas $78\frac{1}{2}$ square inches. Assuming $\pi = \frac{22}{7}$, find the radii of the two circles.

**943.** The diagonal of a rectangle equals 17 feet. If each side was increased by 2 feet, the area of the new rectangle would equal 170 square feet. Find the sides of the rectangle.

**944.** A and B run a race round a two-mile course. In the first heat B reaches the winning post 2 minutes before A. In the second heat A increases his speed 2 miles per hour, and B diminishes his as much; and A then arrives at the winning post 2 minutes before B. Find at what rate each man ran in the first heat.

**945.** The area of a certain rectangle is 2400 square feet; if its length is decreased 10 feet and its breadth increased 10 feet, its area will be increased 100 square feet. Find its length and breadth.

**946.** The area of a certain rectangle is equal to the area of a square whose side is 3 inches longer than one of the sides of the rectangle. If the breadth of the rectangle be decreased by 1 inch and its length increased by 2 inches, the area is unaltered. Find the lengths of the sides of the rectangle.

**947.** The diagonal of a rectangular field is 182 yards, and its perimeter is 476 yards. What is its area?

**948.** A certain number exceeds the product of its two digits by 52 and exceeds twice the sum of its digits by 53. Find the number.

**949.** Find two numbers each of which is the square of the other.

**950.** A number consists of three digits whose sum is 14; the square of the middle digit is equal to the product of the extreme digits, and if 594 be added to the number, the digits are reversed. Find the number.

**951.** Two men can perform a piece of work in a certain time; one takes 4 days longer, and the other 9 days longer to perform the work than if both worked together. Find in what time both will do it.

**952.** The square described on the hypotenuse of a right triangle is 180 square inches, the difference in the lengths of the legs of the triangle is 6. Find the legs of the triangle.

**953.** The sum of the contents of two cubic blocks is 407 cubic feet; the sum of the heights of the blocks is 11 feet. Find an edge of each block.

**954.** Two travelers, A and B, set out from two places, P and Q, at the same time; A starts from P with the design to pass through Q, and B starts from Q and travels in the same direction as A. When A overtook B it was found that they had together traveled 30 miles, that A had passed through Q 4 hours before, and that B, at his rate of traveling, was 9 hours' journey distant from P. Find the distance between P and Q.

**955.** A rectangular lawn whose length is 30 yards and breadth 20 yards is surrounded by a path of uniform width. Find the width of the path if its area is 216 square yards.

**956.** Sum to 32 terms, $4$, $1\frac{3}{4}$, $\frac{7}{2}$ ....

**957.** Sum to 24 terms, $\frac{1}{2}$, $-\frac{3}{4}$, $-2$.

**958.** Sum to 20 terms, $5$, $1\frac{3}{3}$, $1\frac{1}{3}$.

**959.** Find an A. P. such that the sum of the first five terms is one fourth of the sum of the following five terms, the first term being unity.

Find the sums of the series:

**960.** $16 + 24 + 32 + \cdots$, to 7 terms;

**961.** $16 + 24 + 36 + \cdots$, to 7 terms;

**962.** $36 + 24 + 16 + \cdots$, to infinity.

**963.** Given $a = 16$, $d = 4$, $s = 88$. Find $n$.

**964.** How many terms of the series $1 + 3 + 5 + \cdots$ amount to 123,454,321 ?

**965.** Sum to $n$ terms, $1 - 3 + 5 - 7 + \cdots$.

**966.** Sum to $n$ terms, $1 - 2 + 3 - 4 + \cdots$.

**967.** Find the sum of $\frac{4}{3} + 1 + \frac{3}{4} \cdots$, to infinity.

**968.** Sum to infinity, $\frac{3}{2} - \frac{2}{3} + \frac{8}{27} - \cdots$.

**969.** Sum to infinity, $\dfrac{\sqrt{2}+1}{\sqrt{2}-1} + \dfrac{1}{2-\sqrt{2}} + \dfrac{1}{2} + \cdots$.

**970.** Sum to 10 terms, $(x + y) + (x^2 + y^2) + (x^3 + y^3) \cdots$.

**971.** Sum to $n$ terms, $x(x + y) + x^2(x^2 + y^2) + x^3(x^3 + y^3)$.

**972.** Sum to 8 terms, $(x + y) + (2x + y^2) + (3x + y^3) + \cdots$.

**973.** Evaluate $(a)$ $.141414\cdots$; $(b)$ $.3151515\cdots$.

**974.** Find $\dfrac{1}{n} + \dfrac{n-1}{n} + \cdots$, to $n$ terms, the terms being in A. P.

**975.** Find the difference between the sums of the series

$$\frac{n}{n} + \frac{n-1}{n} + \frac{n-2}{n} + \cdots \text{ (to } 2n \text{ terms)},$$

and $\quad \dfrac{n}{n+1} + \dfrac{n}{(n+1)^2} + \dfrac{n}{(n+1)^3} + \cdots \text{ (to infinity)}.$

**976.** The 10th and 18th terms of an A. P. are 29 and 53. Find the first term and the common difference.

**977.** The 9th and 11th terms of an A. P. are 1 and 5. Find the sum of 20 terms.

**978.** Insert 22 arithmetic means between 8 and 54.

**979.** Insert 8 arithmetic means between 1 and 0.

**980.** How many terms of $18 + 17 + 16 + \cdots$, amount to 105?

**981.** The sum of $n$ terms of $7 + 9 + 11 + \cdots$, is 40. Find $n$.

**982.** The sum of $n$ terms of an A. P. is $\frac{n}{2}\left(\frac{n}{3} + 1\right)$. Find the 8th term.

**983.** The 21st term of an A. P. is 225, and the sum of the first nine terms is equal to the square of the sum of the first two. Find the first term, and the common difference.

**984.** Find four numbers in A. P. such that the product of the first and fourth may be 55, and of the second and third 63.

**985.** Find the value of the infinite product $4 \cdot \sqrt[3]{4} \cdot \sqrt[9]{4} \cdot \sqrt[27]{4} \cdots$.

**986.** A *perfect number* is a number which equals the sum of all integers by which it is divisible. If the sum of the series $2^0 + 2^1 + 2^2 \cdots 2^n$ is prime, then this sum multiplied by the last term of the series is a perfect number. (Euclid.) Find four perfect numbers.

**987.** The Arabian Araphad reports that chess was invented by Sessa for the amusement of an Indian rajah, named Sheran, who rewarded the inventor by promising to place 1 grain of wheat on the 1st square of a chess-board, 2 grains on the 2d, 4 grains on the 3d, and so on, doubling the number for each successive square on the board. Find the number of grains which Sessa should have received.

Find the sum of the series:

**988.** $\dfrac{1}{\sqrt{2}} - \dfrac{1}{\sqrt{2} + 1} + \dfrac{2}{4 + 3\sqrt{2}} + \cdots$, to $\infty$.

**989.** $5 + 1 + .2 + .04 + \cdots$, to $\infty$.

**990.** $1.1 + 2.01 + 3.001 + 4.001 + \cdots$, to $n$ terms.

**991.** $\dfrac{1}{8} + \dfrac{4}{8^2} + \dfrac{6}{8^3} + \dfrac{3}{8^4} + \dfrac{1}{8^5} + \dfrac{4}{8^6} + \dfrac{6}{8^7} + \dfrac{3}{8^8} + \cdots$, to $\infty$.

**992.** What value must $a$ have so that the sum of

$$2\,a + a\sqrt{2} + a + \frac{a}{\sqrt{2}} + \cdots, \text{ to infinity may be } 8?$$

**993.** Insert 3 geometric means between 2 and 162.

**994.** Insert 4 geometric means between 243 and 32.

**995.** The fifth term of a G. P. is 4, and the fifth term is 8 times the second; find the series.

**996.** The sum and product of three numbers in G. P. are 28 and 512; find the numbers.

**997.** The sum and sum of squares of four numbers in G. P. are 45 and 765; find the numbers.

**998.** If $a$, $b$, $c$, are unequal, prove that they cannot be in A. P. and G. P. at the same time.

**999.** In a circle whose radius is 1 a square is inscribed, in this square a circle, in this circle a square, and so forth to infinity. Find ($a$) the sum of all circumferences, ($b$) the sum of the perimeters of all squares.

**1000.** The side of an equilateral triangle equals 2. The sides of a second equilateral triangle equal the altitudes of the first, the sides of a third triangle equal the altitudes of the second, and so forth to infinity. Find ($a$) the sum of all perimeters, ($b$) the sum of the areas of all triangles.

**1001.** Each stroke of the piston of an air pump removes $\frac{2}{3}$ of the air contained in the receiver. What fractions of the original amount of air is contained in the receiver, ($a$) after 5 strokes, ($b$) after $n$ strokes?

**1002.** Under the conditions of the preceding example, after how many strokes would the density of the air be $\frac{1}{1000}$ of the original density?

**1003.** In an equilateral triangle $ABC$ a circle is inscribed. A second circle touches the first circle and the sides $AB$ and $AC$. A third circle touches the second circle and the same sides, and so forth to infinity. What is the sum of the areas of all circles, if $AB = n$ inches.

**1004.** Two travelers start on the same road. One of them travels uniformly 10 miles a day. The other travels 8 miles the first day and increases this pace by $\frac{1}{2}$ mile a day each succeeding day. After how many days will the latter overtake the former?

**1005.** Write down the first three and the last three terms of $(a - x)^{18}$.

**1006.** Write down the expansion of $(3 - 2 x^2)^5$.

**1007.** Expand $(1 - 2 y)^7$.

**1008.** Write down the first four terms in the expansion of $(x + 2 y)^n$.

**1009.** Find the 9th term of $(2 ab - cd)^{14}$.

**1010.** Find the middle term of $(a - b)^{16}$.

**1011.** Find the middle term of $(a^{\frac{1}{3}} + b^{\frac{1}{3}})^8$.

**1012.** Find the two middle terms of $(a - b)^{19}$.

**1013.** Find the two middle terms of $(a + x)^{18}$.

**1014.** Find the fifth term of $(1 - x)^9$.

**1015.** Find the middle term of $(a + b)^{10}$.

**1016.** Find the two middle terms of $(a - x)^9$.

**1017.** Find the twenty-ninth term of $\left(2 x - \dfrac{1}{2 x}\right)^{33}$.

**1018.** Find the coefficient $x$ in $\left(x + \dfrac{1}{x}\right)^7$.

**1019.** Find the middle term of $\left(5 a - \dfrac{x}{5}\right)^{16}$.

**1020.** Write down the coefficient of $x^9$ in $(5 a^3 - 7 x^3)^7$.

**1021.** Find the eleventh term of $\left(4 x - \dfrac{1}{2 x}\right)^{15}$.

x

# INDEX

Printed in the United States of America.

# ANSWERS

## TO

# SCHULTZE'S ELEMENTS OF ALGEBRA

COMPILED BY THE AUTHOR

WITH THE ASSISTANCE OF

## WILLIAM P. MANGUSE

New York

THE MACMILLAN COMPANY

1918

*All rights reserved*

COPYRIGHT, 1910,

BY THE MACMILLAN COMPANY.

Set up and electrotyped.   Published September, 1910.
Reprinted April, 1913; December, 1916; August, 1917.

Norwood Press
J. S. Cushing Co. — Berwick & Smith Co.
Norwood, Mass., U.S.A.

# ANSWERS

**Page 2.—1.** 32, 8. **2.** $ 160. **3.** A $9400, B $4700. **4.** South America 45,000,000, Australia 5,000,000. **5.** Seattle 12 ft., Philadelphia 6 ft. **6.** A $96, B $128, C $16. **7.** 48 ft., 8 ft. **8.** 16 in., 16 in., 8 in. **9.** 18°, 18°, 144°. **10.** 150,000,000 negroes, 15,000,000 Indians. **Page 3.—11.** A $40, B $80, C $160. **12.** A $10, B $20, C $60. **13.** 3⅓ in., 16⅔ in. **14.** A 38 mi., B 19 mi. **15.** $100.

**1.** 7. **2.** Not in arithmetic. **3.** 9 is larger than 7. **4.** 7°, 16° − 9° = 7°. **5.** 3° below 0. **6.** − 3°. **7.** − 3.

**Page 4.—1.** His expenditures.

**Page 5.—3.** − sign. **4.** − sign. **5.** 20 B.C., 6 yd. per sec. westerly motion. **6.** $150 loss, −150. **7.** − 1°, − 1. **8.** 13° S., −13. **9.** 37° S., − 37. **10.** − 10. **11.** − 3. **12.** −2. **13.** − 15. **14.** − 14. **15.** −32. **16.** − 7. **17.** −2. **18.** 6. **19.** − 1. **20.** + 1. **21.** − 1. **22.** 1. **23.** − 3. **24.** − 3. **25.** 1. **27.** (*a*) 1, (*b*) − 2. **28.** 5.

**Page 6.— 29.** 73°, 126°, 89°, 131°, 106°, 59°, 115°.

**1.** 5000. **2.** 9, 12. **3.** 12, 2. **4.** 85, − 32. **5.** 13 *d.* **6.** Yes. **7.** 9 *m.* **8.** 14 *b.* **9.** 210. **10.** 9 *c.*

**Page 7.—11.** 7 *m.* **12.** 14, − 14. **13.** − 3 *m*, − 36. **14.** − 2 *p.* **15.** −30 *q.* **16.** 10 *q.* **17.** − 26 *x.* **18.** 2 *x.* **19.** − 22 *x.* **20.** − 20 *p.* **21.** Multiplication.

**Page 8.—1.** $7^2 = 49$, $6^3 = 216$, $3^4 = 81$, $2^5 = 32$. **2.** 49. **3.** 32. **4.** 25. **5.** 512. **6.** 16. **7.** 16. **8.** 1,000,000. **9.** 1. **10.** ⅛. **11.** 0. **12.** 20¼. **13.** 2¼. **14.** 25. **15.** .00000001. **16.** 1.21 **17.** 13. **18.** 27. **19.** 4. **20.** 1. **21.** $\frac{1}{16}$. **22.** 9. **23.** 9. **24.** 8 **25.** 6. **26.** 12. **27.** 16. **28.** 2. **29.** $\frac{1}{4}$. **30.** 3.

**Page 9.—1.** 16. **2.** 3. **3.** 8. **4.** 4. **5.** 12. **6.** 64. **7.** . **8.** 16. **9.** 10. **10.** 576. **11.** 16. **12.** 256. **13.** 1. **14.** 96 **15.** 128. **16.** 192. **17.** 2. **18.** $\frac{16}{3}$. **19.** 2.

**Page 10.—1.** 3. **2.** 3. **3.** 2. **4.** 2. **5.** 1. **6.** 6. **7.** 2. **8.** 3. **9.** 12. **10.** 20. **11.** 6. **12.** 0. **13.** 20. **14.** 24. **15.** 1 **16.** 8. **17.** 12.

1

**Page 11.— 1.** 13.　**2.** 57.　**3.** 13.　**4.** 11.　**5.** 69.　**6.** 104.
**7.** 237.　**8.** 17.　**9.** 58.　**10.** 92.　**11.** 240.　**12.** 143.　**13.** 7.
**14.** 56.　**15.** 27.　**16.** 49.

**Page 12.— 17.** 7.　**18.** 11.　**19.** 35.　**20.** $\frac{11}{12}$.　**21.** 27.　**22.** 8.
**23.** 64.　**24.** 8.　**25.** 1.　**26.** 27.　**27.** 27.　**28.** 8.　**29.** 0.　**30.** 1.
**31.** $6a + 4b$.　**32.** $6a^2 - 3b^3$.　**33.** $8x^3 - 4x^2 + y^2$.　**34.** $6m^3 + 3(a - b)$.
**35.** $(a + b)(a^2 - b^2)$.　**36.** $2a^3 - 5\sqrt{(a-b)^2}$.　**38.** Polynomial, Trinomial, Polynomial, Monomial, Binomial.

**Page 13.— 1.** (a) 16 cm., (b) 135 mi., (c) 2000 m., (d) 50,000 ft.
**2.** (a) 64 ft., (b) 100 ft., (c) $6\frac{1}{25}$ ft.　**3.** (a) 6 sq. ft., (b) 30 sq. in.,
(c) 24 sq. ft.

**Page 14.— 4.** (a) 314 sq. m., (b) 12.56 sq. in., (c) 78.5 sq. mi.
**5.** (a) \$80, (b) \$40.　**6.** (a) 200,960,000 sq. mi., (b) 3.14 sq. in.
(c) 314 sq. ft.　**7.** (a) $523\frac{2}{3}$ cu. ft., (b) 14.13 cu. ft., (c) $267,946,666,666\frac{2}{3}$.
**8.** (a) $50^\circ$, (b) $0^\circ$, (c) $-15^\circ$.

**Page 16.— 1.** 1.　**2.** $-1$.　**3.** $-9$.　**4.** 9.　**5.** $-2$.　**6.** $-15$.
**7.** 8.　**8.** $-14$.　**9.** $-22$.　**10.** 0.　**11.** $-15$.　**12.** $-5$.
**13.** 3.　**14.** $-2$.　**15.** $-6$.　**16.** 0.　**17.** $-31$.　**18.** 6.
**19.** $-12$.　**20.** $-1$.　**21.** 2.　**22.** $-9$.　**23.** 0.　**24.** 5.

**Page 17.— 25.** 14.　**26.** 5.　**27.** 3.　**28.** $-3$.　**29.** 9.　**30.** $-3$.
**31.** 12 yd.　**32.** 12 $a$.　**33.** 6 $a$.　**34.** 14.　**35.** $-4$.　**36.** $-1$.
**37.** 3.　**38.** $-4$.　**39.** $6\frac{1}{2}^\circ$.　**40.** $51\frac{2}{3}^\circ$.　**41.** \$3000 gain.　**42.** $\frac{1}{4}^\circ$.

**Page 18.— 1.** $-3a$.　**2.** $-ab$.　**3.** 0.　**4.** $-9pqr^3$.　**5.** $-2b^3x$.
**6.** $18x^2y^3$.　**7.** $-32a^2bc$.　**8.** $-15a^3x$.　**9.** $4a^2$.　**10.** $-43mp^2$.
**11.** $-22(a + b)$.　**12.** $-26(x + y^2)$.　**13.** $-20\sqrt{m + n}$.

**Page 19. 14.** $-38ab$.　**15.** $39c^3$.　**16.** $-21p^3t$.　**17.** $16xyz$.
**18.** $m + n$.　**19.** $m - n$.　**20.** $-m^2 - n^2$.　**21.** $a - a^3$.　**22.** $b - 1$.
**23.** $c^2 + 1$.　**24.** $b + b^2$.　**25.** $7 - cx^2$.　**26.** $mn + mn^2$.
**27.** $-xyz + xyz^3$.　**28.** $\sqrt{p + q} - \sqrt{p - q}$.　**29.** $b$.　**30.** $a^2 - 15a + 4$.
**31.** $3d - 5e - 12x$.　**32.** $-a^2 + 9ab - 2b^2$.　**33.** $2\sqrt{x + y} + \sqrt{x + y^2}$.
**34.** 173.　**35.** $9^2 = 81$.　**36.** $3^8 = 6561$.

**Page 20.— 1.** $a - 6b - 3c$.　**2.** $-q + 3t - 6s$.　**3.** $8x^2 - 9y^2 - 8z^2$.

**Page 21.— 4.** $e + f - 2g$.　**5.** $-16a^2 + 13b^2 - 32c^2$.　**6.** $8x^2 - 5x + 3$.
**7.** $3m^3 + 5$.　**8.** $-xy + yz$.　**9.** $21a^3 + 3b^3$.　**10.** $3a^3 + a^2$.
**11.** 0.　**12.** $2x^3$.　**13.** $-3(c + a)$.　**14.** $a^4 + a^2 + 1$.　**15.** $(a + b)^2 + 1$.
**16.** $x^3y - x^2y^2 - 3xy^3 - y^4$.　**17.** $6\sqrt{b} - \sqrt{c}$.　**18.** $a^3 + a^2 + a + 1$.
**19.** $8x^3y + 3xy^3$.　**20.** $4m^3 + 6mn^2 - 2n^3$.

**Page 22.** — **21.** $3m^5 + 4m^4 + 8m^3 - 6m$.  **22.** $a + b + c + d + e$.
**23.** $m^3 + 24m^2y - 12my^2 - 17y^3$.  **24.** $8t^3$.  **25.** $-6 - 6x - 6x^2$.
**26.** $3a^3 - a^2 + 7a$.  **27.** $-6x + 12$.

**Page 24.** — **1.** $-2$.  **2.** $-12$.  **3.** $2$.  **4.** $6$.  **5.** $-11$.  **6.** $-14$.
**7.** $-17$.  **8.** $-1$.  **9.** $17$.  **10.** $0$.  **11.** $-24$.  **12.** $31$.
**13.** $7$.  **14.** $-7$.  **15.** $-9$.  **16.** $-1$.  **17.** $1$.  **18.** $-1$.
**19.** $-18$.  **20.** $1$.  **21.** $34$.  **22.** $-n^2$.  **23.** $n^2$.  **24.** $32n^2m$.
**25.** $0$.  **26.** $4mpq^2$.  **27.** $-18p^2st$.  **28.** $-abc$.  **29.** $-10m^3x$.
**30.** $7a^5$.  **31.** $a - b$.  **32.** $a + b$.  **33.** $-a - b$.  **34.** $-a + b$.
**35.** $a^2 - a$.  **36.** $a^2 + a$.  **37.** $1 + a^2$.  **38.** $ab - a$.  **39.** $a^2b + a^2$.
**40.** $11\sqrt{a + b}$.

**Page 25.** — **41.** $-19m$.  **42.** $-4a - 24b + c$.  **43.** $-14x^2y + 2y^3$.
**44.** $-2a^3 + 4$.  **45.** $-6a + 18c - 25d$.  **46.** $4y^2$.  **47.** $2mn + qt - 2mt$.
**48.** $a^3 - a^2 - a - 2$.  **49.** $6a - b - d$.  **50.** $-b^2 + b^3$.  **51.** $6x^4 - 3x^3 - x + 1$.
**52.** $a - c$.  **53.** $-a^3 - a^2 - a + 1$.  **54.** $3x^2 - 5xy + 3y^2$.
**55.** $3a^2 - 4ab + y^2$.  **56.** $-1 - 2a - 2b - 4d$.  **57.** $8 + 2a - a^2$.
**58.** $-8a^3b - 8ab^3$.  **59.** $-(a + b) + 4(b + c) - 8(c + a)$.
**1.** $a + b + 4c$.  **2.** $-9x + 16$.  **3.** $-a - 4$.

**Page 26.** — **4.** $x^2$.  **5.** $-x^2 + x - 1$.  **6.** $-a^2 + 5a - 1$.  **7.** $-3m$.
**8.** $a + 2b + 2c$.  **9.** $2y^2 + 2z^2$.  **10.** $2a + b + c$.  **11.** $2a^3 + a^2 + 2a + 5$.
**12.** $a^3 - 6a - 5$.  **13.** $-5a - 6b + 3c$.  **14.** $3a - b$.  **15.** $4x^2 - 4x - 2$.
**16.** $2x + 2y$.  **17.** $2z$.  **18.** $3x + z$.  **19.** $x - 2y + 3z$.  **20.** $n + 10, 2m$.
**21.** $a + b, a - b + c$.

**Page 28.** — **1.** $x + 2y - z$.  **2.** $a - 3b + 2c$.  **3.** $3a^3 - b^3 + c^3$.
**4.** $-2b^2 + c^2$.  **5.** $-2a^2 + 2b^2 - c^2$.  **6.** $2a - b$.  **7.** $a - 2b$.
**8.** $3a - 3b$.  **9.** $2m - 2n$.  **10.** $0$.  **11.** $-2$.  **12.** $3a + 8b$.
**13.** $10x$.  **14.** $-m + 3n + p$.  **15.** $2m + 2n$.  **16.** $2x$.  **17.** $2x^2 - x - 1$.
**18.** $10a - 3b + c$.  **19.** $-3b$.  **20.** $2x^2 + x$.  **21.** $7 - a + 2b + c$.
**22.** $814$.

**Page 29.** — **Exercise 16.** — **1.** $a + (b - c + d)$.  **2.** $2m - (4n - 2q + 3t)$.
**3.** $5a^2 - (7x^2 - 9x - 2)$.  **4.** $4xy - (2x^2 + 4y^2 + 1)$.  **5.** $m^3 - (2n^2 + 3p^2 - 4q^2)$.  **6.** $x - (y - z - d)$.  **7.** $p - (-q - r + s)$.
**8.** $5x^2 - (5x + 7 - a)$.  **9.** $a^3 - (a^2 + a + 1)$.  **Exercise 17.** — **1.** $m + n$.
**2.** $a - b$.  **3.** $a^2 + b^2$.  **4.** $m^3 - n^3$.  **5.** $n^3 - m^3$.  **6.** $a^4 + b^4$.  **7.** $mn$.
**8.** $m^3n^3$.  **9.** $3m^2n^2$.  **10.** $(mn)^3$.  **11.** $(a - b)^2$.  **12.** $(m + n)(m - n)$.
**13.** $9(a + b)^2 - ab$.

**Page 30.** — **14.** $a^2 + b^2 + \sqrt{x}$.  **15.** $x^3 - (2x^2 - 6x + 6)$.  **16.** $\dfrac{a^3 + b^3 + c^3}{a - d}$.

**17.** $(a+b)(a-b)=a^2-b^2$. **18.** $\dfrac{a^3-b^3}{a-b}=a^2+ab+b^2$. **19.** $\dfrac{a^2-b^2}{a-b}=$ $a+b$.

**Page 31.—1.** $-$. **2.** $+$. **3.** $-$. **4.** 15 lb., $(+3)\times5=+15$. **5.** $-15$ lb., $-15$. **6.** $-15$ lb., $-15$. **7.** 15 lb., $+15$. **8.** 20, $-20$, $-20$, 20, etc.

**Page 33.—1.** $-30$. **2.** 84. **3.** $-12$. **4.** $-42$. **5.** $-18$. **6.** 60. **7.** 15. **8.** $-28$. **9** 13. **10.** 1. **11.** 18. **12.** $14\frac{2}{5}$. **13.** 4. **14.** $-27$. **15.** $-64$. **16.** $-1$. **17.** 10,000. **18.** 120. **21.** 24. **22.** 192. **23.** $-24$. **24.** 0. **25.** $-108$. **26.** $-108$. **27.** 0. **28.** $-64$. **29.** $-36$. **30.** 34. **31.** 29. **32.** $-16$.

**Page 34.—33.** 343. **34.** 216. **35.** $-216$. **36.** 125. **1.** $m^9$. **2.** $a^5$. **3.** $2^9$. **4.** $3^7$. **5.** $5^{17}$. **6.** $127^{23}$. **7.** $(ab)^{11}$. **8.** $(x+y)^8$. **9.** $-a^{12}$. **10.** $7^6$. **11.** $12^{39}$. **12.** $21\,a^2bc$. **13.** 1400. **14.** 210. **15.** 3300. **16.** 770. **17.** 4200.

**Page 35.—18.** 60. **19.** 30. **20.** $20\,a^2b^3$. **21.** $38\,m^6n^6$. **22.** $-15\,x^3y^3z^2$. **23.** $-38\,a^6b^6$. **24.** $-44\,a^4b^4c^3$. **25.** $30\,a^3b^4c$. **26.** $-36\,e^4f^3y$. **27.** $49\,p^4q^5r^4t$. **28.** $16\,abxy$. **29.** $56\,u^{30}v^{30}r^{12}t^{30}$. **30.** $30\,a^2b^2c^2$. **31.** $-18\,a^3x^3y^3$. **32.** $16\,a^4b^6$. **33.** $4\,a^2x^6y^{10}$. **34.** $9\,m^2n^8$. **35.** $-8\,a^6$. **1.** 90. **2.** 360. **3.** 51. **4.** 42. **5.** 13. **6.** 1904. **7.** 25,760.

**Page 36.—8.** $5^2+5^8+5^8$. **9.** $6^3+6^4+6^6$. **10.** $7^6+7^5+7^{18}$. **11.** $4^{22}-4^{18}-4^{14}$. **12.** $11^{19}+11^{20}+11^{21}$. **13.** $2\,m^2+2\,mn+2\,mp$. **14.** $-2\,m^3n-2\,mn^3+2\,mnp^2$. **15.** $-2\,x^3y^3-2\,x^2y^2-2\,xy$. **16.** $-16\,p^3q^4r+20\,p^3q^2r-28\,p^2q^2$. **17.** $25\,x^4+25\,x^3+25\,x^2$. **18.** $-14\,x^3yz-14\,xy^3z+14\,xyz^3$. **19.** $-10\,a^3b^5c^9+15\,a^7b^5c^5-30\,a^3b^4c^4$. **20.** $-35\,a^8b^2c^3+14\,a^3b^4c-21\,a^4b^3c^2$. **21.** $-18\,m^4n^3+10\,m^3n^3-14\,m^2n^2$. **22.** $-32\,x^6y^5z^6-16\,x^6y^4z^5+64\,xy^3z^5$. **23.** $-19\,p^{11}+19\,p^{10}-57\,p^6-190\,p^5$. **24.** $3\,x^2-4\,x+7$. **25.** $x(3\,x^2-4\,x+7)$. **26.** $3(x+y+z)$. **27.** $3\,x^2(2\,x+1+x^2)$. **28.** $2\,x^2(y-2\,xy+4\,x^2)$. **29.** $5\,ab(a-12\,b-2)$. **30.** $3\,xy(2\,x^2y^2-xy+1)$.

**Page 38.—1.** $6\,x^2-5\,xy-6\,y^2$. **2.** $8\,a^2+2\,ab-21\,b^2$. **3.** $10\,c^2-19\,cd+6\,d^2$. **4.** $14\,m^2-5\,mn-24\,n^2$. **5.** $30\,p^2+43\,pq+15\,q^2$. **6.** $4\,a^2-22\,ac+30\,c^2$. **7.** $36\,m^2+55\,mn-14\,n^2$. **8.** $x^2-xy-42\,y^2$. **9.** $66\,l^2-83\,ln+7\,n^2$. **10.** $39\,k^2-19\,k+2$. **11.** $12\,x^2y^2-20\,xyz-8\,z^2$. **12.** $40\,r^2-59\,rt+21\,t^2$. **13.** $132\,r^2+r-1$. **14.** $x^2y^2z^2-11\,xyz-12$. **15.** $2\,a^2b^2c^2+11\,abc-21$. **16.** $a^3-a^2-4\,a-6$. **17.** $2\,x^3+x^2-5\,x-4$. **18.** $4\,m^3+9\,m^2+m-2$. **19.** $2\,n^3-9\,n^2+13\,n-12$. **20.** $3\,a^3-14\,a^2-29\,a+30$. **21.** $-5\,q^3+9\,q^2-12\,q+8$. **22.** $9\,x^3-16\,x^2-9\,x+10$. **23.** $2\,x^3-x^2-3\,x-1$. **24.** $4\,a^3-16\,a^2+32\,a-32$. **25.** $a^3-3\,a^2b+3\,ab^2-b^3$. **26.** $4\,a^3-12\,a^2b+5\,ab^2+6\,b^3$. **27.** $18\,x^3-6\,x^2-25\,x+14$.

**28.** $-2\,m^3 + 4\,m^2 - m - 1$. **29.** $a^2 - 6\,ab + 9\,b^2$. **30.** $16\,a^2 - 8\,a + 1$.
**31.** $36\,a^2 - 84\,a + 49$. **32.** $36\,a^2b^2c^2 - 60\,abc + 25$. **33.** $8\,a^3b^3 - 10\,a^2b^2$
$+ 7\,ab - 2$. **34.** $a^3b^3 - 8\,c^3$. **35.** $m^3n^3p^3 - 8$. **36.** $x^4 - 1$.
**37.** $a^4 + 4\,a^2b^2 + 16\,b^4$. **38.** $x^6 + x^4 - x^3 + x^2 + 1$. **39.** $x^4 - 2\,x^2 + 1$.
**40.** $m^4 - m^2n^2 + n^4$. **41.** $a^3 - 6\,a^2b + 12\,ab^2 - 8\,b^3$.

**Page 39.**—**1.** $a^2 + 5\,a + 6$. **2.** $a^2 - a - 6$. **3.** $a^2 - 7\,a + 12$.
**4.** $x^2 - 4\,x - 21$. **5.** $x^2 - 6\,x + 5$. **6.** $p^2 - p - 132$. **7.** $m^2 + 18\,m + 81$.
**8.** $b^2 + b - 156$. **9.** $y^2 - 21\,y - 100$. **10.** $l^2 + 48\,l - 100$. **11.** $s^2 + 10\,s - 231$.
**12.** $a^2 - ab - 2\,b^2$. **13.** $a^2 - 5\,ab + 6\,b^2$. **14.** $a^2 + ab - 2\,b^2$.
**15.** $a^2 - 81$. **16.** $m^2 - n^2$. **17.** $a^4 - a^2b^2 - 20\,b^4$. **18.** $x^4y^4 - x^2y^2 - 12$.
**19.** $x^4y^4z^4 - x^2y^2z^2 - 132$. **20.** $a^6 - 15\,a^3 + 56$. **21.** $10,506$. **22.** $132$.
**23.** $1,009,020$. **24.** $10,098$. **25.** $1,000,994$. **26.** $10,506$. **27.** $1,009,020$.
**28.** $10,098$. **29.** $(x - 2)(x - 1)$.

**Page 40.**—**30.** $(x - 2)(x - 3)$. **31.** $(a + 4)(a + 2)$. **32.** $(a - 5)$
$(a - 2)$. **33.** $(m - 5)(m + 3)$. **34.** $(m + 5)(m - 3)$. **35.** $(m - 4)(m + 1)$.
**36.** $(n - 8)(n - 2)$. **37.** $(p - 6)(p + 5)$.

**1.** $a^2 + 2\,ab + b^2$. **2.** $a^2 + 4\,a + 4$. **3.** $a^2 - 2\,ax + x^2$. **4.** $a^2 - 4\,a + 4$.
**5.** $p^2 + 6\,p + 9$. **6.** $x^2 - 10\,x + 25$. **7.** $p^2 - 14\,p + 49$. **8.** $a^2 - 4\,ab + 4\,b^2$.
**9.** $x^2 + 6\,xy + 9\,y^2$.

**Page 41.**—**10.** $4\,x^2 - 12\,xy + 9\,y^2$. **11.** $16\,a^2 - 24\,ab + 9\,b^2$.
**12.** $a^4 - 6\,a^2 + 9$. **13.** $p^4 + 10\,p^2q + 25\,q^2$. **14.** $9\,p^4 - 54\,p^2 + 81$.
**15.** $36\,a^4 - 84\,a^2b^2 + 49\,b^4$. **16.** $4\,a^2b^2 - 4\,abc + c^2$. **17.** $4\,x^2y^2 + 12\,xyz + 9\,z^2$.
**18.** $16\,a^4b^4 - 40\,a^2b^2c^2 + 25\,c^4$. **19.** $36\,x^4y^4z^4 - 60\,x^2y^2z^2 + 25$. **20.** $m^2 - n^2$.
**21.** $4\,m^2 - 9$. **22.** $a^4 - 49$. **23.** $c^4d^4 - 25$. **24.** $x^4 - 121\,y^4$.
**25.** $x^4 - 22\,x^2y^2 + 121\,y^4$. **26.** $x^4 + 22\,x^2y^2 + 121\,y^4$. **27.** $25\,r^4 - 4\,t^4$.
**28.** $25\,r^4 - 20\,r^2t^2 + 4\,t^4$. **29.** $a^2 + 10\,a + 25$. **30.** $m^4 - 4\,n^{10}$.
**31.** $10,201$. **32.** $10,404$. **33.** $10,609$. **34.** $10,816$. **35.** $998,001$.
**36.** $996,004$. **37.** $9801$. **38.** $441$. **39.** $484$. **40.** $9996$. **41.** $9999$.
**42.** $990,996$. **43.** $x + y$. **44.** $a - b$. **45.** $m - 1$. **46.** $n + 2$.
**47.** $n - 3$. **48.** $a - 4\,b$. **49.** $a + 5\,b$. **50.** $(x + y)(x - y)$.
**51.** $(a + 3)(a - 3)$. **52.** $(m + 4)(m - 4)$. **53.** $(b + 5\,n)(b - 5\,n)$.
**54.** $(3\,a + 7\,b)(3\,a - 7\,b)$. **55.** $(4\,abc + 5)(4\,abc - 5)$. **56.** $(5\,a^2 + 3)$
$(5\,a^2 - 3)$. **57.** $(3\,a - 5\,b)(3\,a - 5\,b)$.

**Page 42.**—**1.** $2\,a^2 + a - 6$. **2.** $3\,m^2 - m - 2$. **3.** $6\,m^2 - 5\,m - 6$.
**4.** $20\,a^2 - 21\,a + 4$. **5.** $12\,x^2 - 5\,xy - 2\,y^2$. **6.** $25\,a^2b^2 - 35\,ab + 12$.
**7.** $2\,x^4 + 7\,x^2b^2 - 15\,b^4$. **8.** $2\,a^4b^4 + 3\,a^2b^2 - 35$. **9.** $2\,x^4y^4 + 5\,x^2y^2z^2 + 2\,z^4$.
**10.** $6\,x^6 + 13\,x^3 - 15$. **11.** $-x^2 + 5\,x^2y^2 - 6\,y^4$. **12.** $30\,x^4 + 19\,x^3 - 5\,x^2$.
**13.** $156$. **14.** $10,712$.

**Page 43.—Exercise 26.—1.** $m^2 + n^2 + p^2 + 2\,mn + 2\,mp + 2\,np$.
**2.** $x^2 + y^2 + z^2 - 2\,xy + 2\,xz - 2\,yz$. **3.** $a^2 + b^2 + 2\,ab - 10\,a + 25 - 10\,b$.
**4.** $a^2 + 4\,b^2 + c^2 - 4\,ab - 2\,ac + 4\,bc$. **5.** $a^2 + 16\,b^2 + 9\,c^2 - 8\,ab + 6\,ac - 24\,bc$.
**6.** $a^2 + b^2 + c^2 + d^2 + 2\,ab + 2\,ac - 2\,ad + 2\,bc - 2\,bd - 2\,cd$. **7.** $x^4 -$
$8\,x^3 + 20\,x^2 - 16\,x + 4$. **8.** $4\,a^2 + 9\,b^2 + 25\,c^2 - 12\,ab + 20\,ac - 30\,bc$.
**9.** $9\,x^2 + 16\,y^2 + z^2 + n^2 - 24\,xy - 6\,xz + 6\,xn + 8\,yz - 8\,yn - 2\,zn$.
**10.** $4\,a^4 + 9\,b^4 + 16\,c^4 + 12\,a^2b^2 - 16\,a^2c^2 - 24\,b^2c^2$. **11.** $x + y + z$.
**12.** $m + n - p$. **13.** $b - c - 1$. **Exercise 27.—1.** $5\,a - 15$.
**2.** $-2\,a - 8$. **3.** $-x + 23$. **4.** $7\,x - 29$. **5.** $-6\,n$. **6.** $2\,b^3 + b^2 - 3\,b$.
**7.** $-2\,m^2 + 2\,n^2$. **8.** $2$.

**Page 44.—9.** $-5\,xy$. **10.** $4\,pq$. **11.** $-m^2 - 25$. **12.** $4\,x$.
**13.** $-34$. **14.** $-2\,p^2 - p + 77$. **15.** $11\,x^2 + 21\,xy - 8\,y^2$. **16.** $-8\,ab - 4\,b^2$.
**17.** $-5\,x^2 - 9\,xy - 10\,y^2$. **18.** $-y^2 + 27\,y$. **19.** $2\,a^2 + 2\,ab + 2\,ac - 2\,bc$.
**20.** $-4\,xy + 13\,y^2$. **21.** $6\,x^2 + 15\,x - 9$. **22.** $a^2 - a$.

**Page 46.—1.** $5$. **2.** $-5$. **3.** $-13$. **4.** $8$. **5.** $3$. **6.** $-9$.
**7.** $64$. **8.** $-49$. **9.** $-125$. **10.** $12$. **11.** $135$. **12.** $50,000$.
**13.** $12$. **14.** $-3$. **15.** $3\,x$. **16.** $3\,ab^2$. **17.** $-4\,a^2c^2$.

**Page 47.—18.** $-5\,mp^{10}$. **19.** $-12\,y^2z^{14}$. **20.** $1$. **21.** $-2\,mc^{49}$.
**22.** $-4$. **23.** $75\,a^2$. **24.** $14\,x^2$. **25.** $20\,a$. **26.** $a + b$.
**1.** $21$. **2.** $29$. **3.** $-6$. **4.** $15$. **5.** $10$. **6.** $2\,b - 3\,c$.
**7.** $-2\,y^2 + 3\,xy$. **8.** $-9\,a^2b^2 + 3\,ab$. **9.** $-5\,a^4 - 4\,a^2 + 2\,a$.
**10.** $5\,ab - 4\,b^2 + 3$. **11.** $5 - 3\,m + \cdot 7\,m^2$.

**Page 48.—12.** $-2\,m^2 + 3\,mn - 5\,n^2$. **13.** $-3\,x^2y^2z^2 + 2\,xyz - 1$.
**14.** $7\,a^2bc^4 - 4\,c + 2\,a$. **15.** $5\,x^4 + 7\,x^3y - 3\,x^2y^2 - xy^3$.
**16.** $4\,x^2y^2z^2 - 3\,xz^3 - 5\,z^2 - 2$. **17.** $-5\,x^4 + 4\,x^3y - 3\,x^2y^2 + 2\,y^4 - y^5$.

**Page 50.—1.** $x - 4$. **2.** $y + 3$. **3.** $3\,x - 8\,y$. **4.** $5\,m - n$.
**5.** $4\,a - 6\,b$. **6.** $4\,x + 3\,y$. **7.** $k - 8$. **8.** $7\,x + 5$. **9.** $4\,c - 9\,d$.
**10.** $5\,p + 18\,q$. **11.** $6\,x - y$. **12.** $7\,a - 3\,b$. **13.** $5\,a - 6\,b$.

**Page 51.—14.** $3\,ab + 4$. **15.** $4\,x^2y - 5$. **16.** $a^2 - 3\,a + 1$.
**17.** $1 + 8\,m + 7\,m^2$. **18.** $x^2 + 2\,x + 4$. **19.** $x^2 + 2\,x + 1$.
**20.** $9\,m^2 + 6\,m + 1$.
**1.** $m - n$. **2.** $x + 1$. **3.** $c - 3$. **4.** $a + 4\,b$. **5.** $2\,a - 3\,b$.
**6.** $6\,xy - 7\,z$. **7.** $13\,a^2b + 9\,c^3$. **8.** $8\,x^5 + 1$.

**Page 52.—9.** $m^2 + 1,\ m + 1,\ m - 1,\ m^2 - 1$. **10.** $a^2 + b^3,\ a^2 - b^3$.
**11.** $abc + 1,\ abc - 1$. **12.** $x^6 + y^8,\ x^6 - y^8,\ x^3 - y^4,\ x^3 + y^4$.
**13.** $6\,x^2y^2 + 7,\ 6\,x^2y^2 - 7$. **14.** $11\,a^{50} + 3\,b,\ 11\,a^{50} - 3\,b$. **15.** $a^8 + 10\,xy^2$,
$a^8 - 10\,xy^2$. **16.** $1000 + 1,\ 1000 - 1$.

**Page 56.** — **1.** 5. **2.** 9. **3.** 5. **4.** 8. **5.** 7. **6.** 6. **7.** 2. **8.** 2.
**9.** 2. **10.** 1. **11.** 11. **12.** 4. **13.** 5. **14.** 4. **15.** 22.
**16.** 96. **17.** 1. **18.** 7. **19.** 1. **20.** 7. **21.** 2. **22.** 10.
**23.** 5. **24.** 2.

**Page 57.** — **25.** 7. **26.** 3. **27.** 4. **28.** 5. **29.** $-1\frac{3}{7}$. **30.** $-13$.
**31.** $1\frac{1}{5}$. **32.** $-12\frac{1}{2}$. **33.** $\frac{13}{22}$. **34.** 5. **35.** 6. **36.** 0. **37.** $-2\frac{1}{5}$.
**38.** $1\frac{9}{14}$. **39.** 1. **40.** $1\frac{1}{3}$. **41.** $1\frac{3}{25}$. **42.** 6. **43.** 20. **44.** 20.

**Page 58.** — **1.** $a-10$. **2.** $9-x$. **3.** $a+4$. **4.** $m+n$. **5.** $\frac{n}{5}$.

**6.** $\frac{x}{n}$. **7.** $10-\frac{a}{3}$. **8.** $\frac{x}{4}-b$. **9.** $2b-\frac{c}{2}$. **10.** $p+7$. **11.** $a,\ 100-a$.
**12.** $10,\ a-10$. **13.** $b,\ a-b$ **14.** $d+s$. **15.** $g-d$. **16.** $3x^2$.
**17.** $7x^2$.

**Page 59.** — **18.** $2b-a$. **19.** $a+1,\ a+2$. **20.** $x-1,\ x-2$.
**21.** $x-y$ yr. **22.** $y-5$ yr., $y+10$ yr. **23.** $x+y+12$ yr.,
$x+y-10$ yr. **24.** $\$m+6, \$n-6$. **25.** $100\,d$ ct., $10\,x$ ct.
**26.** $100\,a+10\,b+c$ ct. **27.** $100\,a-b$ ct. **28.** $xy$ sq. ft.
**29.** $xy+3x+2y+6$ sq. ft. **30.** $xy+4x-3y-12$ sq. ft. **31.** $2x+$
$2y$ ft. **32.** $10\,x$ ct. **33.** $\frac{20}{x}$ ct. **34.** $\frac{210}{x}$ ct. **35.** $\frac{3\,n}{x}$ ct.

**Page 60.** — **36.** $5\,n$ mi. **37.** $rn$ mi. **38.** $\frac{n}{4}$ mi. **39.** $\frac{n}{r}$ hr.

**40.** $tx$ mi. **41.** $x+20$ yr. **42.** $\frac{1}{x}$. **43.** $\frac{1}{x}+\frac{1}{y}$. **44.** $5\,x$.

**45.** $\frac{3\,x}{50}$. **46.** $10\,a$. **47.** $\frac{x}{25}$. **48.** $\frac{mx}{100}$. **49.** $\frac{m+3}{m}$. **50.** $10\,x+y$.

**Page 61.** — **1.** $2a=10$. **2.** $2x+10=c$. **3.** $a+10=2x$.
**4.** $\frac{x}{3}=c$. **5.** $x-y+7=a$. **6.** $2a+\frac{b}{3}=100$. **7.** $4(a-b)-c=d-9$.
**8.** $(a+b)(a-b)-90=\frac{a^2+b^2}{7}$. **9.** $2a-20=a-7$.

**Page 62.** — **10.** $a-9=17-a$. **11.** $x=\frac{5}{100}\times 450$. **12.** $x=\frac{6}{100}\,m$.

**13.** $100=\frac{x}{100}\times 700$. **14.** $50=\frac{ax}{100}$. **15.** $m=\frac{xn}{100}$. **16.** $(a)\ 2x=2(3x-10)$,
$(b)\ 2x-(3x-10)=4$, $(c)\ 2x-5=x$, $(d)\ 2x+10=n$, $(e)\ 2x+3$
$=3x-10$, $(f)\ (2x-3)+(3x-13)=50$, $(g)\ 2x+3=2(3x-7)$,
$(h)\ (2x+10)+3x+(4x-10)=100$. **17.** $(a)\ 2x-(3x-700)=5$,
$(b)\ 2x+20=3x-740$, $(c)\ (2x+500)+(3x-200)+(x+1700)=12{,}000$,
$(d)\ 2x+(3x-700)=(x+1200)-200$, $(e)\ 3x-800=x+1300$.

**18.** (a) $\dfrac{5x}{100} = 90$, (b) $\dfrac{a}{100}(5x - 30) = 20$, (c) $\dfrac{x^2}{100} = \dfrac{6}{100}(2x + 1)$,

(d) $\dfrac{ax}{100} - \dfrac{b}{100} \cdot (5x - 30) = 900$, (e) $\dfrac{4x}{100} + \dfrac{5}{100}(5x - 30) + \dfrac{6}{100}(2x + 1)$

$= 8000$, (f) $\dfrac{x^2}{100} = \dfrac{2x + 1}{10}$.

**Page 64.—1.** 13.  **2.** 15.  **3.** 25.  **4.** 18.  **5.** 7.  **6.** 9.
**7.** 20 yr.  **8.** 30 yr.  **9.** 90 mi.  **10.** $2\frac{2}{3}$.  **11.** 1250.  **12.** $24\frac{1}{2}$.
**13.** 85 ft.  **14.** 30 mi.

**Page 65.—15.** $250.  **16.** $300.  **17.** $40.  **18.** 80 A.  **19.** 150,000.

**Page 67.—1.** 55, 11.  **2.** 65, 5.  **3.** 36, 6.

**Page 68.—4.** 12, 2.  **5.** 78, 79.  **6.** 52, 13.  **7.** 8, 10.  **8.** 160 lb.,
480 lb.  **9.** 13, 7.  **10.** 40 yr., 10 yr.  **11.** 29,000 ft., 20,000 ft.
**12.** 4 pt., 5 pt.  **13.** 42 yr., 28 yr.  **14.** 45 in., 15 in.  **15.** 7 hr.

**Page 70.—1.** 5, 10, 25.  **2.** 6, 12, 14.

**Page 71.—3.** 12, 8, 24.  **4.** 1, 3, 5.  **5.** 3, 6, 16.  **6.** 20, 21, 22.
**7.** 8 in., 9 in., 11 in.  **8.** 1,000,000 Phil., 2,000,000 Berlin, 4,000,000 N. Y.
**9.** 30°, 50°, 100°.  **10.** 21.  **11.** 20 yr., 10 yr., 25 yr.  **12.** 6, 7, 8.
**13.** $90,000,000 gold, $180,000,000 copper, $480,000,000 pig iron.

**Page 72.—14.** 5 Col., 10 Cal., 16 Mass.

**Page 74.—1.** 15 yd., 20 yd.  **2.** 10 yd. by 12 yd.  **3.** $200.
**4.** $600, $1200.  **5.** $200, $1200.  **6.** 70¢, 210¢.  **7.** 5 lb., 1 lb.
**8.** $4.  **9.** 3 hr., 15 mi.

**Page 75.—10.** 12 mi.  **11.** $5\frac{1}{4}$ hr.  **12.** 82 mi.

**Page 78.—1.** $6x(ab - 2cd)$.  **2.** $3x^2(3x - 2)$.  **3.** $5a^2b(3 + 4b^2)$.
**4.** $7a^2b^2(2a^2b^2 - 1)$.  **5.** $11m(m^2 + m - 1)$.  **6.** $xy(4x^2 + 5xy - 6)$.
**7.** $17x^3(1 - 3x + 2x^2)$.  **8.** $8(a^2b^2 + b^2c^2 - c^2a^2)$.  **9.** $15x^2y(yz^2 - 3xy^2 - 2x^2)$.
**10.** $a^2(a^2 - a + 1)$.  **11.** $16a^2xy(2a^2 - ay + 3y^2)$.  **12.** $3a^6b^7z^6(3a^2 - 2az + z^2)$.
**13.** $6a^2(2ab^3 - 3n^2 - 4a^2p^3)$.  **14.** $17abc(2a^2b^2c^2 - 3abc + 4)$.
**15.** $11p^2q^2(2p^3 - 5p^2q + 7q^3)$.  **16.** $13x^3y^4z^5(5 - 3xyz + x^2y^2z^2)$.
**17.** $q(q^3 - q^2 - q + 1)$.  **18.** $(m + n)(a + b)$.  **19.** $3(a + b)(x^2 - y^2)$.
**20.** $(p + q)(3a - 5b)$.  **21.** $13 \cdot 13$.  **22.** $2 \cdot 3 \cdot 4 \cdot 11$.  **23.** $2 \cdot 3 \cdot 5 \cdot 7$.
**Page 79.—1.** $(a - 4)(a - 3)$.  **2.** $(a + 4)(a + 3)$.  **3.** $(m - 3)(m - 2)$.
**4.** $(x - 5)(x - 2)$.  **5.** $(x + 5)(x + 3)$.  **6.** $(a - 5)(a - 4)$.

**Page 80.—7.** $(x - 4)(x + 2)$.  **8.** $(x + 4)(x - 2)$.  **9.** $(y - 8)(y + 2)$.
**10.** $(y + 8)(y - 2)$.  **11.** $(y - 11)(y - 4)$.  **12.** $(y - 7)(y + 2)$.
**13.** $(y + 7)(y - 3)$.  **14.** $(a + 5)(a + 6)$.  **15.** $(x - 15)(x - 2)$.

**16.** $(p-8)(p+1)$.    **17.** $(q+8)(q-3)$.    **18.** $(ax+9)(ax-2)$.
**19.** $(a-7b)(a-10b)$.    **20.** $(a-11b)(a+2b)$.    **21.** $(a^2+10)(a^2-2)$.
**22.** $(ay-8)(ay-3)$.    **23.** $(m+20)(m+5)$.    **24.** $(y+4)(y-1)$.
**25.** $(a-6b)(a+4b)$.    **26.** $(n^2+12)(n^2+5)$.    **27.** $(a^3+10)(a^3-3)$.
**28.** $(a^4-10)(a^4+3)$.    **29.** $x(x+2)(x+3)$.    **30.** $100(x-3)(x-2)$.
**31.** $6a^2(a-2)(a-1)$.    **32.** $y(x-7)(x+3)$.    **33.** $a^2(m-7)(m+3)$.
**34.** $10x^2(y-9)(y+2)$.    **35.** $200(x+1)(x+1)$.    **36.** $4(a-11b)(a-b)$.

**Page 82.—1.** $(2x-1)(x+5)$. **2.** $(4a-1)(a-2)$. **3.** $(3x-2)(x-2)$.
**4.** $(5m-1)(m-5)$.    **5.** $(3n+4)(2n-1)$.    **6.** $(3x+1)(x+4)$.
**7.** $3(x+2)(x-1)$.    **8.** $(4y-3)(3y+2)$.    **9.** $(2y+3)(y-1)$.
**10.** $(2t+1)(t-9)$.    **11.** $(5a-2)(2a-3)$.    **12.** $(9y-4)(y+4)$.
**13.** $(2m+1)(m+3)$.    **14.** $(5x-7)(2x+1)$.    **15.** $(4x-3)(3x-2)$.
**16.** $(6n+1)(n+2)$.    **17.** $(2y-1)(y+9)$.    **18.** $(7a+4)(2a-1)$.
**19.** $(3x-y)(x+4y)$.    **20.** $(15x-2y)(x-5y)$.    **21.** $(5a-4b)(2a-3b)$.
**22.** $(3x-2y)(2x-3y)$.    **23.** $(4x-5y)(3x+2y)$.    **24.** $2(2x+3)(x+2)$.
**25.** $x(2x+3)(x+4)$.    **26.** $x(5x+4)(x+2)$.    **27.** $10(2x-y)(x-2y)$.
**28.** $100(x-y)^2$.    **29.** $a^4(5a+1)(a-2)$.    **30.** $10y^2(9x+1)(x-3)$.
**31.** $10(3a-5b)^2$.    **32.** $-y^2(2y-3)(2y-1)$.    **33.** $10a^2(4-n^2)(1-2n^2)$.
**34.** $2(9x-8y)(8x-9y)$.    **35.** $(2x^2+3y^2)(2x^2+y^2)$.

**Page 83.—1.** Yes, $(m+n)^2$. **2.** No. **3.** Yes, $(q-5)^2$. **4.** No, $(x-8)(x-2)$. **5.** Yes, $(a-2b)^2$. **6.** No. **7.** Yes, $(m-7n)^2$. **8.** No. **9.** No. **10.** Yes, $(m^2+3n)^2$. **11.** Yes, $(y-8)^2$. **12.** Yes, $(3a-2b)^2$. **13.** Yes, $(5x-2y)^2$. **14.** No.

**Page 84.—15.** Yes, $(4a-3b)^2$. **16.** Yes, $(6x^2+5)^2$. **17.** Yes, $(15xy-2)^2$. **18.** Yes, $10(a-b)^2$. **19.** Yes, $x^2(x-y)^2$. **20.** Yes, $m(m-3)^2$. **21.** $9b^2$. **22.** $16$. **23.** $9$. **24.** $24ab$. **25.** $216ab$. **26.** $140m^2$. **27.** $9$. **28.** $25$. **29.** $9$. **30.** $40x$.

**1.** $(x+y)(x-y)$.    **2.** $(a+3)(a-3)$.    **3.** $(6+b)(6-b)$.
**4.** $(2a+1)(2a-1)$.    **5.** $(1+7a)(1-7a)$.    **6.** $(9+t)(9-t)$.
**7.** $(10a+b)(10a-b)$.    **8.** $(ab+11)(ab-11)$.    **9.** $(3x+4y)(3x-4y)$.

**Page 85. 10.** $(5xy+9z)(5xy-9z)$.    **11.** $(7ay+8)(7ay-8)$.
**12.** $(5a^2+1)(5a^2-1)$.    **13.** $(10ab+c^2)(10ab-c^2)$.
**14.** $(13a^2+10)(13a^2-10)$.    **15.** $(15a+4b^2)(15a-4b^2)$.
**16.** $(a^2+b^2)(a+b)(a-b)$.    **17.** $(a^2b^2+9)(ab+3)(ab-3)$.
**18.** $(1+x^4)(1+x^2)(1+x)(1-x)$.    **19.** $10(a+b)(a-b)$.
**20.** $13x(a+b)(a-b)$.    **21.** $x(x+y)(x-y)$.    **22.** $(x+y^4)(x-y^4)$.
**23.** $3xy^8(5x+4)(5x-4)$.    **24.** $2y(11x^2+1)(11x^2-1)$.
**25.** $x^2(12+y^2)(12-y^2)$.    **26.** $13\times7$.    **27.** $103\times97$.

**1.** $(m+n+p)(m+n-p)$.    **2.** $(m-n+p)(m-n-p)$.
**3.** $(m+n+4p)(m+n-4p)$.    **4.** $(x+y+z)(x-y-z)$.

**5.** $(4x+y+z)(4x-y-z)$.     **6.** $(5a+b-c)(5a-b+c)$.

**7.** $(m+2n+6p)(m+2n-6p)$.    **8.** $(m-3n+a+b)(m-3n-a-b)$.

**9.** $(2a-5b+5c-9d)(2a-5b-5c+9d)$.     **10.** $a(a+2b)$.

**11.** $y(2x-y)$.     **12.** $x(x+6y)$.     **13.** $(5a+1)(9-a)$.

**14.** $(5x+y)(5y-x)$.

**Page 86.—1.** $(a+b)(x+y)$.     **2.** $(m-n)(a+b)$.

**3.** $(2a-3b)(n+q)$.    **4.** $(x+y)(4c-5d)$.    **5.** $(2x-y)(5a-3b)$.

**6.** $(x+1)(x^2+2)$.    **7.** $(x+1)(x+1)(x-1)$.    **8.** $(m^2+n^2)(x+y)(x-y)$.

**9.** $(3a^2-4b^2)(x+y)$.     **10.** $(a-b)(a^2+b^2)$.     **11.** $(c-7)(c^2+2)$.

**12.** $(a-x)(a^4-b)$.

**Page 87. — Exercise 46.—1.** $(x+y+q)(x+y-q)$.     **2.** $(1+a+b)$
$(1-a-b)$.    **3.** $(a-2b+5)(a-2b-5)$.    **4.** $(6x+y-3z)(6x-y+3z)$.

**5.** $(3m-n+ab)(3m-n-ab)$.    **6.** $(a+b+x+y)(a+b-x-y)$.

**7.** $(a-5b+x-2y)(a-5b-x+2y)$.     **8.** $(x^2+x+1)(x^2-x-1)$.

**Exercise 47.—1.** $(m+4)(m-4)$.   **2.** $8(m^2+2)$.   **3.** $(m-2)(m+1)$.

**4.** $(6a^2+1)(a^2+6)$.   **5.** $6(a+1)^2(a-1)^2$.   **6.** $(x+y)^2$.   **7.** $a^2(a-9)$.

**8.** $2(2x-y)(x-2y)$.     **9.** $(m-3+n)(m-3-n)$.

**Page 88.—10.** $x(x+y)(x-y)$.     **11.** $2(5a-b)(a-3b)$.

**12.** $1(a^2+b^2)(a+b)(a-b)$.      **13.** $a^2(7a-3)(7a-3)$.

**14.** $10(2a^2b^2+1)(a^2b^2-5)$.      **15.** $(3+a+6b)(3-a-6b)$.

**16.** $(2x-7)(x^2-2)$.      **17.** $(1+n^4)(1+n^2)(1+n)(1-n)$.

**18.** $n^6(n+y)(n-y)$.    **19.** $5(x-9)(x-1)$.    **20.** $(a^3-13)(a^3+12)$.

**21.** $x(3x^2-2y)(6-x)$.    **22.** $3(a+b+4)(a+b-4)$.    **23.** $10(8x^2+1)$
$(x+2)(x-2)$.    **24.** $a(a^2+1)(a+1)(a-1)$.    **25.** $(a+1)(a^2+1)$.

**26.** $(a^2-a+1)(a^2-a-1)$.      **27.** $(6x-7y)(7x-6y)$.

**28.** $(5m^3+1)(2m^3-9)$.    **29.** $(5ab-3)(5ab+8)$.    **30.** $13(c+3)(c-4)$.

**31.** $a(a+15b)(a-15b)$.    **32.** $2(5n-7s)(2n+3s)$.    **33.** $(a+b+8)$
$(a+b-8)$.    **34.** $(16+x-2y)(16-x+2y)$.    **35.** $2(a^2+8)(a^2-8)$.

**36.** $(xyz-50)(xyz-1)$.    **37.** $17(x+3y)(x-2y)$.    **38.** $(a^2-2)(a^3+b)$.

**39.** $a(a+2b)$.    **40.** $3p^2(p^3-9)(p^3-4)$.    **41.** $3(x^2+3)(x+2)(x-2)$.

**Page 89.—1.** $2a^2b^2$.   **2.** $5a^2b^3c^2$.   **3.** $13x^3y$.   **4.** $12$.   **5.** $450$.
**6.** $7$.

**Page 90.—7.** $6mu^8$.   **8.** $13x^6y^3z^4$.   **9.** $10$.   **10.** $15bcd$.   **11.** $4a^8$.
**12.** $(m+n)^2$.   **13.** $2(m+1)^2$.   **14.** $3x(x-y)^3$.

**1.** $4a^3b^3$.   **2.** $5x^8$.   **3.** $5m^2n$.   **4.** $3x^4$.   **5.** $a-b$.   **6.** $x+y$.

**7.** $a+b$.   **8.** $x-2$.   **9.** $x+3$.   **10.** $x-4$.   **11.** $x+3$.   **12.** $a-4$.

**13.** $a+3$.   **14.** $y-6$.   **15.** $2a+1$.   **16.** $a+3b$.

**Page 92.—1.** $a^3$.   **2.** $x^2y^3$.   **3.** $8x^8$.   **4.** $24x^3y^6$.   **5.** $30a^3b^8$.
**6.** $42a^3x$.   **7.** $40a^5b^5c^5d^5$.   **8.** $9ab(a+b)$.   **9.** $12m^2(m+n)^2$.

**10.** $(a-2)^2(a-3)^2(a-4)^2$. **11.** $6\,a^2b(a-b)$. **12.** $30(x+y)(x-y)$.
**13.** $(a+b)(a-b)$. **14.** $(a-2)(a+2)^2$. **15.** $2(a+b)^2(a-b)$.
**16.** $(x-2)(x-1)(x-3)$. **17.** $(a+2)(a+3)(a+1)$. **18.** $2(2\,a-1)$
$(2\,a+1)$. **19.** $(x+2)^2(x+1)(x+3)$. **20.** $15\,x(x+1)(x-1)$.
**21.** $ab(a+b)(a-3\,b)(a+2\,b)$. **22.** $(x^2+1)(x+1)(x-1)$.
**23.** $(x-5)(x-2)(x-8)(x-3)$ **24.** $6(x+y)(a-b)$.

**Page 94.—1.** $15$. **2.** $\frac{7}{3}$. **3.** $3\,a$. **4.** $\frac{2\,xy}{z}$. **5.** $\frac{9}{c^2}$. **6.** $\frac{2\,c^2}{a^2}$.

**Page 95.—7.** $\frac{1}{2\,abc}$. **8.** $\frac{b^2}{3\,x^2}$. **9.** $\frac{a+b}{x+y}$. **10.** $\frac{7}{9}$. **11.** $\frac{a-2\,b}{3\,a+4\,b}$.

**12.** $\frac{mn}{3}$. **13.** $\frac{a+b}{a-b}$. **14.** $\frac{x-y}{x-2\,y}$. **15.** $\frac{m-4}{m-2}$. **16.** $\frac{b+5}{b+6}$.

**17.** $\frac{2\,x+5\,y}{2}$. **18.** $\frac{x+4}{x+5}$. **19.** $\frac{a-9}{a-10}$. **20.** $\frac{m-4}{m+1}$. **21.** $\frac{n-2}{n+2}$.

**22.** $\frac{a-8\,b}{a+8\,b}$. **23.** $\frac{a+1}{a-4}$. **24.** $\frac{x-3\,y}{3\,x-y}$. **25.** $\frac{a^2+b^2}{6\,b}$. **26.** $\frac{a^2(a+3)}{3\,a-1}$.

**27.** $\frac{4\,m-n}{5\,m-n}$. **28.** $\frac{2\,a^2-6\,an-6\,n^2}{3\,a^2-7\,an+2\,n^2}$. **29.** $\frac{a-m}{b-n}$. **30.** $\frac{a^3+b^3}{2}$.

**31.** $\frac{(a^2+b^2)(a+b)}{a-b}$. **32.** $-\frac{x-4}{x+4}$.

**Page 97.—1.** $\frac{9\,m^2}{12\,n^3},\ \frac{20\,n}{12\,n^3}$. **2.** $\frac{44\,a^2z^2}{6\,x^2y^2z^2},\ \frac{15\,ax^2}{6\,x^2y^2z^2}$. **3.** $\frac{2\,b^2c^2}{b^2c^2}$,

$\frac{a^2}{b^2c^2}$. **4.** $\frac{2\,mx^2}{2\,m^3},\ \frac{2\,m^2}{2\,m^3},\ \frac{5}{2\,m^3}$. **5.** $\frac{yz}{xyz},\ \frac{xz}{xyz},\ \frac{xy}{xyz}$. **6.** $\frac{15\,ac^3}{6\,a^2b^2c^2}$,

$\frac{4\,ab^3}{6\,a^2b^2c^2},\ \frac{5\,b^2c^2}{6\,a^2b^2c^2}$. **7.** $\frac{18\,a^4x}{6\,a^3x},\ \frac{12\,x^2}{6\,a^3x},\ \frac{5\,a^4}{6\,a^3x}$. **8.** $\frac{105\,a^7b^2y}{42\,a^2x^4y^3z^4}$,

$\frac{18\,b^3xz^5}{42\,a^2x^4y^3z^4},\ \frac{28\,a^5x^3y^2z^3}{42\,a^2x^4y^3z^4}$. **9.** $\frac{3}{3(x-y)},\ \frac{4}{3(x-y)}$. **10.** $\frac{3\,b^2}{ab(x-y)}$,

$\frac{2\,a^2}{ab(x-y)}$. **11.** $\frac{2\,a-1}{(3\,a+1)(3\,a-1)},\ \frac{9\,a+3}{(3\,a+1)(3\,a-1)}$. **12.** $\frac{6\,a-6\,b}{a^2-b^2}$,

$\frac{5\,a+5\,b}{a^2-b^2},\ \frac{4\,a}{a^2-b^2}$. **13.** $\frac{3\,x-9}{(x-1)^2(x-3)},\ \frac{5\,x-5}{(x-1)^2(x-3)}$.

**14.** $\frac{6\,a^2+31\,a+5}{a^2-25},\ \frac{6\,a^2-31\,a+5}{a^2-25},\ \frac{2\,a^2}{a^2-25}$. **15.** $\frac{9\,x+9}{6(x+1)(x-1)}$,

$\frac{30}{6(x+1)(x-1)},\ \frac{2\,x-2}{6(x+1)(x-1)}$. **16.** $\frac{a^2-4}{(a-1)(a-2)(a-3)}$,

$\frac{a^2-1}{(a-1)(a-2)(a-3)},\ \frac{a^2-9}{(a-1)(a-2)(a-3)}$.

**Page 99.—1.** $\frac{23\,a+9}{20}$. **2.** $\frac{29\,n}{6}$. **3.** $\frac{16\,y}{45}$. **4.** $\frac{ay+bx}{by}$.

**5.** $\dfrac{n-m}{mn}$.  **6.** $\dfrac{10\,b^2+9\,ab-12\,a^2}{6\,a^2b^2}$.  **7.** $\dfrac{21\,ab-12\,b^2-8\,a^2}{12\,ab}$.

**8.** $\dfrac{24\,ab+15\,b^2-10\,a^2}{20\,ab}$.  **9.** $\dfrac{92\,ab-121\,b^2-195\,a^2}{143\,ab}$.

**10.** $\dfrac{bc+ac+ab}{abc}$.  **11.** $\dfrac{238\,xyz+84\,y^2z-15\,x^2y+30\,x^2z}{105\,x^2yz}$.

**12.** $\dfrac{45\,u-30\,v-50\,u^2-30\,uv^2-30\,uv+48\,v^2}{180\,uv}$  **13.** $\dfrac{ab+2}{2\,b}$.  **14.** $\dfrac{a^3+3\,a+3}{3\,a^2}$.

**15.** $\dfrac{3+2\,q-3\,p}{pq}$.  **16.** $\dfrac{2\,m+5}{(m+2)(m+3)}$.  **17.** $\dfrac{1}{(m-3)(m-2)}$.

**18.** $\dfrac{5\,x+13}{(x+2)(x+3)}$.  **19.** $\dfrac{a^2+b^2}{a^2-b^2}$.  **20.** $\dfrac{3\,a}{a+b}$.  **21.** $\dfrac{4\,m}{1-m^2}$.

**22.** $\dfrac{5\,x^2+8\,x}{(x+1)(x+2)}$.  **23.** $\dfrac{a^2+9\,a-7}{(a-2)(a+3)}$.  **24.** $\dfrac{26}{3(a+1)}$.

**25.** $\dfrac{2}{(x+1)(x+2)(x+3)}$.

**Page 100.—26.** $\dfrac{4\,x-x^2-2}{(x-2)(x-1)}$.  **27.** $\dfrac{18\,y^2-4\,xy}{(x+3\,y)(x-3\,y)}$.

**28.** $\dfrac{37\,x}{(x-9)(x-3)}$.  **29.** $\dfrac{5\,a^2-23\,a+36}{3\,a(a-4)(a-3)}$.  **30.** $\dfrac{a^2+1}{2(a+1)}$.

**31.** $\dfrac{2\,a}{(a-2\,b)^2}$.  **32.** $\dfrac{2\,x-24}{(x-1)(x-3)}$.  **33.** $\dfrac{2}{1-m}$.

**34.** $\dfrac{3\,b(b-1)}{a^2+ab-2\,b^2}$.  **35.** $\dfrac{2\,x^2-x-10}{(2\,x-1)(x-2)(x-3)}$.  **36.** $\dfrac{2\,x^2}{(x-1)^3}$.

**37.** $\dfrac{2\,a^2-2\,ab+2\,b^2}{(a-b)^2(a+b)}$.  **38.** $\dfrac{2\,a^2}{a+1}$.  **39.** $\dfrac{2\,x}{x-1}$.  **40.** $\dfrac{x^2}{x-y}$.

**41.** $\dfrac{-5\,m}{m-3}$.  **42.** $\dfrac{b^3}{a^2+ab+b^2}$.  **43.** $\dfrac{a^3+a+1}{a^2}$.

**44.** $\dfrac{3\,a^2-b^2-2\,a-2\,b-2\,ab}{2\,(a-b)}$.  **45.** $\dfrac{4\,x^3-18\,x^2+57\,x-18}{(x-3)(x-4)(4\,x-1)}$.

**Page 101.—1.** $6\,a-5+\dfrac{7}{a}$.  **2.** $3\,a-2+\dfrac{2}{3\,a}$.  **3.** $m-1+\dfrac{2}{m-4}$.

**4.** $n+4+\dfrac{2}{n+3}$.  **5.** $a+1+\dfrac{5}{a+1}$.  **6.** $x-2+\dfrac{13}{2\,x-1}$.

**7.** $a-1+\dfrac{2}{a+1}$.  **8.** $x^2+5\,x+12+\dfrac{41}{x-3}$.

**Page 103.—1.** $\frac{5}{2}$.  **2.** $\frac{1}{15}$.  **3.** $\dfrac{3\,b}{4\,a}$.  **4.** $\dfrac{2\,dz^2}{x^2}$.  **5.** $\dfrac{3}{4\,bm}$.

**6.** $\dfrac{ay^2}{4\,z^3}$.  **7.** $\dfrac{4\,a^4}{c}$.  **8.** $6$.  **9.** $-\dfrac{15\,a}{4\,c}$.  **10.** $1$.  **11.** $\dfrac{a+2\,b}{7}$.

**12.** $\dfrac{a-b}{a-2b}$. **13.** $\dfrac{x+1}{x+3}$. **14.** $\dfrac{(a-1)^2}{(a+1)^2}$. **15.** $3x-1$. **16.** $\dfrac{x(x-y)^2}{(x+y)^2}$.

**17.** 1. **18.** 1.

**Page 104.—1.** $\dfrac{y}{x}$. **2** $\dfrac{28\,ax}{65\,m}$. **3.** $\dfrac{y^2}{u^2x}$. **4.** $\dfrac{a}{2b}$. **5.** 1.

**6.** $\dfrac{a+b}{a-b}$.

**Page 105.—7.** $\dfrac{x-1}{x+1}$. **8.** $\dfrac{a}{(a-b)(a+b)}$. **9.** $\dfrac{(x-1)^2}{(x+1)^2}$.

**10.** $\dfrac{x+3}{x(x-2)}$. **11.** 1. **12.** $a$. **13.** $\dfrac{2}{x+y}$.

**14.** $\dfrac{4x-1}{2x-1}$. **15.** $\dfrac{1}{a}$.

**Page 106.—1.** $\dfrac{x^3}{y}$. **2.** $\dfrac{x^3}{y^2}$. **3.** $\dfrac{y}{x}$. **4.** $\dfrac{a}{c}$. **5.** $\dfrac{ac+b}{bc+a}$.

**6.** $\dfrac{a^2}{x+y}$. **7.** $\dfrac{a^2}{ac-b}$. **8.** $\dfrac{mp+n}{mp-n}$. **9.** $\dfrac{3x+2y}{2x+3y}$. **10.** $\dfrac{4x+2}{2x+3}$.

**Page 107.—11.** $\dfrac{m+6n}{m-15n}$. **12.** $\dfrac{2a-b}{a+2b}$. **13.** $\dfrac{am+bm}{an+bm}$.

**14.** $y(x+y)$. **15.** $-x$. **16.** $\dfrac{a+b}{a-b}$. **17.** $\dfrac{a+1}{a(a+2)}$. **18.** $\dfrac{2a+1}{a+1}$.

**19.** $\dfrac{1+x}{4+x}$. **20.** $\frac{5}{7}$. **21.** $\dfrac{a^3+a}{a^2+a+1}$. **22.** $\dfrac{x^2+1}{2x}$.

**Page 109.—1.** 23. **2.** 11. **3.** $6\frac{3}{7}$. **4.** 5. **5.** 1. **6.** 21.
**7.** 8. **8.** 5. **9.** 12. **10.** 5. **11.** $\frac{1}{2}$. **12.** $\frac{1}{4}$. **13.** $\frac{3}{5}$. **14.** $\frac{1}{2}$.
**15.** 3. **16.** 5. **17.** 6. **18.** 6. **19.** $\frac{2}{11}$. **20.** 7. **21.** 6.
**22.** 5. **23.** $-5\frac{5}{13}$.

**Page 110.—24.** 9. **25.** 4. **26.** $\frac{1}{4}$. **27.** $-3$. **28.** 9. **29.** $-3$.
**30.** 4. **31.** 3. **32.** 4. **33.** 3. **34.** $1\frac{4}{25}$. **35.** 4. **36.** 7.
**37.** 1. **38.** 11. **39.** 0. **40.** $-1$.

**Page 111.—41.** $-9\frac{2}{5}$. **42.** 4. **43.** $\frac{7}{15}$. **44.** $-2\frac{1}{5}$. **45.** $-21\frac{1}{15}$.
**46.** 7. **47.** 6.

**Page 113.—1.** $\dfrac{n-m}{2}$. **2.** $\dfrac{8-m}{7}$. **3.** $a-b$. **4.** $\dfrac{n}{m}$.

**5.** $\dfrac{p+n}{m}$. **6.** $\dfrac{3a+2b}{3}$. **7.** $\dfrac{ab-c}{a}$. **8.** $4a-3b$. **9.** 0

**10.** $\dfrac{c}{a+b}$. **11.** $\dfrac{n-m}{m-a}$. **12.** $\dfrac{5}{a+b}$. **13.** $\dfrac{6}{a+b+c}$. **14.** $\dfrac{a}{n}$

**15.** $\dfrac{q}{m-n-p}$. **16.** $mn$. **17.** $\dfrac{m}{n}$. **18.** $a+b$. **19.** $\dfrac{b-a}{2}$

**20.** $\dfrac{-m}{m+n}$. **21.** $\dfrac{abc}{a+b}$. **22.** $\dfrac{5\,ab-a-b}{a}$. **23.** $\dfrac{b-a}{b}$. **24.** 0.

**25.** $\dfrac{m+n}{m-n}$. **26.** $n$. **27.** $\dfrac{a+2\,b-4}{a+b-3}$. **28.** $\dfrac{s}{t}$. **29.** $\dfrac{s}{v}$. **30.** $\dfrac{2\,s}{t^2}$.

**31.** $\dfrac{b-c}{b+c}$. **32.** $\dfrac{pq}{p+q}$. **33.** $\dfrac{fq}{q-f}$. **34.** $(a)$ $\dfrac{100\,i}{rn}$, $(b)$ $\dfrac{100\,i}{pn}$,

$(c)$ $\dfrac{100\,i}{pr}$.

**Page 114.—35.** $(a)$ $\dfrac{9\,c+160}{5}$; $(b)$ 104°, 212°, −4°. **36.** $\dfrac{C}{2\,\pi}$.

**Page 116.—1.** 36. **2.** 60. **3.** 8, 2. **4.** 21, 9. **5.** 15, 16.
**6.** 18, 15. **7.** 30 yrs. **8.** 40 yrs., 10 yrs. **9.** 30 ft. **10.** $30,000.
**11.** $6,000. **12.** $60.

**Page 117.—13.** 40 mi./hr. **14.** 30 mi./hr. **15.** $21\frac{9}{11}$ min. after 4.
**16.** $38\frac{2}{11}$ min. after 7. **17.** $5\frac{5}{11}$ min. after 7. **18.** $12,000.
**19.** $40,000 = A's, $45,000 = B's. **20.** $9\frac{1}{2}$ oz. gold, $10\frac{1}{2}$ oz. silver.
**21.** $1\frac{5}{7}$ da. **22.** $1\frac{1}{2}$ da. **23.** 3 da.

**Page 118.—24.** $(a)$ $2\frac{1}{2}$ da., $(b)$ 5 da., $(c)$ $3\frac{1}{5}$ da., $(d)$ 4 da. **25.** 13,
14, 15. **26.** 18, 19, 20. **27.** $\dfrac{m-3}{3}$, $\dfrac{m}{3}$, $\dfrac{m+3}{3}$; 7, 8, 9 ; 10,002, 10,003,
10,004 ; 306,137, 306,138, 306,139.

**Page 119.—28.** 5, 6. **29.** 10, 11. **30.** 9 ft. **31.** $(a)$ 25, 26 ;
$(b)$ 74, 75 ; $(c)$ 8360, 8361 ; $(d)$ 500,000, 500,001. **32.** 11 hrs., 33, 55 mi.

**33.** $\dfrac{d}{m+n}$ hr., $\dfrac{dm}{m+n}$, $\dfrac{dn}{m+n}$ mi. $(a)$ 12 hr., 36, 24 mi.; $(b)$ 5 hr.,
10, 25 mi. ; $(c)$ 8 hr., 28, 36 mi. **34.** $\dfrac{mn}{m+n}$ min. ; $(a)$ 4 min. ;
$(b)$ 7 min.; $(c)$ 2 hr.

**Page 121.—1.** 4. **2.** 3. **3.** $\frac{2}{5}$. **4.** $\frac{7}{8}$. **5.** 3. **6.** $\frac{2}{3}$.
**7.** 7:9. **8.** 2:1. **9.** 275:168. **10.** 3:2. **11.** $2x:3y$.
**12.** $4x^2:3y^2$. **13.** $1:3z$. **14.** $x-y:x+y$. **15.** 1:4. **16.** $1:\frac{5}{8}$.
**17.** 1:2. **18.** $1:\dfrac{3\,b}{2\,a}$.

**Page 124.—1.** Yes. **2.** Yes. **3.** No. **4.** Yes. **5.** Yes.
**6.** $1:1 = 1:1$, Yes. **7.** $1:1 = 1:1$, Yes. **8.** $3:19 = 4:25$, No.
**9.** $1:1 = 1:1$, Yes. **10.** $1:1 = 1:1$, Yes.

**Page 125.—11.** $10\frac{1}{2}$. **12.** $3\frac{3}{4}$. **13.** $9\frac{1}{4}$. **14.** $1\frac{3}{4}$. **15.** 7.
**16.** 20. **17.** $7\frac{1}{2}$ b. **18.** $40\,mn$. **19.** 15. **20.** 12. **21.** $\frac{2}{3}$.

**22.** $\frac{np}{m}$. **23.** $pq$. **24.** 16. **25.** 49. **26.** $31\frac{1}{2}$. **27.** $b^2$. **28.** $a^2$. **29.** $\frac{1}{a}$.

**30.** 8. **31.** $\frac{10}{3}$. **32.** $6a$. **33.** $4ab$. **34.** $\sqrt{m^2-1}$. **35.** $5:6 = 10:12$ ;
$5:10 = 6:12$. **36.** $b:x = y:a$ ; $b:y = x:a$. **37.** $x:y = 7:6$.
**38.** $x:y = 2:9$. **39.** $x:y = 1:6$. **40.** $x:y = n:m$. **41.** $x:y = c:a+b$.
**42.** $x:y = 5:2$. **43.** $x:y = m:n$. **44.** $x:y = 3:2$. **45.** $x:y = 2:7$.
**46.** $x:y = a:b$. **47.** $x:y = a^2:1$. **48.** $5:3 = 4:x$. **49.** $11:5 = 15:x$.
**50.** $a+2:2 = 5:x$. **51.** $10:3 = 2:x$. **52.** $1:18 = 3:x$. **53.** $3:2 = 3:x$.

**Page 126.** — **54.** $(a)$ $T:T' = b:b'$. $\qquad\qquad$ $(b)$ $C:C' = R:R'$.
$(c)$ $V:V' = P':P$. $\qquad$ $(d)$ $A:A' = R^2:R'^2$. $\qquad$ $(e)$ $m:m' = d':d$.
**55.** $(a)$ Directly. $\quad$ $(b)$ Inversely. $\quad$ $(c)$ Directly. $\quad$ $(d)$ Inversely.
$(e)$ Directly. **56.** 15 mi. **57.** $24\frac{1}{2}$ sq. in. **58.** 20 cu. ft.

**Page 127.** — **59.** 200 mi., $32\frac{1}{2}$+ mi., 174+ mi.

**Page 128.** — **1.** 8, 36. **2.** $13\frac{1}{2}$, $31\frac{1}{2}$. **3.** $6\frac{1}{2}$, $32\frac{1}{2}$. **4.** 9, 15.
**5.** $6\frac{2}{3}$, $3\frac{1}{3}$. **6.** 19.8 oz. copper, 2.2 oz. tin. **7.** 945 cu. ft.
**8.** 55,160,000 sq. mi. land, 141,840,000 sq. mi. water. **9.** $11\frac{1}{9}$ gms.
**10.** $\frac{10\,a}{a+b}$, $\frac{10\,b}{a+b}$. **11.** $\frac{20}{1+m}$, $\frac{20\,m}{1+m}$. **12.** $\frac{3\,a}{10}$, $\frac{7\,a}{10}$.
**13.** $\frac{mx}{x+y}$, $\frac{my}{x+y}$. **14.** $7\frac{4}{23}$, $7\frac{19}{23}$. **15.** $\frac{ac}{a+b}$, $\frac{bc}{a+b}$.

**Page 131.** — **1.** 30, 17. **2.** 7, 12. **3.** 2, 3. **4.** 1, 3.

**Page 132.** — **5.** 5, 4. **6.** 3, 3. **7.** $-7, 4$. **8.** $2, -3$. **9.** $4, -2$.
**10.** 7, 9. **11.** 5, 4. **12.** 5, 7. **13.** 9, 4. **14.** 5, 5. **15.** 5, $6\frac{2}{3}$.
**16.** 7, 5. **17.** 1, $1\frac{1}{2}$. **18.** 1, 15. **19.** 28, 22. **20.** $4\frac{1}{2}$, $\frac{1}{2}$. **21.** 19, 57.
**22.** 27, 20. **23.** 5, $-5$. **24.** 9, $6\frac{1}{2}$. **25.** 3, 4.

**Page 133.** — **1.** 14, 1. **2.** $-1, -2$. **3.** $2, -\frac{1}{2}$. **4.** 2, 5. **5.** 3, 3.
**6.** $2\frac{7}{8}$, $3\frac{1}{4}$. **7.** $4, -1$. **8.** $2, -3$. **9.** 4, 5. **10.** 2, 1. **11.** 2, 2.
**12.** 1, $-19$.

**Page 134.** — **1.** 2, 3. **2.** 9, 7. **3.** 2, 3.

**Page 135.** — **4.** 4, 5. **5.** $-3, 9$. **6.** 7, 4. **7.** 2, 3. **8.** 5, 7.
**9.** 7, 5. **10.** 16, 12. **11.** $4\frac{1}{2}$, 2. **12.** 36, 3. **13.** $6, -4$. **14.** 7, 46.
**15.** 11, 7. **16.** 17, 13. **17.** 6, 8. **18.** 4, 3.

**Page 136.** — **19.** 2, 1. **20.** 2, 3. **21.** 10, 5. **22.** 1, 1. **23.** 4, 6.
**24.** 2, 3. **25.** $-7, -7$. **26.** 2, 3.

**Page 137.** — **1.** 1, 2. **2.** 2, 3. **3.** 2, 5. **4.** $\frac{1}{2}$, $\frac{1}{3}$.

**Page 138.** — **5.** $2, -2$. **6.** 3, 4. **7.** $4, -5$. **8.** $1, -1$.
**9.** $-3, -2$. **10.** $\frac{1}{4}$, $\frac{1}{2}$. **11.** $\frac{1}{5}$, $\frac{1}{3}$. **12.** $\frac{1}{8}$, $\frac{1}{7}$. **13.** $\frac{1}{2}$, $\frac{1}{3}$. **14.** $\frac{1}{3}$, $\frac{1}{4}$

**Page 139.—1.** $\dfrac{m+n}{2}$, $\dfrac{m-n}{2}$.   **2.** $\dfrac{m-n}{a-b}$, $\dfrac{an-bm}{a-b}$.   **3.** 0, 1.

**4.** $\dfrac{a-b}{m-1}$, $\dfrac{mb-a}{m-1}$.   **5.** $\dfrac{n-m}{m+n}$, $\dfrac{2}{m+n}$.   **6.** $\dfrac{d-b}{ad-bc}$, $\dfrac{a-c}{ad-bc}$.

**7.** $\dfrac{ce-bf}{ae-bd}$, $\dfrac{af-cd}{ae-bd}$.   **8.** $b$, $a$.   **9.** $2\,b$, $3\,a$.   **10.** $\dfrac{cq}{p+q}$, $\dfrac{cp}{p+q}$.

**11.** $2\,a$, 1.   **12.** $a = l - (n-1)\,d$, $s = ln - \dfrac{n\,(n-1)\,d}{2}$.

**13.** $d = \dfrac{l-a}{n-1}$.   **14.** $s = \dfrac{l^2 - a^2 + ad + ld}{2\,d}$.

**Page 141.—1.** 1, 2, 3.   **2.** 1, 5, 6.   **3.** 4, 5, 6.   **4.** 3, 2, 1.   **5.** 2, 3, 4.
**6.** 5, 3, 1.   **7.** 1, 2, 3.   **8.** 2, 3, 4.   **9.** $-4, 4, 4$.   **10.** $2, -3, -\frac{1}{2}$.
**11.** 9, 7, 3.   **12.** 20, 6, 4.   **13.** $-3, 5, 7$.   **14.** 11, 13, 17.

**Page 142.—15.** $\frac{1}{2}, \frac{1}{3}, \frac{1}{4}$.   **16.** $9, 7, -7$.   **17.** 3, 5, 7.   **18.** 2, 3, 4.
**19.** 2, 3, 4.   **20.** $2, 3, -4$.   **21.** 11, 8, 7.   **22.** 8, 6, 2.   **23.** 11, 33, 55.
**24.** 24, 30, 40.   **25.** 5, 7, 1.   **26.** 1, 2, 3.   **27.** 18, 32, 10.   **28.** $-9, 72,$
$-90$.

**Page 143.—29.** 2, 2, 2.   **30.** 2, 3, 4.   **31.** $m+n-p$, $m-n+p$,
$-m+n+p$.

**Page 145.—1.** 3, 7.   **2.** 4, 3.   **3.** 6, 2.   **4.** $\frac{5}{12}$.   **5.** $\frac{4}{11}$.   **6.** $\frac{1}{5}$.   **7.** $\frac{2}{11}$.
**8.** 24.   **9.** 25.   **10.** 423.

**Page 146.—11.** A's 50 yrs., B's 40 yrs., C's 30 yrs.   **12.** A's 30 yrs.,
B's 15 yrs., C's 10 yrs.   **13.** $1000, $4000.   **14.** $500, $250.   **15.** $5000,
$3000, $2000.   **16.** $6500 at $3\frac{11}{13}$%.   **17.** $900 at 5%.   **18.** 6%, 5%.
**19.** 19 gms., $10\frac{1}{2}$ gms.

**Page 147.—20.** 2 horses, 6 cows, 16 sheep.   **21.** 100°, 60°, 20°.
**22.** 2, 4, 3.   **23.** 5, 4, 3.   **24.** 20°, 40°, 30°.   **25.** 3, 4 mi./hr.

**Page 149.—5.** About $12\frac{3}{4}$.   **6.** 5.   **8.** On a parallel to the $x$ axis.
**9.** On the $y$ axis.   **10.** On the $x$ axis.   **11.** A parallel to the $x$ axis
through point (0, 3).   **12.** The ordinate.   **13.** 0, 0.

**Page 151.—1.** (a) 12°; (b) $23\frac{1}{4}$°; (c) $-1\frac{1}{4}$°; (d) 5°.

**Page 152.—2.** (a) Apr. 1, Nov. 15; (b) May 20, Oct. 1; (c) Jan. 1,
Feb. 1; (d) Apr. 15, Nov. 5.   **3.** July 20, $23\frac{1}{4}$°.   **4.** Jan. 15, $-1\frac{1}{4}$°.
**5.** June, July, Aug., & part of Sept.   **6.** Jan. & part of Feb.   **7.** Jan. 15
to July 20.   **8.** Apr. 20 & Oct. 25.   **9.** 18°.   **10.** 4°.   **11.** Apr. & May.
**12.** Nov.   **13.** Jan.   **14.** Jan.   **15.** July.   **16.** 10°.   **17.** Apr. 20 to Oct. 25.

**Page 153.—18.** Nov. 15.

**Page 157.** — **21.** (*a*) 12.25; (*b*) 2.25; (*c*) 7.84; (*d*) $3\frac{1}{16}$; (*e*) 2.5; (*f*) 3.5; (*g*) 2.24; (*h*) 3.25. **22.** (*a*) 4.25, $-1.75$, $-1.75$; (*b*) 2, 3.73 and .27, 3.87 and .13; (*c*) $-2$; (*d*) 2; (*e*) 3.41 and .59; (*f*) 3.41 and .59; (*g*) 3 and 1; (*h*) 0 and 4. **23.** (*a*) 2.75, $-3.25$, 1.5; (*b*) 3.24; $-1.24$; (*c*) 2.73, $-.7$; (*d*) 2.73, $-.73$; (*e*) 2.4, $-4$.

**Page 158.** — **24.** (*b*) $-18\frac{1}{3}°$ C., $-13°$ C., $-10°$ C., 0°C.; (*c*) 14° F.; 32° F., 34° F.

**Page 159.** — **1.** 1.75. **2.** $-2.5$. **3.** 6. **4.** 2.67. **5.** 2. **6.** 3. **7.** $-1$. **8.** $1\frac{1}{5}$. **9.** 3, $-2$. **10.** 2.79, $-1.79$. **11.** 3.83, $-1.83$. **12.** (*a*) 3, 3; (*b*) 5. 83, .17; (*c*) 1, 5. **13.** (*a*) 5. 1, $-1.2$; (*b*) 5, $-1$. **14.** (*a*) 1.64, $-3.64$; (*b*) $-4$, 2.

**Page 163.** — **9.** 2, 1. **10.** 2, 3. **11.** 3, $-1$. **12.** $1\frac{1}{2}$, 1. **13.** $-1$, $-2$. **14.** 3, $1\frac{1}{2}$. **15.** 3, $-1$. **16.** 5, 5. **17.** $\frac{1}{2}$, 0. **18.** Inconsistent. **19.** 4, 3. **20.** Inconsistent. **21.** $\frac{3}{2}$, $\frac{5}{4}$. **22.** $4\frac{1}{4}$, $2\frac{5}{8}$. **23.** 3, 2.

**Page 164.** — **24.** $\frac{3}{2}$, 1. **25.** Indeterminate. **26.** 3, 2. **27.** 2, $\frac{5}{2}$. **28.** Indeterminate. **29.** $-2$, 5 and 2, $-3$. **30.** 3, 9 and $-2$, 4.

**Page 166.** — **1.** $x^{15}$. **2.** $x^8$. **3.** $-a^{10}$. **4.** $a^{11}b^{22}$. **5.** $a^{15}b^{10}c^{20}$. **6.** $-8\,m^3n^{12}$. **7.** $-125\,a^3$. **8.** $64\,x^{12}y^{12}z^{12}$. **9.** $-x^{18}y^{27}$. **10.** $16\,x^{12}y^8$. **11.** $81\,x^4y^8z^{12}$. **12.** $-a^{60}b^{185}c^{150}$. **13.** $121\,a^4b^6c^2$. **14.** $-27\,a^3x^{60}y^3$.

**15.** $\frac{1}{8}\,m^9n^9p^9$. **16.** $\frac{4}{9}\,x^4y^4z^2$. **17.** $\frac{16\,m^4}{81}$. **18.** $-\frac{8\,m^3}{27}$. **19.** $\frac{16\,x^4}{81\,y^4}$.

**20.** $\frac{121\,x^4y^4z^6}{25\,a^8}$. **21.** $\frac{a^{70}b^{80}c^{90}}{x^{100}y^{110}}$. **22.** $\frac{16\,m^4n^4}{81\,x^8y^4}$. **23.** $\frac{64\,a^6b^6c^6}{27\,x^3y^3}$.

**24.** $-\frac{343\,x^{30}}{125\,a^{12}b^9}$. **25.** $\frac{-1}{125\,x^9y^9z^9}$. **26.** $\frac{-1}{a^{121}}$. **27.** $-\frac{64}{a^{12}b^{27}}$.

**28.** $\frac{1}{a^{40}b^{44}}$. **29.** $a^{4m}$. **30.** $a^{6m}$. **31.** $\frac{64}{81}$.

**Page 167.** — **1.** $a^3 + 3\,a^2b + 3\,ab^2 + b^3$. **2.** $x^3 - 3\,x^2y + 3\,xy^2 - y^3$. **3.** $a^3 + 3\,a^2 + 3\,a + 1$. **4.** $m^3 - 6\,m^2 + 12\,m - 8$. **5.** $m^3 + 3\,m^2n + 3\,mn^2 + n^3$. **6.** $a^3 + 21\,a^2 + 147\,a + 343$. **7.** $125 - 75\,x + 15\,x^2 - x^3$. **8.** $1 + 6\,x + 12\,x^2 + 8\,x^3$. **9.** $27\,a^3 - 27\,a^2 + 9\,a - 1$. **10.** $1 + 12\,x + 48\,x^2 + 64\,x^3$. **11.** $343\,a^6 - 147\,a^4 + 21\,a^2 - 1$. **12.** $1 + 15\,a^3 + 75\,a^6 + 125\,a^9$. **13.** $27\,a^3 + 54\,a^2b + 36\,ab^2 + 8\,b^3$. **14.** $125\,m^3 + 150\,m^2n + 60\,mn^2 + 8\,n^3$. **15.** $27\,a^3 - 135\,a^2b + 225\,ab^2 - 125\,b^3$. **16.** $27\,a^6b^6 - 27\,a^4b^4 + 9\,a^2b^2 - 1$. **17.** $a^3m^3 + 3\,a^2m^2bn + 3\,amb^2n^2 + b^3n^3$. **18.** $64\,x^6 - 240\,x^4y^2 + 300\,x^2y^4 - 125\,y^6$. **19.** $a + b$. **20.** $x - y$. **21.** $a - 1$. **22.** $1 + m$. **23.** $2\,b - 1$.

**Page 168.** — **1.** $p^4 + 4\,p^3q + 6\,p^2q^2 + 4\,pq^3 + q^4$. **2.** $m^4 - 4\,m^3n + 6\,m^2n^2 - 4\,mn^3 + n^4$. **3.** $x^4 + 4\,x^3 + 6\,x^2 + 4\,x + 1$. **4.** $1 + 4\,y + 6\,y^2 + 4\,y^3 + y^4$. **5.** $m^4 - 4\,m^3 + 6\,m^2 - 4\,m + 1$. **6.** $1 - 4\,ab + 6\,a^2b^2 - 4\,a^3b^3 + a^4b^4$

7. $1+4\,x^2+6\,x^4+4\,x^6+x^8$.　　**8.** $a^5-5\,a^4b+10\,a^3b^2-10\,a^2b^3+5\,ab^4-b^5$.
9. $c^5+5\,c^4d+10\,c^3d^2+10\,c^2d^3+5\,cd^4+d^5$.　　**10.** $m^6-6\,m^5n+15\,m^4n^2-20\,m^3n^3+15\,m^2n^4-6\,mn^5+n^6$.　　**11.** $a^7+7\,a^6b+21\,a^5b^2+35\,a^4b^3+35\,a^3b^4+21\,a^2b^5+7\,ab^6+b^7$.
12. $a^5-5\,a^4+10\,a^3-10\,a^2+5\,a-1$.
13. $32+80\,a+80\,a^2+40\,a^3+10\,a^4+a^5$.　　**14.** $m^{12}+4\,m^9+6\,m^6+4\,m^3+1$.
15. $1-3\,m^2n^2+3\,m^4n^4-m^6n^6$.　　**16.** $m^{15}-5\,m^{12}+10\,m^9-10\,m^6+5\,m^3-1$.
17. $m^8+8\,m^7n+28\,m^6n^2+56\,m^5n^3+70\,m^4n^4+56\,m^3n^5+28\,m^2n^6+8\,mn^7+n^8$.
18. $m^5n^5+5\,m^4n^4c+10\,m^3n^3c^2+10\,m^2n^2c^3+5\,mnc^4+c^5$.　　**19.** $m^5n^5p^5-5\,m^4n^4p^4+10\,m^3n^3p^3-10\,m^2n^2p^2+5\,mnp-1$.　　**20.** $32\,m^{10}+80\,m^8+80\,m^6+40\,m^4+10\,m^2+1$.　　**21.** $81\,a^8+540\,a^6+1350\,a^4+1500\,a^2+625$.
22. $16\,x^8-160\,x^6+600\,x^4-1000\,x^2+625$.　　**23.** $16\,a^4-160\,a^3c+600\,a^2c^2-1000\,ac^3+625\,c^4$.　　**24.** $1+8\,x+24\,x^2+32\,x^3+16\,x^4$.
25. $1+5\,a^2b^2+10\,a^4b^4+10\,a^6b^6+5\,a^8b^8+a^{10}b^{10}$.

**Page 170.—1.** 2.　**2.** $\pm 9$.　**3.** $-5$.　**4.** $\pm 4$.　**5.** $\pm 35$.　**6.** 30.
**7.** $\pm 60$.　**8.** $-20$.　**9.** $\pm 180\,a$.　**10.** $\pm 90$.

**Page 171.—11.** $\pm x^2$.　**12.** $m^4$.　**13.** $\pm 25\,a$.　**14.** $49\,a$.　**15.** $\pm 10\,x$.
**16.** $-10$.　**17.** $3\,a^2$.　**18.** $\pm 2\,xy^2$.　**19.** $\pm a^2b^3$.　**20.** $\pm 2\,mn^2$.
**21.** $\dfrac{-2\,a}{b}$.　**22.** $\pm\dfrac{a^4}{3\,b}$.　**23.** 6.　**24.** $\dfrac{a^2b^7}{c^{10}}$.　**25.** $-x$.　**26.** $\pm(x+y)$.
**27.** $2(a+b)$.　**28.** $\pm(a+b)$.　**29.** 420.　**30.** 90.　**31.** 72.　**32.** 96.
**33.** 70.　**34.** 300.　**35.** 40.　**36.** 23.

**1.** $\pm(a+1)$.　**2.** $\pm(1-y)$.　**3.** $\pm(x-2\,y)$.　**4.** $\pm(3\,x+y)$.
**5.** $\pm(x^2-1)$.　**6.** $\pm(1+8\,x^2)$.

**Page 172.—7.** $\pm(2\,a-11\,b)$.　**8.** $\pm(2\,a+b)$.　**9.** $\pm(mn-7\,p)$.
**10.** $\pm(a^2-b^4)$.　**11.** $\pm(7\,a^4-3\,b^4)$.　**12.** $\pm(4\,a^2-9\,b^2)$.　**13.** $\pm(x+y+z)$.
**14.** $\pm(x+y+1)$.　**15.** $\pm(a-b+c)$.　**16.** $\pm(a+b-c)$.

**Page 174.—1.** $\pm(x^2-3\,x+2)$.　**2.** $\pm(a^2-a+1)$.　**3.** $\pm(x^2+x-1)$.
**4.** $\pm(4\,a^2+3\,a+9)$.　**5.** $\pm(5\,m^2-2\,m+3)$.　**6.** $\pm(2-3\,ab+7\,a^2b^2)$.
**7.** $\pm(5\,x^2+4\,xy+3\,y^2)$.　**8.** $\pm(4\,x^3+5\,x^2+6\,x)$.　**9.** $\pm(1+x+x^2)$.
**10.** $\pm(1+2\,x+3\,x^2+4\,x^3)$.　**11.** $\pm(6\,a^3+5\,a^2+4\,a)$.　**12.** $\pm(6\,x^3+3\,x^2y+5\,xy^2)$.　**13.** $\pm(2\,m^3+3\,m^2-5)$.　**14.** $\pm(7\,a^2-3\,ab+2\,b^2)$.
**15.** $\pm(x^3+2\,x^2-2\,x+4)$.　**16.** $\pm(2\,x^3-3\,x^2+x-2)$.　**17.** $\pm(x^2+xy-y^2)$.　**18.** $\pm(27+3\,a^2-a^3)$.　**19.** $\pm(6\,a^6+5\,a^4+4\,a^2)$.
**20.** $\pm(5\,a^3+4\,a^2+3\,a+2)$.　**21.** $\pm\left(\dfrac{x}{3}+\dfrac{y}{4}+\dfrac{z}{5}\right)$.　**22.** $\pm\left(4-\dfrac{3}{x}+\dfrac{2}{x^2}\right)$.
**23.** $\pm\left(1+\dfrac{2}{x}+\dfrac{3}{x^2}+\dfrac{4}{x^3}\right)$.　　**24.** $\pm\left(\dfrac{6}{x}+3+5\,x\right)$.

**Page 176.—1.** 75.　**2.** 64.　**3.** 57.　**4.** 71.　**5.** 84.　**6.** 98.　**7.** 99.
**8.** 119.　**9.** 101.　**10.** 237.　**11.** 309.　**12.** 247.　**13.** 763.　**14.** 978.
**15.** 2.83.　**16.** 6.5.　**17.** 8.5.　**18.** .94.　**19.** .037.　**20.** 1247.　**21.** 2038.

**22.** 7563. **23.** 5083. **24.** 6561. **25.** 15,367. **26.** $6\frac{1}{4}$. **27.** $1\frac{1}{5}$. **28.** $\frac{25}{64}$.

**Page 177.**—**29.** 2.236. **30.** 3.60. **31.** .469. **32.** 1.237. **33.** 1.005. **34.** .935. **35.** .645. **36.** .243. **37.** 7.522 ft. **38.** 9.798 yds. **39.** 3.925 ft. **40.** 4.690.

**Page 179.**—**1.** $\pm 13$. **2.** $\pm .5$. **3.** $\pm 17$. **4.** $\pm 5$. **5.** $\pm 2$. **6.** $\pm 1$. **7.** $\pm 3$. **8.** $\pm 4$. **9.** $\pm\sqrt{2}$. **10.** $\pm\sqrt{18}$. **11.** $\pm 5$. **12.** $\pm\frac{3}{4}$. **13.** $\pm\frac{3}{5}$. **14.** $\pm 5$. **15.** $\pm 5$. **16.** $\pm\frac{1}{3}$. **17.** $\pm\sqrt{\frac{b}{a}}$. **18.** $\pm\sqrt{ab}$. **19.** $\pm(a+b)$. **20.** $\pm\sqrt{ab}$.

**Page 180.**—**21.** $\pm\sqrt{\frac{a}{c}}$. **22.** $\pm(a+1)$. **23.** $\pm\sqrt{c^2-b^2}$. **24.** $\pm\sqrt{\frac{2s}{g}}$. **25.** $\pm\sqrt{\frac{s}{\pi}}$. **26.** $\pm\sqrt{\frac{s}{4\pi}}$. **27.** $\pm\sqrt{\frac{2a^2+2b^2-c^2}{4}}$. **28.** $\pm\sqrt{\frac{2E}{m}}$. **29.** $\pm\sqrt{\frac{mm'}{g}}$.

**1.** 1. **2.** $\pm 15$. **3.** $\pm 10, \pm 15$. **4.** $\pm 2, \pm 4, \pm 6$. **5.** 9 ft., 15 ft. **6.** 21 yds., 6 yds.

**Page 181.**—**7.** 21 in., 28 in. **8.** 39 in., 36 in., 270 sq. in. **9.** $\sqrt{2}$. **10.** 6, 8. **11.** 2 sec., $2\frac{1}{2}$ sec. **12.** (a) 7 in.; (b) $\sqrt{14}$ or 3.742 ft. **13.** 21 ft., 28 ft. **14.** $\sqrt{35}$ or 5.916 yds.

**Page 183.**—**1.** 10, −2. **2.** −7, 5. **3.** 3, −9. **4.** 7, −1. **5.** 12, −5. **6.** 7, 4. **7.** $1\pm\sqrt{3}$. **8.** $6\pm\sqrt{21}$. **9.** −7, −10. **10.** $6\frac{1}{2}, \frac{1}{2}$. **11.** 8, $-2\frac{1}{4}$. **12.** 5, $-4\frac{1}{3}$. **13.** 1, $-\frac{4}{3}$. **14.** $\frac{-1\pm\sqrt{5}}{2}$. **15.** $\frac{3}{2}, -\frac{1}{2}$. **16.** $\frac{5}{3}, -\frac{1}{4}$. **17.** $7\frac{1}{2}, -2$. **18.** 11, 1. **19.** 7, −3. **20.** $6\pm\sqrt{21}$. **21.** $-\frac{5}{12}, -6$. **22.** 7, −12. **23.** 11, −3. **24.** 2, $-\frac{23}{33}$. **25.** 14, 6. **26.** 7, $-\frac{3}{2}$. **27.** 3, $\frac{1}{3}$. **28.** 3, −4. **29.** 4, −7.

**Page 184.**—**30.** 7, $-4\frac{1}{3}$. **31.** 3, −4. **32.** 12, $-\frac{3}{4}$. **33.** 14, −10. **34.** 3, −1. **35.** 9, −1. **36.** −4, 13. **37.** $1\pm\sqrt{13}$. **38.** 9, $1\frac{2}{13}$. **39.** 10, 18. **40.** 4, $\frac{1}{4}$. **41.** 3, 6. **42.** $5\frac{2}{3}, 5$. **43.** 5, $-3\frac{1}{3}$. **44.** $-6\frac{3}{7}, 2$. **45.** 18, 6. **46.** 2, $1\frac{5}{24}$. **47.** $1\pm\sqrt{17}$. **48.** $-a, 4a$. **49.** $7m, m$. **50.** $a, b$.

**Page 185.**—**1.** 2, $\frac{1}{2}$. **2.** 2, $\frac{4}{3}$. **3.** 3, $\frac{5}{2}$. **4.** −5, $3\frac{1}{2}$. **5.** −5, $-\frac{3}{2}$. **6.** $\frac{1}{5}, -\frac{1}{5}$. **7.** 4, −16. **8.** 10, 1. **9.** 12, 5. **10.** 5, $\frac{7}{3}$. **11.** $\frac{7}{3}, \frac{3}{5}$. **12.** $\frac{3\pm\sqrt{7}}{5}$. **13.** $\frac{3}{2}, -\frac{5}{3}$. **14.** $\frac{8}{7}, -4$. **15.** $\frac{5}{3}, -\frac{7}{2}$. **16.** $-\frac{39}{7}, 3$. **17.** $7m, -m$. **18.** $4n, -16n$. **19.** $-1\pm\sqrt{-3}$. **20.** $-n, -\frac{1}{n}$. **21.** 1, $a+b-1$.

**Page 186.—22.** .62, $-1.62$. **23.** 1.37, $-1.70$. **24.** 9.48, $-1.48$
**25.** 1.23, $-3.23$. **26.** 3.41, .59. **27.** 2.74, $-3.11$. **28.** .231, $-.088$.

**Page 187.—1.** $-6, -1$. **2.** 3, 7. **3.** 6, 2. **4.** $-2, 12$.
**5.** $-6, -4$. **6.** 2, $-12$. **7.** 6, 4. **8.** $-1, -7$. **9.** $\frac{1}{2}, -5$.

**10.** 7, $1\frac{1}{3}$. **11.** 0, $-4, -5$. **12.** 0, $-4, -\frac{1}{2}$. **13.** $\dfrac{1 \pm \sqrt{61}}{6}$.

**14.** 0, 3, $\frac{2}{3}$. **15.** 7, 2. **16.** 0, $-2, 1\frac{1}{3}$.

**Page 188.—17.** 0, $-3, \frac{2}{3}$. **18.** 1, $-2$. **19.** 0, 3, $-2$. **20.** $a, b - a$.
**21.** $-1, 3$. **22.** 6, $-2$. **23.** 0, $-1, 2$. **24.** 0, $\pm \sqrt{7}$. **25.** 0, 2, 3.
**26.** 0, $a + b + c$. **27.** 1, 2, $-3$. **28.** $3\frac{2}{3}$, 7. **29.** $6 \pm \sqrt{-64}$.
**30.** $\frac{1}{2}, -1$. **31.** 28, $-3$. **32.** 6, 3. **33.** $-4, -21$. **34.** $-3, 2$.
**35.** 2, $-3\frac{1}{2}$. **36.** $-1\frac{2}{3}$, 2. **37.** 4, 7. **38.** 4, 7. **39.** 9, $3\frac{3}{5}$.
**40.** $\frac{3}{2}$. **41.** $\frac{4}{3}$, 3. **42.** $-\frac{1}{2}$, 1. **43.** 0, $\dfrac{c - b}{a}$. **44.** $a + b, \dfrac{a + b}{2}$.
**45.** $\pm 3\,a$. **46.** $a + b \pm \sqrt{a^2 - ab + b^2}$. **47.** $-a, b$. **48.** $a - 2\,b$,
$b - 2\,a$. **49.** $-5, 2, -2$. **50.** 3, $\pm 3$.

**Page 189.—51.** $x^2 - 4x + 3 = 0$. **52.** $x^2 + x - 12 = 0$.
**53.** $x^2 + 7x + 10 = 0$. **54.** $x^2 - 9x = 0$. **55.** $x^3 - 2x^2 - 5x + 6 = 0$.
**56.** $x^3 - x^2 - 6x = 0$. **57.** $x^3 - 6x^2 + 11x - 6 = 0$. **58.** $x^3 - 4x = 0$.
**1.** 6 or $\frac{1}{2}$. **2.** 25, 35. **3.** 2, 6. **4.** 12, 24. **5.** 6, 7. **6.** 14, 15.
**7.** $\frac{1}{4}$ or $\frac{2}{3}$. **8.** 4, 10. **9.** 10 or 19. **10.** 10 in., 19 in. **11.** 70 ft., 120 ft.

**Page 190.—12.** $AB = 204$ ft., $AD = 85$ ft. **13.** $8\sqrt{2}$ in., $6\sqrt{2}$ in.
**14.** \$40 or \$60. **15.** \$30 or \$70. **16.** \$80. **17.** 10 mi./hr.,
$10\frac{1}{2}$ mi./hr. **18.** 20 mi./hr. **19.** 8 or 12 mi./hr. **20.** 5 ¢. **21.** \$120.

**Page 191.—22.** $AB = 26$, $BC = 3$. **23.** 5 ft. **24.** 15 ft. **25.** 4 da.
**26.** $2\sqrt{3}$ in. **27.** 20 eggs.

**Page 192.—1.** $\pm 3, \pm 1$. **2.** $\pm 3, \pm 2$. **3.** $\pm 2, \pm 1$. **4.** $\pm \sqrt{-4}$,
$\pm 5$. **5.** $\pm 2, \pm \sqrt{-2}$. **6.** 2, $-1, -1 \pm \sqrt{-3}, \dfrac{1 \pm \sqrt{-3}}{2}$. **7.** $\pm \sqrt{11}$,
$\pm \sqrt{1\frac{1}{3}}$. **8.** $\pm \dfrac{a}{2}, \pm \dfrac{3\,a}{2}$. **9.** $\pm \frac{1}{2}, \pm 3$. **10.** 2, $-3, 3, -4$.
**11.** $-1, 2, -2, 3$. **12.** $\pm 1, -2, -4$. **13.** 0, 1, $1 \pm \sqrt{2}$.
**14.** 1, 1, $\dfrac{1 \pm \sqrt{-15}}{4}$. **15.** $\pm \sqrt{-4}, \pm \sqrt{2}$. **16.** 1, 2, 4, 5. **17.** 1, 2, 2, 3.
**18.** 0, 1, $-1, -2$. **19.** $-1, \pm 2, -5$. **20.** $-2, -1, 0, 1$.

**Page 194.—1.** Real, unequal, rational. **2.** Real, unequal, rational.
**3.** Real, unequal, irrational. **4.** Imaginary, unequal. **5.** Real,
equal, rational. **6.** Imaginary, unequal. **7.** Real, equal, rational.
**8.** Imaginary, unequal. **9.** Real, unequal, irrational. **10.** Real,
unequal, rational, **11.** Imaginary, unequal. **12.** Real, equal,

rational. **13.** 5, 2. **14.** 9, − 3. **15.** 2, − $\frac{5}{2}$. **16.** − 3, − $\frac{2}{3}$.
**17.** $m$, $p$. **18.** $\frac{1}{3}$, $\frac{1}{5}$. **19.** 4, 15. **20.** − 1 ± $\sqrt{3}$. **21.** − 5, 3.
**22.** − 2, 6. **23.** − 1 ± $\sqrt{-1}$. **24.** $\frac{7}{2} \pm \frac{1}{2}\sqrt{37}$.

**Page 196.** — **1.** ± 3. **2.** ± 2. **3.** ± 4. **4.** 2. **5.** 3. **6.** 4.
**7.** 2. **8.** 4. **9.** 8. **10.** 9. **11.** 5. **12.** 4. **13.** 1. **14.** − 2.
**15.** ± 8. **16.** 0. **17.** ± $\frac{1}{3}$. **18.** ± $\frac{1}{8}$. **19.** $\frac{1}{4}$. **20.** ±$(m - n)$.
**21.** $x + y$. **22.** $\sqrt[3]{m^7}$. **23.** $\sqrt[4]{x}$. **24.** $\sqrt[3]{a^2}$. **25.** $\sqrt{a^i}$. **26.** $\sqrt[5]{xy}$.
**27.** $\sqrt[3]{3}$. **28.** $\sqrt[5]{b^4c^4d^4}$. **29.** $\sqrt[3]{a^m}$. **30.** $\sqrt{bx}$. **31.** $\sqrt[5]{m^a}$.

**Page 197.** — **32.** $x^{\frac{2}{3}}$. **33.** $a^{\frac{1}{5}}b^{\frac{1}{5}}c^{\frac{1}{5}}$. **34.** $y^{\frac{3}{8}}$. **35.** $m^{\frac{1}{3}}$. **36.** $a^{\frac{1}{2}}$.
**37.** $x^{\frac{1}{4}}y^{\frac{1}{4}}$. **38.** $m^{\frac{1}{6}}n^{\frac{1}{5}}$. **39.** 2. **40.** $\frac{1}{2}$. **41.** 9. **42.** 8. **43.** 3.
**44.** $\frac{1}{3}$. **45.** 16. **46.** 49. **47.** 20. **48.** 10. **49.** 29. **50.** $\frac{1}{32}$.

**Page 199.** — **1.** $\frac{1}{49}$. **2.** $\frac{1}{5}$. **3.** $\frac{1}{27}$. **4.** 1. **5.** 1. **6.** $\frac{1}{32}$.
**7.** 1. **8.** $\frac{1}{225}$. **9.** $\frac{1}{217}$. **10.** 4. **11.** 8. **12.** $\frac{2}{3}$. **13.** $\frac{16}{9}$.
**14.** ± $\frac{1}{2}$. **15.** $\frac{1}{4}$. **16.** ± $\frac{1}{8}$. **17.** 32. **18.** 49. **19.** ± $\frac{1}{4}$.
**20.** − 4$\frac{1}{3}$. **21.** $\dfrac{1}{x^5}$. **22.** $\dfrac{6}{x^2}$. **23.** $\dfrac{9\,c^5}{a^2b}$. **24.** $\dfrac{a^2b^2}{7}$. **25.** $\dfrac{3\,c^2x^5}{b^3}$.
**26.** $\dfrac{b^8}{25\,a^2}$. **27.** $\dfrac{30\,a^2z^2}{b^3c^4xy}$. **28.** $\dfrac{180\,x^4}{a^8b^7}$. **29.** $6\,ab^{-1}$. **30.** $5\,a^2b^2c^{-2}$.
**31.** $a^3b^3c^3x^{-3}y^{-3}$. **32.** $abxy^2$. **33.** $m^{-2}x^{\frac{1}{2}}n^2$. **34.** $7^{-1}m^{-1}n^{-1}p^{-1}$.
**35.** $\sqrt{m}$. **36.** $\dfrac{1}{\sqrt{m}}$. **37.** $3\sqrt[3]{x}$. **38.** $\dfrac{3}{\sqrt[3]{x}}$. **39.** $\dfrac{2}{\sqrt[5]{m}}$. **40.** $\dfrac{1}{\sqrt[5]{2\,m}}$.
**41.** $\dfrac{\sqrt{x}}{2}$. **42.** $\sqrt{m^n}$. **43.** $\dfrac{1}{\sqrt[3]{n^a}}$. **44.** $\dfrac{\sqrt{abc}}{6}$. **45.** $\sqrt[n]{m}$.

**Page 200.** — **46.** 5. **47.** 2. **48.** − 1. **49.** − 2. **50.** 0.
**51.** 125. **52.** 16. **53.** − 3. **54.** − 1. **55.** − 3. **56.** − 3.
**57.** 33. **58.** − 1. **59.** 25. **60.** − 17. **61.** 6.

**Page 201.** — **1.** 8. **2.** 3. **3.** 1. **4.** $\frac{1}{4}$. **5.** $a$. **6.** 12$x^2$.
**7.** 30$a$. **8.** 5. **9.** 7$\sqrt{7}$. **10.** $\sqrt[3]{x^5}$. **11.** 5$\sqrt[12]{5}$. **12.** 243.
**13.** $\frac{1}{125}$. **14.** 3. **15.** $a^{18}$. **16.** 2$\sqrt{a}$. **17.** $\frac{4}{3}x^{\frac{7}{2}} - \frac{1}{2}x^{\frac{5}{2}}$. **18.** $a^2$.
**19.** 11. **20.** $\sqrt[3]{x}$. **21.** $\sqrt[3]{25}$. **22.** $\sqrt{3}$. **23.** $\sqrt[3]{49}$. **24.** $\sqrt{a}$.
**25.** $x^4$. **26.** $\sqrt[5]{m^4}$. **27.** $x$. **28.** $m$. **29.** 1. **30.** $\dfrac{2}{n\sqrt[15]{n^4}}$. **31.** $xy\sqrt[6]{z}$.

**Page 202.** — **32.** $\dfrac{1}{m}$. **33.** $\sqrt[8]{x}$. **34.** $\sqrt[12]{a}$. **35.** $\dfrac{5\,x}{4\,y}$. **36.** $\dfrac{b^3}{27\,a^3}$.
**37.** 1. **38.** $\sqrt[12]{\dfrac{a^7}{x^7}}$. **39.** $\dfrac{1}{a^4b^{17}}$. **40.** 1.

**Page 203.** — 1. $a^2 - a\sqrt{b} - 2b$.　　2. $x - \sqrt{xy} - 6y$.　　3. $\dfrac{16}{r} + \dfrac{34s}{r} - 15s^2$.　　4. $63r^{\frac{3}{2}} - 121r + 101\sqrt{r} - 35$.　　5. $a^4 + a^{-4} + 1$.

6. $x + 2\sqrt{xy} + y - 1$.　　7. $x - 1$.　　8. $a^{2m} + a^{-2m} + 1$.　　9. $4x^2 - 13x^{\frac{3}{2}} + 40x - 12x^{\frac{1}{2}} + 9$.　　10. $x^{\frac{2}{3}} - x^{\frac{1}{3}}y^{\frac{1}{3}} + y^{\frac{2}{3}}$.　　11. $4a^{-2} + 9a^{-1} + 3$.

12. $7p + 8\sqrt{p} + 1$.　　13. $a + 2a^{\frac{1}{2}}b^{\frac{1}{2}} + b$.　　14. $3a^{-3} - 5a^{-2}b^{-1} + 2a^{-1}b^{-2}$.　　15. $\sqrt[3]{a} - \sqrt[3]{b}$.　　16. $(\sqrt{a} - \sqrt{b} + \sqrt{c})$.　　17. $\pm(x - 3 + 2x^{-1})$.

18. $\pm(x^{-1} - 1 + x)$.　　19. $\pm(5x^{-1} - 2 + 3x)$.　　20. $\pm(x^{\frac{2}{3}} - 3x^{\frac{1}{3}} + 2)$.

21. $\pm(1 + \sqrt{x} + x)$.　　22. $1 + 2\sqrt[3]{x} + 3\sqrt[3]{x^2} + 4x$.　　23. $\sqrt{2} + 2$.

24. $-2$.　　25. $19 - 5\sqrt{3}$.　　26. $-13 - 5\sqrt{5}$.　　27. $17 - 4\sqrt{22}$.

**Page 204.** — 28. $x + 5x^{\frac{1}{2}} + 6$.　　29. $x - 2x^{\frac{1}{2}} - 15$.　　30. $x - y$.

31. $2$.　　32. $5 - 2\sqrt{6}$.　　33. $x - 25$.　　34. $x^{\frac{2}{3}} + 2x^{\frac{1}{3}} + 1$.

35. $x^{\frac{3}{2}} - 3xy^{\frac{1}{2}} + 3x^{\frac{1}{2}}y - y^{\frac{3}{2}}$.　　36. $7 - 2\sqrt{10}$.　　37. $13 + 2\sqrt{22}$.

38. $\pm(\sqrt{x} + \sqrt{y})$.　　39. $\pm(a^{\frac{1}{2}} + 2)$.　　40. $m^{\frac{1}{2}} - n^{\frac{1}{2}}$.

**Page 207.** — 1. $3\sqrt{3}$.　　2. $3\sqrt{5}$.　　3. $4\sqrt{2}$.　　4. $2\sqrt{7}$.　　5. $2\sqrt{6}$.

6. $9\sqrt{3}$.　　7. $11\sqrt{3}$.　　8. $x\sqrt{x}$.　　9. $a^3\sqrt{a}$.　　10. $ab\sqrt{ab}$.　　11. $2a\sqrt{7}$.

12. $8ab\sqrt{5}$.　　13. $2ab\sqrt{2a}$.　　14. $20b\sqrt{5}$.　　15. $10a\sqrt{10c}$.

16. $28ax\sqrt{3x}$.　　17. $\dfrac{5c^3}{2}\sqrt{5}$.　　18. $2\sqrt[3]{2}$.　　19. $3\sqrt[3]{2}$.　　20. $10a\sqrt[3]{ab}$.

21. $5a^2b^2\sqrt[3]{2a}$.　　22. $2ab\sqrt[3]{6}$.　　23. $-3x^2y^2\sqrt[3]{4y}$.　　24. $(a+b)\sqrt{2}$.

25. $(a+b)\sqrt{3}$.　　26. $5xy\sqrt{5y}$.　　27. $\frac{1}{2}\sqrt{2}$.　　28. $\frac{1}{5}\sqrt{5}$.　　29. $\frac{1}{3}\sqrt{3r}$.

30. $\dfrac{ab}{7}\sqrt{35}$.　　31. $\dfrac{ab}{c}\sqrt{2c}$.　　32. $\sqrt{ax}$.　　33. $n\sqrt{m}$.　　34. $a\sqrt{2t}$.

35. $\frac{1}{2}\sqrt[3]{4}$.　　36. $\frac{a}{2}\sqrt[3]{6}$.

**Page 208.** — 37. $\dfrac{ab^2}{2c}\sqrt[3]{9bc}$.　　38. $3\sqrt[3]{4m^2n}$.　　39. $\sqrt{x^2 - y^2}$.　　40. $b\sqrt[3]{2b}$.

41. $3\sqrt{-1}$.　　42. $4a\sqrt{-1}$.　　43. $\dfrac{15ab}{c}\sqrt{-1}$.　　44. $5xy\sqrt{-xy}$.

45. $3ab\sqrt{-2}$.　　46. $\sqrt{-5a}$.　　47. $.577$.　　48. $.707$.　　49. $.632$.

50. $.548$.　　51. $.592$.

1. $\sqrt{80}$.　　2. $\sqrt{63}$.　　3. $\sqrt[3]{88}$.　　4. $\sqrt[3]{135}$.　　5. $\sqrt{\frac{3}{2}}$.　　6. $\sqrt{a^2b}$.

7. $\sqrt{\dfrac{112a}{5}}$.　　8. $\sqrt[3]{\dfrac{81ab}{2}}$.　　9. $\sqrt{\dfrac{b}{a}}$.　　10. $\sqrt[3]{\dfrac{2m^3}{n}}$.　　11. $\sqrt[3]{4z}$.

**Page 209.—1.** $\sqrt[6]{x^3}$.  **2.** $\sqrt[6]{m^2n^2}$.  **3.** $\sqrt[6]{\dfrac{x^3y^3}{z^3}}$.  **4.** $\sqrt[6]{a}$.  **5.** $\sqrt[6]{\dfrac{8\,x^3y^3}{27}}$.

**6.** $\sqrt[6]{m^6n^6}$.  **7.** $\sqrt[12]{64\,a^6}$.  **8.** $\sqrt[12]{m^8n^4}$.  **9.** $\sqrt[12]{a^{16}b^8c^4}$.  **10.** $\sqrt[12]{27\,a^3x^3}$.

**11.** $\sqrt[12]{a^{-10}b^2}$.  **12.** $\sqrt[12]{25\,x^4z^{14}}$.  **13.** $\sqrt[12]{abc}$.  **14.** $\sqrt[12]{a^{12}}$.  **15.** $\sqrt[3]{a}$.

**16.** $\sqrt{abc}$.  **17.** $\sqrt{2}\,a$.  **18.** $ab^2\sqrt{ac}$.  **19.** $\dfrac{a^5}{b^2}$.  **20.** $\dfrac{n}{2}\sqrt{2\,m}$.

**21.** $\sqrt[3]{2\,m^2}$.  **22.** $3\,n\sqrt{m}$.  **23.** $\dfrac{(a+b)^2}{c^4}$.  **24.** $\sqrt{8}$.

**Page 210.—25.** $\sqrt[6]{27}$, $\sqrt[6]{4}$.  **26.** $\sqrt[12]{8}$, $\sqrt[12]{81}$.  **27.** $\sqrt[12]{27}$, $\sqrt[12]{16}$.

**28.** $\sqrt[10]{49}$, $\sqrt[10]{32}$.  **29.** $\sqrt[14]{128}$, $\sqrt[14]{9}$.  **30.** $\sqrt[6]{8}$, $\sqrt[6]{9}$, $\sqrt[6]{5}$.  **31.** $\sqrt[12]{81}$, $\sqrt[12]{125}$, $\sqrt[12]{49}$.

**32.** $\sqrt[30]{a^{30}}$, $\sqrt[30]{a^5}$, $\sqrt[30]{a^3b^3}$.  **33.** $\sqrt[6]{27}$, $\sqrt[6]{9}$, $\sqrt[6]{8}$.  **34.** $\sqrt[12]{16}$, $\sqrt[12]{16}$, $\sqrt[12]{8000}$.

**35.** $\sqrt[3]{3}$, $\sqrt{2}$.  **36.** $\sqrt[3]{4}$, $\sqrt[4]{5}$.  **37.** $\sqrt{5}$, $\sqrt[3]{7}$.  **38.** $\sqrt[6]{126}$, $\sqrt{5}$, $\sqrt[3]{11}$.

**39.** $5\sqrt{2}$, $4\sqrt[3]{4}$.  **40.** $\sqrt[3]{2}$, $\sqrt[15]{30}$, $\sqrt[5]{3}$.

**Page 211.—1.** $4\sqrt{6}$.  **2.** $8\sqrt{2}$.  **3.** $-13\sqrt{3}$.  **4.** $0$.  **5.** $2\sqrt{7}$.

**6.** $6\sqrt{2}$.  **7.** $\sqrt{5}$.  **8.** $3\sqrt{2}$.  **9.** $\tfrac{29}{2}\sqrt{3}$.  **10.** $\tfrac{28}{15}\sqrt{15}$.  **11.** $8\sqrt[3]{2}$.

**12.** $8\sqrt{7}-\sqrt[3]{3}$.  **13.** $a\sqrt{5\,c}$.  **14.** $ab\sqrt{ab}$.  **15.** $\sqrt{3\,m}$.  **16.** $\dfrac{2}{x}\sqrt{21}\,x$.

**17.** $\tfrac{2}{3}\sqrt{10}$.  **18.** $\tfrac{1}{8}\sqrt{3}$.  **19.** $a\sqrt{x}$.  **20.** $\sqrt{1+x^2}$.

**Page 212.—21.** $\dfrac{(x^2-x+1)\sqrt{x}}{x^3}$.  **22.** $\dfrac{b-a+ab\sqrt{11\,ab}}{ab}$.

**23.** $5\,a\sqrt{2}-\sqrt[3]{a^2}$.  **24.** $0$.  **25.** $-\dfrac{1}{2\,b}\sqrt[3]{4\,a^2b^2x}$.

**Page 213.—1.** $6$.  **2.** $10$.  **3.** $3\sqrt{2}$.  **4.** $5\sqrt{2}$.  **5.** $7\sqrt{6}$.

**6.** $5\sqrt{6}$.  **7.** $7\sqrt{10}$.  **8.** $10\sqrt{6}$.  **9.** $2$.  **10.** $3$.  **11.** $3\sqrt[3]{2}$.  **12.** $5\sqrt[3]{2}$.

**13.** $\sqrt{ax}$.  **14.** $a\sqrt{5}$.  **15.** $2\,y\sqrt{2\,y}$.  **16.** $2\,abx^2$.  **17.** $4\,a^2$.

**18.** $14\,c^4\sqrt{5}$.  **19.** $\sqrt[x]{abc}$.  **20.** $\tfrac{1}{3}\sqrt{3\,ax}$.  **21.** $\dfrac{n}{2}$.  **22.** $\dfrac{7\,a}{20}\sqrt{3}$.

**23.** $\dfrac{5\,a}{3\,b}$.  **24.** $2\,a^2b^3\sqrt[5]{b}$.  **25.** $\sqrt{6}+3+2\sqrt{3}$.  **26.** $3\sqrt{15}+20$.

**27.** $5\sqrt{6}-6-6\sqrt{15}$.  **28.** $1-\sqrt{5}$.  **29.** $a^2-b$.  **30.** $m-n$.

**31.** $5\,x-2\,y$.  **32.** $32\,m-27\,n$.  **33.** $1$.  **34.** $b$.  **35.** $n$.

**36.** $45$.  **37.** $-3$.  **38.** $m+n-2\sqrt{mn}$.  **39.** $5-2\sqrt{6}$.

**40.** $6+2\sqrt{5}$.  **41.** $7-4\sqrt{3}$.

**Page 214.—42.** $120-30\sqrt{15}$.  **43.** $\dfrac{(a-b)^2}{ab}$.  **44.** $8-\sqrt{15}$.

**45.** $3\sqrt{15}-4$.  **46.** $98-39\sqrt{14}$.  **47.** $9\sqrt{10}+5\sqrt{2}$.  **48.** $2\sqrt{2}$.

**49.** $3\sqrt{15}-6$.  **50.** $2+\sqrt{10}-\sqrt{6}$.  **51.** $4\sqrt{3}$.  **52.** $\sqrt[6]{a^5}$.  **53.** $a\sqrt[6]{a}$.

**Page 216.—1.** $\sqrt{2}$.  **2.** $\sqrt{3}$.  **3.** $\sqrt{3}$.  **4.** $2$.  **5.** $\sqrt{n}$.  **6.** $3$.

**7.** $2\sqrt{3}$. **8.** $\frac{5}{4}$. **9.** $\frac{1}{2}\sqrt{2}$. **10.** $\sqrt{3}$. **11.** $\sqrt{a}$. **12.** $p^{\frac{2}{3}}$. **13.** $n\sqrt{11}$.

**14.** $\frac{1}{2}\sqrt{5}$. **15.** $\frac{1}{2}\sqrt{3}$. **16.** $\sqrt{a+b}$. **17.** $\frac{5a}{4}\sqrt{2}$. **18.** $\frac{3\sqrt{2}}{5}$. **19.** $.7071$.

**20.** $1.732$. **21.** $.3535$. **22.** $4.4722$. **23.** $1.0606$. **24.** $1.1547$. **25.** $.2828$.

**Page 217.** — **1.** $2-\sqrt{3}$. **2.** $3\sqrt{2}-3$. **3.** $15+3\sqrt{21}$. **4.** $\frac{1}{2}(\sqrt{2}+1)$.

**5.** $\sqrt{15}+\frac{3}{2}\sqrt{6}$. **6.** $\sqrt{3}-\sqrt{2}$. **7.** $\frac{1}{5}(\sqrt{7}+\sqrt{2})$. **8.** $4\sqrt{3}+6$.

**9.** $8+5\sqrt{2}$. **10.** $3(7+3\sqrt{5})$. **11.** $\frac{1}{2}(4+3\sqrt{2})$. **12.** $6\sqrt{3}-13$.

**Page 218.** — **13.** $\frac{3\sqrt{5}-9}{4}$. **14.** $5+2\sqrt{6}$. **15.** $\frac{18+5\sqrt{10}}{2}$.

**16.** $6\sqrt{35}+12\sqrt{7}-3\sqrt{15}-6\sqrt{3}$. **17.** $\sqrt{35}$. **18.** $\frac{(m+\sqrt{m})}{m-1}$.

**19.** $\frac{1+\sqrt{x}}{1-x}$. **20.** $\frac{x-2\sqrt{xy}+y}{x-y}$. **21.** $\sqrt{ab}$. **22.** $\frac{\sqrt{ab}-\sqrt{ac}}{b-c}$.

**23.** $2.4142$. **24.** $.732$. **25.** $9.7083$. **26.** $.691$. **27.** $6.464$.

**28.** $5.5536$. **29.** $1.1805$. **30.** $-26.389$. **31.** $7+5\sqrt{2}$.

**Page 219.** — **1.** $9mn$. **2.** $2x\sqrt[3]{2x}$. **3.** $2x\sqrt{2x}$. **4.** $125$. **5.** $512$.

**6.** $4$. **7.** $8$. **8.** $25$. **9.** $15,625\,n^8$. **10.** $\sqrt{abc}$. **11.** $\sqrt[3]{a}$. **12.** $\sqrt[3]{5n}$.

**Page 220.** — **1.** $\pm(\sqrt{5}+\sqrt{3})$. **2.** $\pm(\sqrt{2}-1)$. **3.** $\pm(\sqrt{5}-1)$.

**4.** $\pm(3+\sqrt{2})$. **5.** $\pm(\sqrt{3}+\sqrt{2})$. **6.** $\pm(\sqrt{5}-\sqrt{2})$. **7.** $\pm\frac{1}{2}(\sqrt{6}+\sqrt{2})$.

**8.** $\pm(\sqrt{3}+1)$. **9.** $\pm(2-\sqrt{3})$. **10.** $\pm\frac{1}{2}(\sqrt{10}-\sqrt{2})$. **11.** $\pm(2-\sqrt{2})$.

**12.** $\pm(2+\sqrt{5})$. **13.** $\pm(2-\sqrt{11})$. **14.** $\pm\frac{1}{2}(\sqrt{22}+\sqrt{10})$.

**15.** $\pm(\sqrt{6}+2\sqrt{2})$. **16.** $\pm(\sqrt{a}+\sqrt{b})$. **17.** $\pm(\sqrt{a+b}-\sqrt{a})$.

**Page 221.** — **18.** $\pm(\sqrt{11}-\sqrt{2})$. **19.** $\frac{\sqrt{6}+1}{5}$. **20.** $\sqrt{5}-1$.

**21.** $2\sqrt{2}$. **22.** $4$. **23.** $2-\sqrt{3}$.

**Page 223.** — **1.** $4$. **2.** $\frac{25}{4}$. **3.** $(a+b)^2$. **4.** $27$. **5.** $7$. **6.** $5$. **7.** $4$.

**8.** $3$. **9.** $10$. **10.** $9$. **11.** $10$. **12.** $4$. **13.** $2$. **14.** $-\frac{3}{4}$. **15.** $16$. **16.** $5$.

**17.** $\pm 24$. **18.** $5$. **19.** $5$. **20.** $1$.

**Page 224.** — **21.** $7$. **22.** $5$. **23.** $4$. **24.** $\frac{1}{2}$. **25.** $\frac{1}{4}, \frac{1}{3}$. **26.** $25$. **27.** $25$.

**28.** $\frac{36}{5}$. **29.** $9$. **30.** $64$. **31.** $100$. **32.** $\frac{12a}{5}$. **33.** $4$. **34.** $\frac{1}{4}$. **35.** $\frac{mn}{m+n}$.

**36.** $\frac{5}{4}$. **37.** $\pm\frac{7}{8}$.

**Page 225.** — **1.** $4$. **2.** $9$. **3.** $16$. **4.** $25, 81$. **5.** $216, -64$. **6.** $16$.

**7.** $4$. **8.** $\pm 4$. **9.** $5$. **10.** $\pm 3$.

**Page 226.** — **11.** $-1, 9$. **12.** $8, -3$. **13.** $-3, 2$. **14.** $-4, 5$.

**15.** $2, -\frac{1}{2}$. **16.** $-2, -3$. **17.** $3, -1\frac{2}{3}$. **18.** $3, -6$. **19.** $9, -2$.

**20.** $7, 4$. **21.** $\frac{17}{5}, \frac{7}{10}$.

**Page 228.** — **1.** $-7$.   **2.** $36$.   **3.** $56$.   **4** $1$.   **5.** $0$.   **6.** $2\,b^6$.
**7.** $0$.   **8.** $160$.

**Page 229.** — **4.** $(m-1)(m-2)(m-3)$.   **5.** $(a+2)(a^2-2\,a+2)$.
**6.** $(p-1)(p-2)(p-2)$.   **7.** $(p-1)(p-3)(p-5)$.   **8.** $(x+1)(x-2)$
$(x+3)$.   **9.** $(m-2)(m-3)(2\,m+5)$.   **10.** $(a-1)(a-3)(a-4)$.
**11.** $(b-3)(b^2-5\,b-1)$.   **12.** $(m-p)(m-2\,p)(m-3\,p)(m+2\,p)$.
**13.** $(m-n)(m-4\,n)(m^2+6\,mn+n^2)$.   **14.** $(a+2)(a^4-2\,a^3+$
$4\,a^2-8\,a+16)$.   **15.** $-1,-2,-3$.   **16.** $3, 1\pm\sqrt{6}$.   **17.** $1, 4, 5$.
**18.** $3, 3, -1$.   **19.** $1, 3, 4$.   **20.** $1, 2, -3$.   **21.** $-3, -3, -3$.
**22.** $2, 2, 3$.   **23.** $2, -4, -5$.   **24.** $-1, 1, 1, 3$.   **25.** $-2, -\frac{1}{4}, 3$.

**Page 231.** — **1.** $(a-1)(a^2+a+1)$.     **2.** $(x+1)(x^2-x+1)$.
**3.** $(1-ab)(1+ab+a^2b^2)$.   **4.** $(x-2)(x^2+2x+4)$.   **5.** $(x+2)(x^2-2x+4)$.
**6.** $(2\,a+1)(4\,a^2-2\,a+1)$.     **7.** $(2\,a-1)(4\,a^2+2\,a+1)$.
**8.** $(a-4\,b)(a^2+4\,ab+16\,b^2)$.   **9.** $27(2\,a+b)(4\,a^2-2\,ab+b^2)$.
**10.** $(10-x^2)(100+10\,x^2+x^4)$.     **11.** $x(x-1)(x^2+x+1)$.
**12.** $a(1+a)(1-a+a^2)$.   **13.** $(a+b)(a^2-ab+b^2)(a-b)(a^2+ab+b^2)$.
**14.** $(a^2+b^2)(a^4-a^2b^2+b^4)$.     **15.** $(xy-2)(x^2y^2+2\,xy+4)(xy+2)$
$(x^2y^2-2\,xy+4)$.      **16.** $(a-1)(a^4+a^3+a^2+a+1)$.
**17.** $(a+1)(a^4-a^3+a^2-a+1)$.    **18.** $(m^4+1)(m^8-m^4+1)$.
**19.** $(1+a^2b^2)(1-a^2b^2+a^4b^4)$.     **20.** $(8-a)(64+8\,a+a^2)$.
**21.** $(xy+5)(x^2y^2-5\,xy+25)$.   **22.** $(4\,mn-y^2)(16\,m^2n^2+4\,mny^2+y^4)$.
**23.** $(3\,m^3-7)(9\,m^6+21\,m^3+49)$.     **24.** $(a-b)(a+b)(a^4+a^3b+a^2b^2+$
$ab^3+b^4)(a^4-a^3b+a^2b^2-ab^3+b^4)$.   **25.** $2, -1\pm\sqrt{-3}$.   **26.** $-2, 1\pm\sqrt{-3}$.
**27.** $3, \dfrac{-3\pm 3\sqrt{-3}}{2}$.   **28.** $1, \dfrac{-1\pm\sqrt{-3}}{2}$.

**Page 233.** — **1.** $4, 2 ; 2, 4$.   **2.** $5, 3 ; -3, -5$.   **3.** $5, -3 ; -3, 5$.
**4.** $10, 30 ; 30, 10$.    **5.** $22, 3 ; -3, -22$.      **6.** $25, 4 ; 4, 25$.
**7.** $73, 12 ; -12, -73$.   **8.** $3, 4 ; 4, 3$.   **9.** $7, 2 ; 2, 7$.   **10.** $11, 10 ; -11, -10$.
**11.** $13, 3 ; -3, -13$.    **12.** $20, 6 ; -6, -20$.

**Page 234.** — **13.** $4, 4 ; 4, 4$.   **14.** $6, 1 ; 1, 6$.   **15.** $4, 1 ; -1, -4$.
**16.** $5, 2 ; 2, 5$.      **17.** $3, 4 ; 4, 3$.       **18.** $a-b, b ; b, a-b$.
**19.** $6, 30 ; 30, 6$.

    **1.** $4, 1 ; -4, -1$.     **2.** $3, 1 ; -3, -1$.     **3.** $2, 1 ; -\frac{3}{2}, -\frac{5}{2}$.
**4.** $3, 2 ; \frac{11}{7}, -\frac{9}{7}$.    **5.** $2, 3 ; \frac{14}{3}, \frac{1}{3}$.    **6.** $1, 2 ; 1, 2$.

**Page 235.** — **7.** $2, 1 ; -\frac{19}{6}, -\frac{25}{6}$.   **8.** $6, 5 ; 0, 0$.   **9.** $-24, 12 ;$
$\frac{6}{5}, \frac{3}{5}$.   **10.** $1, 3 ; 5, -1$.   **11.** $3, 4 ; 4, 3$.   **12.** $5, 2 ; -\frac{43}{9}, -\frac{26}{9}$.
**13.** $5, 2 ; -5, -2$.   **14.** $-2, -3 ; \frac{4}{3}, 2$.

**Page 236.** — **1.** $\pm 1, \pm 2 ; \pm 2, \pm 3$.     **2.** $\pm 2, \pm 3 ; \pm 3, \pm 4$.
**3.** $\pm 1, \pm 2 ; \pm 3, \pm 5$.   **4.** $\pm 1, \pm 2 ; \pm 2, \mp 1$.   **5.** $\pm 1, \pm 5 ; \pm 7, \pm 2$.
**6.** $\pm 6, \pm 2$.   **7.** $\pm 2, \pm 4 ; \pm 3, \pm 5$.   **8.** $\pm 2, \pm 3 ; \pm 3, \pm 4$.

**Page 237.—9.** $\pm 2$, $\pm 5$; $\pm 2\sqrt{3}$, $\pm\sqrt{3}$. **10.** $\pm 3$, $\pm 2$; $\pm\sqrt{5}$, $\pm 3\sqrt{5}$. **11.** $\pm 5$, $\pm 2$; $\pm\sqrt{7}$, $\pm 2\sqrt{7}$. **12.** $\pm\frac{2}{5}$, $\pm\frac{3}{5}$; $\pm\frac{3}{5}$, $\pm\frac{4}{5}$. **13.** $\pm\frac{1}{3}$, $\pm\frac{2}{3}$; $\pm\frac{2}{3}$, $\pm 1$. **14.** $\pm 7$, $\pm 5$; $\pm 5$, $\pm 7$. **15.** $\pm 2$, $\pm 1$. **16.** $\pm 3$, $\pm 2$; $\pm\frac{7}{3}\sqrt{6}$, $\pm\frac{19}{18}\sqrt{6}$. **17.** $\pm 4$, $\pm 3$; $\pm 1$, $\pm 12$.

**Page 238.—1.** $6, 4$; $-4, -6$. **2.** $3, 1$. **3.** $5, 4$; $4, 5$. **4.** $2, 3$. **5.** $5, 2$; $2, 5$. **6.** $4, 1$; $1, 4$. **7.** $\pm 3$, $\pm 2$. **8.** $4, 3$; $-3, -4$. **9.** $\pm 2$, $\pm 1$; $\pm 1$, $\pm 2$.

**Page 239.—1.** $1, 2$; $1, -2$; $-1, 2$; $-1, -2$. **2.** $1, 3$; $2, 1$. **3.** $4, 1$; $1, 4$. **4.** $6, 3$; $3, 6$. **5.** $2, 3$; $3, 2$. **6.** $\frac{1}{3}, \frac{1}{2}$; $\frac{1}{2}, \frac{1}{3}$. **7.** $512, 8$; $8, 512$. **8.** $1, 3$; $4, -3$.

**Page 240.—9.** $\pm 7$, $\pm 11$. **10.** $5, 2$; $-\frac{95}{28}$, $-\frac{33}{7}$. **11.** $4, 3$; $3, 4$. **12.** $3, 2$; $\frac{11}{6}, \frac{17}{6}$. **13.** $\pm 1$, $\pm 2$; $\pm 3$, $\pm 5$. **14.** $\pm 1$, $\pm 2$; $\pm 2$, $\pm 1$. **15.** $\pm 9$, $\pm 12$; $\pm 12$, $\pm 9$. **16.** $4, 3$; $-\frac{48}{13}$, $-\frac{41}{4}$. **17.** $6, 3$; $-3, -6$. **18.** $\pm 2$, $\pm 1$. **19.** $1, 125$; $125, 1$. **20.** $\pm\frac{5}{2}$, $\pm 1$; $\pm\frac{3}{2}$, $\pm\frac{5}{3}$. **21.** $\pm 1$, $\pm 2$. **22.** $1, 2$; $3, 4$. **23.** $15, 5$; $-15, -5$. **24.** $m+n$, $m-n$; $m-n$, $m+n$. **25.** $1, 4$; $3, 2$. **26.** $5, 4$; $4, 5$. **27.** $\frac{1}{4}, \frac{1}{3}$; $\frac{1}{3}, \frac{1}{4}$. **28.** $\pm 2$, $\pm 1$. **29.** $5, 3$; $3, 5$.

**Page 241.—30.** $2, 4$; $4, 2$. **31.** $\pm\frac{8}{3}$, $\pm\frac{7}{2}$. **32.** $\pm 3$, $\pm 1$; $\pm 1$, $\pm 3$. **33.** $4, 1$; $-1, -4$. **34.** $\pm 8$, $\pm 6$. **35.** $5, 3$; $5, -3$; $-5, 3$; $-5, -3$. **36.** $1, 3$; $3, 1$. **37.** $\pm 2$, $\pm 3$. **38.** $2, 8$; $8, 2$. **39.** $\dfrac{a}{2}, \dfrac{b}{2}$. **40.** $1, 3$; $-1\frac{3}{4}, 7\frac{1}{4}$. **41.** $3, 4$; $4, 3$.

**Page 243, Exercise 113.—1.** Indeterminate. **2.** $x = \infty$, *i.e.* no solution. **3.** $\infty$. **4.** Indeterminate. **5.** $\infty, \infty$; $-4, -4$. **6.** $\infty, \infty$. **7.** $\infty$.

**Exercise 114.—1.** $37, 39$. **2.** $29, 13$.

**Page 244.—3.** $18, 1$. **4.** $\pm 17$, $\pm 15$. **5.** 8 ft., 12 ft. **6.** $55, 48$. **7.** 12 ft., 35 ft. **8.** 49 ft., 25 ft. **9.** 40 ft., 9 ft. **10.** 28 yd., 45 yd. **11.** $2, 18$. **12.** 40 in., 30 in. **13.** $1\frac{1}{2}$ in., 3 in. **14.** 5 cm., 3 cm.

**Page 245.—15.** 35 in., 12 in. **16.** 29 in., 21 in. **17.** $36$.

**Page 247.—1.** $a, c$, and $d$. **2.** ($a$) $5, 8, 11, 14, 17, 20$; ($b$) $2, -1, -4, -7, -10, -13$; ($c$) $-1, -3, -5, -7, -9, -11$. **3.** $14$. **4.** $35$. **5.** $4\frac{1}{2}$. **6.** $-30$. **7.** $-37$. **8.** $201$. **9.** $2n$.

**Page 248.—10.** $31, 136$. **11.** $-14, -56$. **12.** $84, 920$. **13.** $-3, 0$. **14.** $900$. **15.** $288$. **16.** $-460$. **17.** $50$. **18.** $-10$. **19.** $69.6$. **20.** $ax + \dfrac{a(a+1)}{2}$. **21.** $5050$. **22.** $\dfrac{n(n+1)}{2}$. **23.** $n^2$. **24.** $78$. **25.** \$237. **26.** ($a$) \$3400; ($b$) \$46,200.

**Page 250.** — **1.** $a$. **2.** $x + 2y$. **3.** $\dfrac{m+n}{2mn}$. **4.** $\dfrac{a}{a^2 - b^2}$. **5.** $5, 6, 7$

**6.** $9\frac{1}{2}, 9, 8\frac{1}{2}, 8, 7\frac{1}{2}, 7, 6\frac{1}{2}$. **7.** 16. **8.** 20. **9.** 5, 50. **10.** 5.

**11.** 4. **12.** 4. **13.** 3, 476. **14.** 35. **15.** 5. **16.** $7\frac{3}{4}$. **17.** $\dfrac{2s - an}{n}$.

**18.** \$50. **19.** 25, 35, 45, 55, 65, 75.

**Page 252.** — **1.** $a$, $c$, and $d$. **2.** 3, 12, 48, 192, 768. **3.** 16, 8, 4, 2,

1, $\frac{1}{2}$. **4.** $\frac{8}{27}$. **5.** 1. **6.** $1\frac{17}{512}$. **7.** 327,680. **8.** 16,384.

**9.** $-\frac{1}{96}$. **10.** $\frac{16}{5}$.

**Page 253.** — **11.** 665. **12.** 364. **13.** 45,920. **14.** 43. **15.** $6\frac{6}{3}5$.

**16.** $\frac{7}{4}$. **17.** 910. **18.** $\dfrac{x^5(x^5 - 1)}{x - 1}$. **19.** 10, 210. **20.** 81, 120.

**21.** 10, 160. **22.** 5, 405. **23.** 45.

**Page 254.** — **1.** 2. **2.** $1\frac{1}{2}$. **3.** 64. **4.** $2\frac{1}{4}$. **5.** $6\frac{1}{4}$. **6.** $416\frac{2}{3}$. **7.** 27.

**8.** 70. **9.** $\frac{5}{9}$. **10.** $\frac{71}{99}$. **11.** $\frac{19}{99}$. **12.** $\frac{3}{11}$. **13.** $\frac{5}{18}$. **14.** $\frac{103}{330}$. **15.** 6.

**16.** $\frac{1}{2}$. **17.** $(a)$, 8 sq. in. ; $(b)$ $8(2 + \sqrt{2})$.

**Page 257.** — **1.** $a^5 + 5\,a^4 b + 10\,a^3 b^2 + 10\,a^2 b^3 + 5\,ab^4 + b^5$. **2.** $x^6 -$
$6\,x^5 y + 15\,x^4 y^2 - 20\,x^3 y^3 + 15\,x^2 y^4 - 6\,xy^5 + y^6$. **3.** $1 + 8\,x + 28\,x^2 +$
$56\,x^3 + 70\,x^4 + 56\,x^5 + 28\,x^6 + 8\,x^7 + x^8$. **4.** $x^7 - 14\,x^6 + 84\,x^5 - 280\,x^4$
$+ 560\,x^3 - 672\,x^2 + 448\,x - 128$. **5.** $x^5 + \frac{5}{2}\,x^4 + \frac{5}{2}\,x^3 + \frac{5}{4}\,x^2 + \frac{5}{16}\,x + \frac{1}{32}$.
**6.** $16\,a^4 + \frac{32}{3}\,a^3 b + \frac{8}{3}\,a^2 b^3 + \frac{8}{27}\,ab^3 + \frac{1}{81}\,b^4$. **7.** $1 - 2\,m + \frac{5}{3}\,m^2 - \frac{20}{27}\,m^3$
$+ \frac{5}{27}\,m^4 - \frac{2}{81}\,m^5 + \frac{1}{729}\,m^6$. **8.** $x^{10} - 5\,x^6 + 10\,x^2 - 10\,x^{-2} + 5\,x^{-6} - x^{-10}$.
**9.** $2 + 12\,x + 2\,x^2$. **10.** $10\,a^2 b^{\frac{1}{2}} + 20\,ab^{\frac{3}{2}} + 2\,b^{\frac{5}{2}}$. **11.** $126\,a^5 b^4$.
**12.** $45\,a^8 b^2$. **13.** $220\,m^9 n^3$. **14.** $330\,x^4$. **15.** $280\,a^4 b^3$. **16.** $-53,130\,x^{36}$.
**17.** 70. **18.** $189\,a^4$. **19.** 105. **20.** 1820. **21.** $-15,504$. **22.** $-161,700$.
**23.** 495. **24.** $6\,x^2 y^2$. **25.** $-20\,a^3 b^3$. **26.** 70. **27.** $12,870\,m^8 n^8$.
**28.** $4950\,a^2 b^{98}$. **29.** $1000\,ab^{999}$.

## REVIEW EXERCISE

**Page 258.** — **1.** 1, 0, 125, 64, 27, 8, 343, 125, 343. **2.** $-8, -53, 8,$
$-53, -192, 27, -176, -419$. **3.** 1, 1, 1, 4, 1, 9, 4, 4. **4.** 0, 8, 0, 3,
0, 32, $-1$, 0. **5.** 6, 7, 7, 10, 8, 12, 12, 18. **6.** 6, 3, 8, 2, 2, 4, 6, 6.
**7.** 6, $-4$, 3, $-6$, 4, 5, $-5$, 2.

**Page 259.** — **8.** 4, 5, 2, 4, 6, 2, 3, 1. **9.** 1, 4, 3, 3, 6, 4. **10.** $a^2 x +$
$3\,a^3$. **11.** $5\,x^2$. **12.** $x^4$. **13.** $x^3 + y^3 + z^3 - 3\,xyz$. **14.** 0. **15.** $7\,x^4 + x^3$
$- x^2 + 15\,x + 5$. **16.** $7\,x^4$. **17.** $3\,a^4 + 15\,a^3 + 11\,a^2 + 14\,a - 13$.
**18.** $16\,x^3 + 12\,x^2 y - 2\,y^3 - 4\,y^2 z + xyz$.

**Page 260.** — 19. $\sqrt{1+x} + \sqrt{1+y} + 1$. 20. $-3x^2$. 21. $-5a^2x - 8x^3$.
22. $14x - 7$. 23. $x^5 + x^7y^2$. 24. $1 + x - x^2$. 25. $2a$. 26. $x^3 + x^2 + x$.
27. 0. 28. $12x$. 29. 0. 30. $(a)$ $2x + 2y$; $(b)$ $2y - 2z$; $(c)$ $3x + y$
$+z$; $(d)$ $x - y + 3z$; $(e)$ $2x + 2y + 2z$; $(f)$ $2x - 2y + 2z$. 31. $11x$
$-15y$. 32. $3a - 2c$. 33. $-8x^3 - 8x$. 34. $13 - 12a$. 35. $-9x$.
36. $-5b + c - 2$.

**Page 261.** — 37. $3a^2 + 5a + 7$. 38. $5a + 8$. 39. $-5x + 2y - z$.
40. $3a - 3b - c$. 41. $x^2 - 9x - 4$. 42. $7a - 9b - c$. 43. $6c - e + f$.
44. $3a - 5b - 7c$. 45. $a - 6b$. 46. $-3b + 9c - 9d$. 47. $x^3 + 3x^2$
$+3x + 2$. 48. $x^3 - 3x^2 + 3x - 1$. 49. $x^3 + x^2 - 7x - 15$. 50. $2 - 15x$
$+28x^2 - 15x^3$. 51. $1 - x + x^3 - x^4$. 52. $x^4 + 2x^2 + 9$. 53. $6x^4 -$
$13x^3 + 11x^2 - 5x + 1$. 54. $b^8 - 2b^4 + 1$. 55. $a^8 - a^6 - 13a^4 + 11a^2 - 2$.
56. $a^3 + b^3 + c^3 - 3abc$. 57. $a^3 - b^3 - c^3 - 3abc$. 58. $x^4 + 2x^2y^2 -$
$8y^4 - 18y^2z^2 - 9z^4$. 59. $a^3 - b^3 + 27 + 9ab$. 60. $8x^3 + 27y^3 + z^3 - 18xyz$.
61. $x^3 + 15x^2 + 71x + 105$. 62. $x^3 - 15x^2 + 71x - 105$. 63. $x^3 - 15x^2$
$+62x - 72$. 64. $x^8 + x^4 + 1$. 65. $x^8 - a^8$. 66. $x^8 + x^4y^4 + y^8$.
67. $1 - x^8$. 68. $x^6 - y^6$.

**Page 262.** — 69. $a^{24} - x^{24}$. 70. $a^6 - a^4 - a^2 + 1$. 71. $a^4 - 2a^2b^2 + b^4$.
72. $4x^3 - 3xy^2 + y^3$. 73. $a^4 - 4a^3b + 16ab^3 - 16b^4$. 74. $x^4 - 36x^2y^2$
$+108xy^3 - 81y^4$. 75. $a^6 - a^4 - a^2 + 1$. 76. $x^{3n} - y^{3m}$. 77. $a^{3p+3} + b^{3q}$.
78. $a^{3m} + b^{3n} + c^3 + 3a^{2m}b^n + 3a^mb^{2n}$. 79. $m^{3a} + n^{3b} + p^{3c} - 3m^an^bp^c$.
80. $24a^2b - 12ab^2$. 81. $p^3 + 3p^2q$. 82. $-4a^3 + 3ab^2 - b^3$.
83. $3x^3 + 4x^2y - 4xy^2 - 3y^3$. 84. $6p^2q - 54q^3$. 85. $2x^2y^2 + 2y^4$.
86. 0. 87. $4bc + 4ac$. 88. $xy^2 + y^2z + 2xyz$. 89. $-x^4 - y^4 - z^4$
$+2y^2z^2 + 2x^2z^2 + 2x^2y^2$. 90. $a^2b - ab^2 + b^2c - bc^2 + c^2a - a^2c$. 91. 0.
92. $6abc$. 93. $10a - 12b$. 94. $6a^2 - 15ab + 9ac + 6bc$.

**Page 263.** — 96. $2(a+b)^2 - (a+b)$. 97. $(a+b)^n - 2$. 98. $2a^{2x} -$
$a^x + 5$. 99. $2^{2a} - 2^a + 2^{3a} - 2$. 100. $3^{m-n} + 3^m \cdot 3^{2n} - 1$. 101. $5^4 + 5^3$
$-5^2 + 5 = 730$. 102. $2x^2 + 5x + 4$. 103. $5x^2 - 2x + 3$. 104. $2y^2 -$
$6y - 4$. 105. $x^2 - 2xy + 4y^2$. 106. $a^2 + 2ab + b^2$. 107. $2x^2 - 5ax -$
$2a^2$. 108. $x^2 - 3xy - y^2$. 109. $x^2 - 5ax - 5a^2$. 110. $2y^2 - 6y - 4$.
111. $3a^2 - 5a - 5$. 112. $2a^2 - 4ab - 5b^2$. 113. $-3x^2 - 2xy + 3y^2$.
114. $2x^2 - 3xy - y^2$. 115. $y - 3$. 116. $x^2 + xy + y^2$. 117. $a^2 - 4a +$
16. 118. $a^4 + a^2b^2 + b^4$. 119. $a^2 + 4b^2 + c^2 + 2bc - ac + 2ab$.
120. $x^2 + 9y^2 + 1 + 3xy + x - 3y$. 121. $x^4 - x^3y + xy^3 - y^4$.
122. $x^7 + x^4 + x$. 123. $a^{n+1} + a^n - a^{n-1}$. 124. $a^x + a^{x-1}b + a^{x-2}b^2$.

**Page 264.** — 125. $1 + a + a^2 - 2ax + 2x + 4x^2$. 126. $x^2 + y^2 + z^2 -$
$xy - xz - yz$. 127. $x^{2a} - x^a + x^{a-1}$. 128. $1 + 9a^{2m} + 4b^{2n} + 3a^m + 2b^n$
$-6a^mb^n$. 129. $2ab^3 + 3b^4$. 130. $x^6 - 3x^5 + 9x^4 - 27x^3 + 81x^2 -$
$243x + 729$. 131. $a^2 - 2ab - 2b^2$. 132. $x + 1$. 133. $1\frac{1}{2}$. 134. 0.

**135.** 2. **136.** $4\frac{27}{66}$. **137.** 1. **138.** 12. **139.** 24. **140.** $6\frac{1}{4}$. **141.** 7.
**142.** $-3$. **143.** $2\frac{6}{17}$. **144.** $-2\frac{1}{2}$. **145.** 2. **146.** $5\frac{1}{4}$. **147.** $-\frac{7}{8}$.
**148.** 1. **149.** 2. **150.** $1\frac{3\cdot 4}{6\cdot 6}$.

**Page 265.** — **151.** $-1$. **152.** $-1$. **153.** 1. **154.** 3. **155.** 15.
**156.** 1. **157.** 2. **158.** 3. **159.** 4. **160.** 6. **161.** $-1$. **162.** 50.
**163.** 20. **164.** $y-3, y-2, y-1, y$. **165.** $15-a$. **166.** $23-k$.
**167.** $30+x$ yr. **168.** 18, 19, 20, 21, 22. **169.** 15, 18, 30.
**170.** $37\frac{1}{2}°, 75°, 67\frac{1}{2}°$. **171.** 10 in. **172.** $(a)\ 59°; \ (b)\ -40°; \ (c)\ 160°$ C.

**Page 266.** — **173.** $1\frac{1}{2}$. **174.** $4\frac{1}{2}$. **175.** 147 mi. **176.** 36 ft.
**177.** 36 yr. **178.** 12, 10, 6 panes. **179.** 40 yr., 12 yr. **180.** 32 h. p.,
48 h. p. **181.** 12 ft., 10 ft. **182.** 12 yr., 6 yr. **183.** 42 yr.
**184.** $(x+2)(x-1)$. **185.** $(y-7)(y+6)$. **186.** $(x-12)(x+3)$.
**187.** $(a+11)(a-10)$. **188.** $(y-17)(y+6)$. **189.** $(ab-11)$
$(ab+22)$. **190.** $(4a+1)(a+3)$. **191.** $(5x-2)(2x+3)$.
**192.** $(x+8)(x-7)$.

**Page 267.** — **193.** $(y-75)(y-2)$. **194.** $(2a+1)(a-10)$.
**195.** $(a+7)(a-4)$. **196.** $(3x+2y)(2x-3y)$. **197.** $(3y-1)(y-4)$.
**198.** $(x^2+3)(x^2+5)$. **199.** $(2x+5y)(x-2y)$. **200.** $(x-16)$
$(x+4)$. **201.** $(y-24)(y-5)$. **202.** $(x-y+3)(x-y-3)$.
**203.** $x^3(x-7y)(x-12y)$. **204.** $(5x-3y)(x-y)$. **205.** $(7x-2y)$
$(2x-3y)$. **206.** $(x+2)(x-2)(3x-1)$. **207.** $(4x^2+9)(2x+3)$
$(2x-3)$. **208.** $2a(a+2b)(a-2b)$. **209.** $(y-b)(y+b-1)$.
**210.** $x(x-10)(x-1)$. **211.** $5x^2(4x^2-4x-1)$. **212.** $3(x-9)(x+2)$.
**213.** $a^2(15-b)(4+b)$. **214.** $2xy(3x-5)(2x+1)$. **215.** $x(x+6)$
$(x-1)$. **216.** $2(x-8)(x-3)$. **217.** $2y(x-11)(x-3)$.
**218.** $x(x+3)(x+10)$. **219.** $a(a-1)(a^2+a+2x^2)$. **220.** $xy(x+2)$
$(y+2)$. **221.** $(3a-2)(2a+3)$. **222.** $x^2y(x+6y+9xy^2)$.
**223.** $(3x-2y)(2x+3y)$. **224.** $(7x-y)(x+7y)$. **225.** $(a-c)$
$(a+b+c)(a^2+ab-2ac+bc+c^2)$. **226.** $(x+z)(x^2-xz+z^2)(x+y)$
or $(x+y)(x^3+z^3)$. **227.** $(7x-y)(x-3y)$. **228.** $y(x-y)$
$(2x^2+xy+y^2)$. **229.** $(x+y+2)(x+y-2)$. **230.** $(5x+2)(3x+4)$.
**231.** $a(3a+2b)(3a-2b)$. **232.** $b(2a+b-2c)$. **233.** $(y+1)$
$(y+1)(y-1)$. **234.** $(3a+4b+c)(5c-a)$. **235.** $(x^3+x-1)$
$(x^3-x+1)$. **236.** $(8x+3)(3x-4)$. **237.** $(x-1)(x^3+x^2-1)$.
**238.** $(x+1)(x-1)(y+1)(y-1)$. **239.** $(5x+y)(5x-y)(x+3y)(x-3y)$.
**240.** $(a^2b^2+2ab-c^2)(-a^2b^2+2ab+c^2)$. **241.** $2(a-d)(a+b+c+d)$.
**242.** $(x-y)(x^m+y^m)$. **243.** $(4x^2-y)(x-3y)$. **244.** $(2x-3y)$
$(x+b)(x-b)$. **245.** $3(a-c)(p-2q)$. **246.** $3xy(x-1)(y+1)$.
**247.** $(a^2+1)(a^4+1)$. **248.** $(l-m)(x-y+a)$.

**Page 268.—249.** $(7x-3y+1)(a+2b)$. **250.** $(abc-1)(a^2bc^2-3)$.

**251.** $2x(x-a)(x+b)$. **252.** $(a+b)(2x+y+z)$. **253.** $3(x+1)$.

**254.** $3x+4$. **255.** $5x+2$. **256.** $3xy+5$. **257.** $a-1$. **258.** $ab-1$.

**259.** $2x-3$. **260.** $7x-5$. **261.** $3xy-7$. **262.** $x+5$. **263.** $x-7$.

**264.** $x-10$. **265.** $x-12$. **266.** $(x+1)(x+2)(x+3)$. **267.** $(x+1)$ $(x+3)(x-4)$. **268.** $(2x-5)(x-1)(3x-4)$. **269.** $(x+5)$ $(2x+3)(4x-1)$. **270.** $(x-16y)(x-7y)(x-2y)$. **271.** $\dfrac{2a-3b}{2a+3b}$.

**272.** $\dfrac{x-7}{x+7}$. **273.** $\dfrac{p-1}{p-3}$. **274.** $\dfrac{x-3y}{x+8y}$. **275.** $\dfrac{5x+3}{5x-7}$.

**276.** $\dfrac{3x+y}{7x-y}$.

**Page 269.—277.** $\dfrac{4x+3}{3x^2-4x}$. **278.** $\dfrac{m^2-n^2}{m^2+n^2}$. **279.** $\dfrac{x^2+2}{x+3}$.

**280.** $\dfrac{m+n-p}{m-n+p}$. **281.** $\dfrac{a-x}{3}$. **282.** $\dfrac{x-b}{x-c}$. **283.** $\dfrac{3x-y}{4x-y}$.

**284.** $\dfrac{2x-3}{2x+3}$. **285.** $\dfrac{x-1}{x+1}$. **286.** $\dfrac{1-2x}{1+3x}$. **287.** $\dfrac{x+z}{x^2-yz}$.

**288.** $\dfrac{3a+5b}{3c-1}$. **289.** $\dfrac{x+y-z}{x-y+z}$. **290.** $\dfrac{x+y-z}{x-y+z}$. **291.** $\dfrac{x+y+z}{x-y-z}$.

**292.** $\dfrac{2x-y}{x^2-1}$. **293.** $\dfrac{cx+d}{ax-b}$. **294.** $(b+c-a)(c+a-b)$. **295.** $1$.

**296.** $(a+b)(c+d)$. **297.** $\dfrac{1+2x^2}{1-2x^2}$.

**Page 270.—298.** $4$. **299.** $\dfrac{3x^3-20x^2-32x-235}{(x+4)(x-3)(y+7)}$.

**300.** $\dfrac{3x^3-24x^2+60x-46}{(x-2)(x-3)(x-4)}$. **301.** $\dfrac{x}{(x+1)(x+2)(x+3)}$. **302.** $\dfrac{3x^2}{x^2-1}$.

**303.** $\dfrac{e-d}{(a+c)(a+d)(a+e)}$. **304.** $2$. **305.** $0$. **306.** $\dfrac{2x+11}{(x+4)(x+5)(x+7)}$.

**307.** $\dfrac{2x-17}{(x-4)(x+11)(x-13)}$. **308.** $\dfrac{m^3+4m^2n+mn^2}{n(m+n)^2}$. **309.** $0$.

**310.** $\dfrac{x^2-x+1}{x^2-5x+6}$.

**Page 271.—311.** $0$. **312.** $\dfrac{1}{x^3-1}$. **313.** $\dfrac{a^2}{a^3-b^3}$.

**314.** $\dfrac{1}{(x-3)(x-4)(x-5)}$. **315.** $\dfrac{1}{1+2x}$. **316.** $\dfrac{2}{(x^2-1)(x^2-4)}$.

**317.** $\dfrac{18}{(x-1)(x+2)(x+5)}$. **318.** $\dfrac{4x^3y}{x^4-y^4}$. **319.** $\dfrac{1}{x(1-x^4)(1+x)}$.

**320.** $1$. **321.** $\dfrac{3x-a-b-c}{x+a+b+c}$. **322.** $0$. **323.** $0$.

**Page 272.—324.** $\dfrac{5\,x(x+5)}{(x+1)(x+2)}$.     **325.** $\dfrac{6}{5(x-7)}$.

**326.** $\dfrac{a^2(a-5)(a+2)(a-2)}{(a+7)^2(a+1)^2}$.     **327.** $\dfrac{2(a-3)(a-4)}{3\,a^2(a+2)(3\,a-2)}$.

**328.** $\dfrac{2\,(x-3)}{3\,x(x+6)}$.     **329.** $\dfrac{2\,x^2-3\,xy+9\,y^2}{2(x-3y)(2\,x-3\,y)}$.     **330.** $\dfrac{y-5}{6\,y-5}$.

**331.** $\dfrac{2\,y+3}{5\,y+3}$.    **332.** 1.    **333.** $\dfrac{x-3\,y}{x+3\,y}$.    **334.** $\dfrac{x-4}{x+4}$.    **335.** $\dfrac{x^4}{x^2+y^2}$.

**Page 273.—336.** $(a+b+c)$.     **337.** $a^2+4\,ab+4\,b^2$.

**338.** $\dfrac{a^4+2\,a^2+1}{a^2}$.     **339.** $\dfrac{y^2-2\,xy+x^2}{x^2y^2}$.     **340.** $\dfrac{27\,x^3}{125\,y^3}$.

**341.** $\dfrac{m^2+2\,mn+n^2}{mn}$.     **342.** $\dfrac{a^2-b^2}{ab}$.     **343.** $\dfrac{x^4+2\,x^3+3\,x^2+2\,x+1}{x^2}$.

**344.** $\dfrac{m^4+2\,m^2n^2+n^4}{m^2n^2}$.     **345.** $\dfrac{64\,x^4-144\,x^2y^2+81\,y^4}{144\,x^2y^2}$.     **346.** $\dfrac{x^3-1}{x}$.

**347.** $x^2-4$.    **348.** $\dfrac{m-n}{m+n}$.    **349.** $\dfrac{x-8}{x-1}$.    **350.** $\dfrac{121}{343}$.    **351.** $1\tfrac{3}{4}$.

**352.** $1\tfrac{5}{11}$.    **353.** $\dfrac{10\,x-113}{60}$.    **354.** $\dfrac{2\,x^4+2\,y^4}{x^4-y^4}$.

**Page 274.—355.** $\dfrac{1}{x^2-1}$.     **356.** $-\dfrac{1+2\,x}{1+x}$.     **357.** $4\,xy$.

**358.** $\dfrac{1}{2\,a^2-1}$.    **359.** $a^2$.    **360.** 1.    **361.** 0.    **362.** $\dfrac{3\,x-y}{2\,y}$.    **363.** 1.

**364.** $\dfrac{1}{36}$.    **365.** $\dfrac{1}{2\,x^2-1}$.    **366.** $\dfrac{a^2+b^2}{ab(a-b)^2}$.

**Page 275.—367.** $b^4+c^4$.    **368.** $\dfrac{a-c}{(1+ac)}$.    **369.** $a^4+b^4$.    **370.** 1.

**371.** $yz$.    **372.** $\dfrac{3\,n-m}{2}$.    **373.** $1+x$.    **374.** $\dfrac{x^2+bx-a^2}{x^2-ab-a^2}$.

**Page 276.—375.** $\dfrac{x(x+1)}{x^2+4\,x+1}$    **376.** 12.    **377.** $4\tfrac{1}{24}$.    **378.** $-4$.

**379.** $-7\tfrac{1}{2}$.    **380.** $-\tfrac{1}{5}$.    **381.** 2.    **382.** $\tfrac{9}{7}$.    **383.** $2\tfrac{3}{16}$.    **384.** 9.

**385.** $-\tfrac{1}{2}$.    **386.** $9\tfrac{7}{17}$.    **387.** 1.    **388.** $1\tfrac{1}{2}$.    **389.** 0.    **390.** 3.

**Page 277.—391.** 7.    **392.** 4.    **393.** $-1$.    **394.** $-23$.    **395.** 3.

**396.** $\tfrac{4}{13}$.    **397.** 0.    **398.** 20.    **399.** 3.    **400.** 11.    **401.** $\dfrac{ab}{c}$.    **402.** $\dfrac{1}{a}$.

**403.** $\dfrac{nc}{m(b-a)}$.    **404.** $a-b$.    **405.** $b-a$.    **406.** $\dfrac{2\,ab}{a+b}$.    **407.** $2(a+b)$.

**408.** $\dfrac{c}{2(a-b)}$.    **409.** $a+b$.

**Page 278.** —**410.** $ab$.　**411.** $\dfrac{a^2 + ab + b^2}{a + b}$.　**412.** $\dfrac{ab}{a + b - c}$.

**413.** $\dfrac{ab(a + b - 2\,c)}{c(a+b) - a^2 - b^2}$.　**414.** $\dfrac{2\,ab}{a + b}$.　**415.** $\dfrac{a+b}{2}$.　**416.** $\dfrac{a+b+c+d}{m + n}$.

**417.** $c$.　**418.** 0.　**419.** $a + b + c$.　**420.** A 5 mi , B 4 mi.　**421.** 42.

**422.** $15\frac{1}{3}$ mi., $18\frac{3}{5}$ mi.　**423.** 14 miles.

**Page 279.** —**424.** 24 days.　**425.** 54, 55.　**426.** $x^3 + y^3$.　**427.** $\frac{2}{3}$.

**428.** $a^2 - b^2$.　**429.** $2\frac{1}{4}$.　**430.** $xz - \dfrac{1}{xz}$.　**431.** $7:5$, $m - 1 : m$,

$(d - b) : (a - c)$.　**432.** $\dfrac{10\,m}{m + n}$ in., $\dfrac{10\,n}{m + n}$ in.　**433.** $40^\circ$, $50^\circ$, $90^\circ$.

**434.** $\frac{2}{3}$.　**435.** $m$.　**436.** $\frac{151}{208}$.　**438.** $(a)$ not true, $(b)$ true, $(c)$ not true, $(d)$ true.

**Page 280.** —**439.** $(a)$ 1, $(b)$ 2 $ab$, $(c)$ $3a - 4\,b$.　**440.** $40\frac{1}{2}$ oz.
**441.** 3, 10.　**442.** 2, 3.　**443.** $-\frac{1}{11}, \frac{1}{5}$.　**444.** $16\frac{1}{4}, 10\frac{3}{4}$.　**445.** $\frac{2}{3}, 1\frac{2}{3}$.
**446.** $1\frac{1}{2}, 2\frac{1}{2}$.　**447.** 39, 2.　**448.** $\dfrac{1}{a}, \dfrac{1}{b}$.　**449.** $\frac{3}{11}, \frac{1}{11}$.　**450.** $\dfrac{1}{a}, \dfrac{1}{b}$.
**451.** 0, $-7$.　**452.** 7, $-5$.　**453.** $-7, 8$.　**454.** 0, $-2$.　**455.** $-2$, $-7$.　**456.** $-1, 1$.　**457.** 6, $-1$.　**458.** 10, 7.　**459.** 10, $-2$.

**Page 281.** —**460.** 8, 0.　**461.** $-10, 7$.　**462.** 6, 12.　**463.** 5, $-3$.
**464.** 6, 7.　**465.** 11, $-1$.　**466.** $\frac{3}{2}, 2$.　**467.** 0, $-5$.　**468.** 1, 6.
**469.** $-1, -1$.　**470.** 4, 0.　**471.** $\dfrac{ce + bf}{bd + ae}, \dfrac{cd - af}{bd + ae}$.　**472.** $\dfrac{em + bn}{ae + bc}$,

$\dfrac{an - cm}{ae + bc}$.　**473.** $\dfrac{de}{c + d}, \dfrac{ce}{c + d}$.　**474.** $\dfrac{c(f - bc)}{af - bd}, \dfrac{c(ac - d)}{af - bd}$.

**475.** $\dfrac{2\,b^2 - 6\,a^2 + d}{3\,a}, \dfrac{3\,a^2 - b^2 + d}{3\,b}$.

**Page 282.** —**476.** $3\frac{1}{3}$.　**477.** $\frac{2}{3}$.　**478.** $\frac{3}{8}$.　**479.** 9, 7, 5.　**480.** 12, 4.
**481.** 20 yr., 32 yr.　**482.** 53 yr., 28 yr.　**483.** A \$ 3500, B \$ 2500.
**484.** 17, 4.　**485.** \$2000 at 6%.　**486.** \$ 250 at 6%.　**487.** 84.

**Page 283.** —**488.** 63.　**489.** $\dfrac{s + d}{2}, \dfrac{s - d}{2}$.　**490.** $\dfrac{a^2 + b}{2\,a}, \dfrac{a^2 - b}{2\,a}$.
**491.** 16, $7\frac{3}{4}, 5\frac{1}{2}$.　**492.** 18, 32, 10.　**493.** 2, 2, 2.　**494.** 5, 6, 8.
**495.** 1, 4, 6.　**496.** 5, 6, 7.　**497.** 20, 10, 5.　**498.** 5, 6, $-7$.　**499.** 17, 22, 45.　**500.** $\dfrac{2}{a - b + c}, \dfrac{2}{a + b - c}, \dfrac{2}{b + c - a}$.　**501.** 10, 2, 3.
**502.** 2, 1, 4.　**503.** $\frac{1}{2}, 2, 2\frac{1}{4}$.　**504.** 2, $-3$, 0.　**505.** 4, $-1$, $-2$.
**506.** 7, 2, 0.　**507.** $-3$, 2, 5.　**508.** 8, $-4$, 5.

**Page 284.—509.** 10, 6, 0. **510.** 5, 3, 8. **511.** 3, 0, 5. **512.** 8, − 1,

− 2. **513.** − 7, 6, 14. **514.** $1\frac{1}{3}$, 4, $\frac{1}{3}$. **515.** $\frac{2}{3}$, $\frac{3}{4}$, $\frac{2}{5}$. **516.** $\dfrac{2\,a}{a+c}$,

$\dfrac{2\,b}{b-c}$, $\dfrac{2\,c}{a-b}$. **517.** $\dfrac{2}{a+b-c}$, $\dfrac{2}{b-a-c}$, $\dfrac{2}{a+b+c}$. **518.** 2, 3, 4.

**519.** 1, 1, 1. **520.** $-\frac{1}{2}(2\,a+b+c)$, $-\frac{1}{2}(a+2\,b+c)$, $-\frac{1}{2}(a+b+2\,c)$.

**521.** $\dfrac{b^2+c^2-a^2}{2\,bc}$, $\dfrac{c^2+a^2-b^2}{2\,ac}$, $\dfrac{a^2+b^2-c^2}{2\,ab}$. **522.** $a$, $b$, $c$. **523.** 315,

3465, 9009, 6435. **524.** $a+b+c$, $a+b-c$, $a-b+c$, $-a+b+c$.

**Page 285.— 525.** (a) 74 lb. tin, 46 lb. lead. (b) 111 lb. tin, 115 lb.
lead. **526.** $4\frac{4}{5}$ da., $3\frac{3}{7}$ da., 24 da. **527.** $3\frac{3}{5}$, $4\frac{4}{5}$, 24. **528.** $17\frac{1}{2}$ min.
**529.** $\frac{3}{2}(a+b-c)$, $a+b$. **530.** 9 da., $7\frac{1}{5}$ da. **531.** 232. **532.** 3, 4, 9.
**533.** 8 mi. per hour, 4 mi. per hr.

**Page 286.— 536.** $2\frac{2}{11}$ da. **537.** 1.56 sec. **550.** (a) 2.75, − 3.25, 1.5.
(b) 3.24, − 1.24. (c) 3. (d) 1. (e) 2.4, − .4.

**Page 287.— 551.** (b) 31.25 m. (c) 2.24 sec. **552.** 3.83, − 1.83.
**553.** 3, 3. **554.** .7, − 5.7. **555.** 5.54, − .54. **556.** 4.37, − 1.37.
**557.** − 2.16, 4.16. **558.** − 1.93, 2.93. **559.** − 1.92, 3.92. **560.** − 5.62,
.62. **561.** − 1.31, 3.31. **562.** − 1.53, − .35, 1.88. **563.** 1.78, 2 imag.
**564.** − 1.94, .55, 1.39. **565.** ± .94, ± 3.02. **566.** − 2.5, 1.73, 2 imag.
**567.** − 2.99, − 1.15, .21, 1.9, 3.05. **568.** 1.38. **569.** (a) − 4.51,
− 1.75, 1.26. (b) −4.12, −2.4, 1.52. (c) −4.78, −1.14, .92. (d) − 5.19,
2 imag. (e) 3. (f) − 10 to 8.5 +. (g) − 10 or 8.5 +. (h) 8.5. (i) −3.33.
**570.** $1\frac{1}{2}$, $2\frac{1}{2}$. **571.** $\frac{2}{5}$, $1\frac{2}{5}$. **572.** 3.6, − 1.6 ; − 1.6, 3.6. **573.** 3, 2 ; 2, 3.
**574.** 4, 3 ; − 3, − 4. **575.** 4, 2 ; − 2, − 4.

**Page 288.—576.** 4.3, 1.4 ; − 1.8, − 3.4. **577.** 2.3, 1.15 ; − 2.3,
− 1.15. **578.** ± 4.8, ± 1.3 ; ± 1.3, ± 4.8. **579.** ± 3, ± 1 ; ± ∞, ∓∞.
**580.** Roots imaginary. **581.** − 7, − 1 ; 1, 7. **582.** 2, 4 ; 0, 0.
**583.** 3, 0 ; 1, − 2. **584.** $-\dfrac{64\,x^3}{27\,y^6}$. **585.** $\dfrac{x^{12}}{y^8 z^8}$. **586.** $a^7+7\,a^6b+21\,a^5b^2$
$+ 35\,a^4b^3 + 35\,a^3b^4 + 21\,a^2b^5 + 7\,ab^6 + b^7$. **587.** $a^7 - 7\,a^6b + 21\,a^5b^2 -$
$35\,a^4b^3 + 35\,a^3b^4 - 21\,a^2b^5 + 7\,ab^6 - b^7$. **588.** $a^6 - 3\,a^4b^2 + 3\,a^2b^4 - b^6$.
**589.** $1 - 3\,x + 3\,x^2 - x^3$. **590.** $8 + 12\,x + 6\,x^2 + x^3$. **591.** $27 - 54\,x$
$+36\,x^2-8\,x^3$. **592.** $1+4\,x+6\,x^2+4\,x^3+x^4$. **593.** $x^4-8\,x^3+24\,x^2-32\,x+16$.
**594.** $2\,a^3x^3 + 6\,axb^2y^2$. **595.** $2\,a^4x^4 + 12\,a^2x^2b^2y^2 + 2\,b^4y^4$.
**596.** $2(5\,x + 10\,x^3 + x^5)$. **597.** $1 - 4\,x^2 + 6\,x^4 - 4\,x^6 + x^8$.
**598.** $1 + 2\,x + 3\,x^2 + 2\,x^3 + x^4$. **599.** $1 - 2\,x + 3\,x^2 - 2\,x^3 + x^4$.
**600.** $2(4 + 25\,x^2 + 16\,x^4)$. **601.** $1 + 3\,x + 6\,x^2 + 7\,x^3 + 6\,x^4 + 3\,x^5 + x^6$.
**602.** $8\,a^6 - 8\,a^4b + 6\,a^2b^2 - 4\,b^3$. **603.** $2 - 2\,xy^2 + 8\,x^2y^4 - 8\,x^3y^6$.
**604.** $x^8 - x^6y + x^4y^2 - x^2y^3 + y^4$. **605.** $a^4 - 2\,a^3b + 3\,ab^3 - b^4$.

**606.** $3 - x + 2x^2$.    **607.** $2a^2b + a - b^2$.    **608.** $x^3 + 3x - \dfrac{2}{x} + \dfrac{1}{x^3}$.

**609.** $2x^2 + 4xy^2 + 3xy + y^3$.    **610.** $\pm(ax - x + b)$.    **611.** $\pm(a - 3b + 5c)$.

**612.** $\pm\left(x + \dfrac{a}{3} - \dfrac{b}{2}\right)$.

**Page 289.—613.** $a + b$.    **614.** $2x + 3$.    **615.** $3x^2 - y$.    **616.** $2a - b^2$.
**617.** $a - b$.    **618.** $x - 2$.    **619.** 971.    **620.** 5062.    **621.** 78.04.    **622.** 2092.
**623.** 7003.    **624.** 210.6.    **625.** 1010.    **626.** 898.    **627.** 7.062.    **628.** 609.3.
**629.** 25.203.    **630.** 14.702.    **631.** 703.001.    **632.** $1\frac{1}{2}$.    **633.** $1\frac{2}{3}$.    **634.** $1\frac{1}{5}$.
**635.** 2.049.    **636.** 3.061.    **637.** 6.363.    **638.** $8\frac{1}{4}$.    **639.** 4330 da.    **640.** 5, $-1\frac{1}{4}$.
**641.** 25, $-4$.    **642.** 12, $-13$.    **643.** 50, 3.    **644.** 1, $-4$.    **645.** 4, $-25$.
**646.** 2, $-8$.    **647.** 2, $-11$.    **648.** 11, $-6$.    **649.** 9, $-9\frac{2}{3}$.    **650.** 2, $-2\frac{1}{3}$.
**651.** $-2$, 10.

**Page 290.—652.** 2, $-1\frac{1}{2}$.    **653.** 2, 3.    **654.** $-3$, $\frac{2}{3}$.    **655.** $1\frac{1}{3}$, $-2\frac{1}{2}$.
**656.** 7, $-1\frac{1}{3}$.    **657.** $-1$, $1\frac{1}{4}$.    **658.** 3, $-4\frac{2}{3}$.    **659.** 3, $1\frac{10}{11}$.    **660.** 2, $4\frac{6}{13}$.

**661.** $-a$, $-b$.    **662.** $a + b$, $a - b$.    **663.** $\dfrac{a}{b}$, $-\dfrac{2a}{b}$.    **664.** 4, 0.    **665.** $1\frac{1}{3}$, 0.

**666.** 13, $\frac{5}{7}$.    **667.** 6, $-3\frac{1}{3}$.    **668.** 5, $-1\frac{5}{13}$.    **669.** 5, $1\frac{1}{5}$.    **670.** 5, $-1\frac{1}{4}$.

**671.** $2\frac{2}{3}$, 0.    **672.** $a \pm \dfrac{1}{a}$.    **673.** $(a \pm b)^2$.    **674.** $\pm\sqrt{ab}$.    **675.** $a$, $-\dfrac{b(a+b)}{2a+b}$.

**676.** $m$, $-\dfrac{a^2 + b^2 + m(a+b)}{2m + a + b}$.    **677.** $\dfrac{a}{2}(-3 \pm \sqrt{3})$.    **678.** $\pm\sqrt{a^2 + b^2}$

**679.** $\pm\sqrt{a^2 + 10\,ab + b^2}$.    **680.** $1\frac{1}{2}$.    **681.** 3.    **682.** $a^2 + a$, $-1$.

**683.** $\frac{1}{5}(2a + 3b)$, $\frac{1}{5}(3a + 2b)$.    **684.** $\dfrac{a^2}{b}$, $\dfrac{b^2}{a}$.    **685.** 0, $\dfrac{-(a^2 + b^2)}{a + b}$.

**686.** $\frac{3}{2}a$, $3a$.    **687.** $a + b$, $\dfrac{2ab}{a+b}$.

**Page 291.—688.** $c$, $\dfrac{-ab(a+c)}{ab + bc - ac}$.    **689.** $-a$, $-b$.    **690.** $\pm\frac{1}{2}$, $\pm\frac{1}{2}$.

**691.** $2a - b$, $2b - a$.    **692.** $\dfrac{a^2 + b^2}{2b}$, $\dfrac{2ab + b^2 - a^2}{2a}$.    **693.** $\dfrac{a+b}{a-b}$,

$\dfrac{a-b}{a+b}$, $-1$.    **694.** $\dfrac{b-a}{1-ab}$, $\dfrac{1-ab}{b-a}$.    **695.** $\dfrac{a+b}{1-ab}$, $\dfrac{a-b}{1+ab}$.    **696.** 1, $\dfrac{-c}{a+b+c}$.

**697.** 1, $\dfrac{-(a+b+c)}{a}$.    **698.** $\sqrt{3} \pm \sqrt{2}$.    **699.** $\sqrt{5} \pm 1$.

**700.** $-1$, $-2\frac{1}{3}$, $-4$, $-2\frac{2}{3}$.    **701.** $\pm 1$, $-2$, $-4$.    **702.** 1, $1\frac{1}{4}$, $-\frac{1}{2}$, $-\frac{1}{4}$.
**703.** 3, 2, $\frac{1}{2}(5 \pm \sqrt{-3})$.    **704.** 2, $-3$, $\frac{1}{2}(-1 \pm 3\sqrt{-3})$.
**705.** $-2 \pm \sqrt{2}$, $1 \pm \sqrt{2}$.    **706.** $\dfrac{\pm a \pm \sqrt{a^2 - b^2}}{b}$.    **707.** $\frac{1}{2}(-5 \pm \sqrt{41})$,
$\frac{1}{2}(-5 \pm \sqrt{-15})$.    **708.** $1 \pm \sqrt{7}$, $1 \pm \sqrt{-6}$.

**Page 292.**— **709.** $\pm 3, \pm 2.$  **710.** $\pm \dfrac{a}{2}, \dfrac{3\,a}{2}.$  **711.** $\sqrt[n]{\pm b - a}.$

**712.** $\pm \sqrt{25 \pm 4\sqrt{-34}}.$  **713.** $\pm \sqrt{11}, \perp \sqrt{1\frac{1}{3}}.$  **714.** $\pm \sqrt{3}, \pm \sqrt{1\frac{1}{6}}.$

**715.** $1, 1, \dfrac{2 \pm \sqrt{-5}}{3}.$  **716.** $24, 25.$  **717.** $\frac{5}{3}.$  **718.** $\dfrac{a \pm \sqrt{a^2 - 4\,b}}{2}.$

**719.** $45$ da.  **720.** $17$ ft., $13$ ft.  **721.** $6\,\not{e}.$  **722.** $8, 9.$  **723.** $8, 9, 10, 11.$

**724.** $\dfrac{a}{2}\sqrt{3}.$  **725.** $15$ shares.  **726.** $39, 8.$  **727.** $30\,\not{e}.$  **728.** $\frac{1}{12}.$

**729.** $13\frac{1}{2}.$

**Page 293.**— **730.** $y - 1.$  **731.** $a^2 - x^2.$  **732.** $a + b + c - 3\,a^{\frac{1}{3}}b^{\frac{1}{3}}c^{\frac{1}{3}}.$

**733.** $m - n.$  **734.** $m^{\frac{4}{3}} + 4\,d^{\frac{1}{2}}m^{\frac{2}{3}} + 16\,d.$  **735.** $x^4 + 1 + x^{-4}.$

**736.** $a^{-4} - b^{-4}.$  **737.** $a^{-2} + 2\,a^{-1}c^{-1} - b^{-2} + c^{-2}.$  **738.** $1 + a^2b^{-2} + a^4b^{-4}.$

**739.** $4\,x^{-5} - x^{-4} + 3\,x^{-3} + 2\,x^{-2} + x^{-1} + 1.$  **740.** $16\,x^{-\frac{2}{3}} - 12\,x^{-\frac{1}{3}}y^{-\frac{2}{3}} + 9\,y^{-\frac{4}{3}}.$

**741.** $x + y.$  **742.** $a^{\frac{1}{3}} - a^{\frac{1}{6}}b^{\frac{1}{6}} + b^{\frac{1}{3}}.$  **743.** $a^{\frac{1}{3}} + b^{\frac{1}{3}} - c^{\frac{1}{3}}.$  **744.** $x^{\frac{1}{2}} + 2\,x^{\frac{5}{8}}a^{\frac{1}{2}}$

$+ 3\,x^{\frac{1}{4}}a + 2\,x^{\frac{1}{8}}a^{\frac{3}{2}} + a^2.$  **745.** $x - 2 + \dfrac{3}{x}.$  **746.** $2\,x - 3\,x^{\frac{1}{3}} + 7.$

**747.** $a^3 + 2\,a^2b^{-1} - ab^{-2} - b^{-3}.$  **748.** $x^5, x^{\frac{2n}{3}}, 1.$  **749.** $x^{-1}, x, a^2b^{\frac{1}{2}}.$

**750.** $2\,a\sqrt{2\,a}, 24\,a^4, x^{-\frac{1}{6}}y^{\frac{1}{2}}z^{1\frac{3}{2}}.$  **751.** $a^{\frac{61}{2}}, b^{-\frac{1}{7}}x^{-\frac{5}{7}}y^{-\frac{1}{5}}.$  **752.** $1.$

**Page 294.**— **753.** $x.$  **754.** $\dfrac{x^{\frac{1}{3}}}{x^{\frac{1}{3}} + y^{\frac{1}{3}}}.$  **755.** $\dfrac{1}{1 - x^{\frac{1}{2}}}.$  **756.** $34.$

**757.** $6 - 6\sqrt[3]{6}.$  **758.** $5.$  **759.** $29\sqrt{3}.$  **760.** $30\sqrt{10} + 164\sqrt{2}.$

**761.** $5\sqrt[3]{2}.$  **762.** $(a^2 + b^2 + c^2)\sqrt{x}.$  **763.** $33\sqrt[3]{2}.$  **764.** $59{,}257\frac{3}{8}.$

**765.** $\sqrt[60]{7}.$  **766.** $\sqrt[xy]{a^3}.$  **767.** $\sqrt[24]{a^{-19}b^7c^{13}}.$  **768.** $-3\frac{11}{30}.$  **769.** $\frac{88}{120}.$

**770.** $\left(\dfrac{1}{a} + \dfrac{1}{n}\right)\sqrt{m - n}.$  **771.** $\dfrac{2\,ab}{c}\sqrt{2}.$  **772.** $\dfrac{a\sqrt{ab^2c^3d^4}}{bcd}.$  **773.** $a^{7x-8y}.$

**774.** $x^{5m-6}.$  **775.** $a^{12x+14y}.$  **776.** $2 + \sqrt{2}.$  **777.** $3 + 2\sqrt{3}.$  **778.** $3 - 2\sqrt{2}.$

**779.** $\dfrac{a + x + 2\sqrt{ax}}{a - x}.$

**Page 295.**— **780.** $m^2 - \sqrt{(m^4 - 1)}.$  **781.** $\dfrac{1 + x + 2\sqrt{x}}{1 - x}.$

**782.** $\dfrac{a + \sqrt{(a^2 - x^2)}}{x}.$  **783.** $\dfrac{2\,a^2 - x^2 + 2\,a\sqrt{(a^2 - x^2)}}{x^2}.$  **784.** $\sqrt{7} + \sqrt{3}.$

**785.** $\sqrt{11} + \sqrt{5}.$  **786.** $\sqrt{7} - \sqrt{2}.$  **787.** $7 - 3\sqrt{5}.$  **788.** $\sqrt{10} - \sqrt{3}.$

**789.** $2\sqrt{5} - 3\sqrt{2}.$  **790.** $2\sqrt{3} - \sqrt{2}.$  **791.** $3\sqrt{11} - 2.$  **792.** $3\sqrt{7} - 2\sqrt{3}.$

**793.** $3\sqrt{7} - 2\sqrt{6}.$  **794.** $\frac{1}{2}(\sqrt{10} - 2).$  **795.** $3\sqrt{5} - 2\sqrt{3}.$  **796.** $3 - \sqrt{7}.$

**797.** $6 + \sqrt{7}$.     **798.** $\sqrt{a + b} + \sqrt{a - b}$.     **799.** $\sqrt{a - b} + \sqrt{b - c}$

**800.** $\sqrt{\dfrac{a}{b} - \dfrac{1}{2}} + \sqrt{\dfrac{b}{a} + \dfrac{1}{2}}$.     **801.** $\frac{1}{5}\sqrt{5}$.     **802.** $\frac{1}{2}\sqrt{6}$.     **803.** $0$.     **804.** $\sqrt{2}$.

**805.** $\sqrt{\dfrac{x}{y} + \dfrac{1}{2}} + \sqrt{\dfrac{y}{x}}$.     **806.** $48$.     **807.** $\dfrac{2\,x}{a^2}$.     **808.** $\dfrac{2\sqrt{a^2 - x^2}}{x}$.     **809.** $x$.

**Page 296.** — **810.** $8(y^2 + 2\,x^2 y - x^4),\ 16\,xy\sqrt{2\,y - x^2}$.     **811.** $16$.

**812.** $1$.     **813.** $19$.     **814.** $\dfrac{a^4 - 2\,a^2 b^2 + b^4}{4\,b^2}$.     **815.** $17$.     **816.** $23$.     **817.** $10$.

**818.** $4$.     **819.** $11$.     **820.** $13$.     **821.** $5$.     **822.** $3$.     **823.** $7$.     **824.** $7$.

**825.** $7$.     **826.** $9$.     **827.** $\pm\,\dfrac{2\,a}{3}\sqrt{3}$.     **828.** $-1$.     **829.** $17$.     **830.** Roots are extraneous.     **831.** Roots are extraneous.     **832.** $9$.     **833.** $-\frac{3}{4}$. **834.** $5$.     **835.** $0, 3$.     **836.** $(c + d)^2$.     **837.** $25$.

**Page 297.** — **838.** $-4, 5, -7\frac{1}{5}, 1\frac{4}{5}$.     **839.** $2, 3, -\frac{2}{3}, \frac{1}{3}$.     **840.** $-4, 1$. **841.** $0, 2\frac{1}{2}$.     **842.** $-11, 14$.     **843.** $(x + 2)(x^3 + 5\,x + 8)$.     **844.** $(x + 1)$ $(x^2 + x + 1)(x^2 - x + 1)$.     **845.** $(a + 2)(a - 2)(5\,a^2 + 7\,a + 20)$. **846.** $(x + 2\,y)(x - 2\,y)(4\,x^2 - xy + 16\,y^2)$.     **847.** $(x - 1)(x - 1)(x + 2)$. **848.** $(x - 1)(x + 2)(x + 2)$.     **849.** $(a - b)(a^2 + ab - 3\,b^2)$.     **850.** $(x + 3)$ $(x^2 + x - 1)$.     **851.** $(x + y)(x + y)(x + y)(x - y)$.     **852.** $(x^2 + x - 5)$ $(x - 3)$.     **853.** $(x + 9)(x^2 + 2\,x - 5)$.     **854.** $(x - 1)(x - 3)(x + 4)$. **855.** $(x - 1)(x - 3)(x - 4)$.     **856.** $(x + 1)(x + 2)(x - 7)$.     **857.** $(x - 1)$ $(2\,x + 3)(2\,x + 3)$.     **858.** $(x + 1)(4\,x - 7)(4\,x - 7)$.     **859.** $(2\,x + 3\,y)$ $(4\,x^2 - 6\,xy + 9\,y^2)$.     **860.** $(2 + ax)(4 - 2\,ax + a^2 x^2)$.     **861.** $(3\,x - 4)$ $(9\,x^2 + 12\,x + 16)$.     **862.** $(1 - 4\,a)(1 + 4\,a + 16\,a^2)$.     **863.** $a(xy + z)$ $(x^2 y^2 - xyz + z^2)$.     **864.** $(3\,b - 1)(9\,b^2 + 3\,b + 1)$.     **865.** $(2\,a - 5\,b)$ $(4\,c^2 + 10\,ab + 25\,b^2)$.     **866.** $(9\,x + 8\,y)(81\,x^2 - 72\,xy + 64\,y^2)$. **867.** $(2\,x - 3\,yz)(4\,x^2 + 6\,xyz + 9\,y^2 z^2)$.     **868.** $(a + 2\,bc)(a^2 - 2\,abc + 4\,b^2 c^2)$. **869.** $(a^m - b^m)(a^{2m} + a^m b^m + b^{2m})$.     **870.** $(a^{2m} + b^{2n})(a^{4m} - a^{2m} b^{2n} + b^{4n})$. **871.** $(a^4 + 1)(a^8 - a^4 + 1)$.     **872.** $(a + 6\,n^3)(a^2 - 6\,an^3 + 36\,n^6)$. **873.** $b(a - b)(a^2 + ab + b^2)$.     **874.** $a(a - b)(a^2 + ab + b^2)(a^6 + a^3 b^3 + b^6)$. **875.** $a(a^4 + 1)(a^8 - a^4 + 1)$.     **876.** $(a^m + 1)(a^m - 1)(a^{4m} + a^{3m} + a^{2m}$ $+ a^m + 1)(a^{4m} - a^{3m} + a^{2m} - a^m + 1)$.

**Page 298.** — **879.** $4$.     **880.** $m = 2,\ n = -20$.     **881.** $4, 4\ ;\ -4, 8$. **882.** $3, 1\ ;\ 1, 3$.     **883.** $5, 2\ ;\ 2, 5$.     **884.** $\pm 2, \pm 3$.     **885.** $\pm 4, \pm 3$, $\pm 3, \pm 4$.     **886.** $2, \frac{2}{3}\ ;\ 1, \frac{1}{4}$.     **887.** $\pm 5, \pm 3$.     **888.** $2, 1\ ;\ 1, 2$; $-1, -2\ ;\ -2, -1$.     **889.** $3, 6\ ;\ -\frac{1}{3}, -\frac{2}{3}$.     **890.** $5, 2\ ;\ -\frac{5}{3}, -\frac{2}{3}$. **891.** $2, 4\ ;\ 4, 2$.     **892.** $\frac{1}{3}, \frac{1}{2}\ ;\ \frac{1}{2}, \frac{1}{3}$.     **893.** $\frac{1}{4}, 2$.     **894.** $4, 3\ ;\ -3, -4$. **895.** $5, 3\ ;\ 3, 5$.     **896.** $\pm 7, \pm 2$.     **897.** $\pm 7, \pm 11$.     **898.** $4, 3\ ;\ \pm \frac{7}{2}\sqrt{3}$, $\mp \frac{5}{2}\sqrt{3}$.     **899.** $\pm 3, \pm 2\ ;\ \pm \frac{7}{3}\sqrt{\frac{2}{3}}, \pm \frac{19}{6}\sqrt{\frac{2}{3}}$.     **900.** $0, 0\ ;\ \pm 8\sqrt{\frac{5}{41}}$,

$\pm 10\sqrt{\frac{5}{41}}$; $\mp 8\sqrt{\dfrac{-5}{41}}$, $\mp 10\sqrt{\dfrac{-5}{41}}$. **901.** $\dfrac{b^2}{b-a}$, $\dfrac{a^2}{b-a}$; $\dfrac{-b^2}{a+b}$, $\dfrac{-a^2}{a+b}$.

**Page 299.—902.** 2, 1; $-1$, 1; $\frac{1}{2}(1 \pm \sqrt{-3})$, $1 \pm \sqrt{3}$. **903.** $\pm 3$, $\mp 3$. **904.** 8, 2; 2, 8. **905.** $\pm 4$, $\pm 3$; $\pm 1$, $\pm 12$. **906.** 115, 329; 333, 111. **907.** 0, 0; $\sqrt[3]{(a-1)(b-1)^2}$, $\sqrt[3]{(a-1)^2(b-1)}$. **908.** 3, 6; 6, 3. **909.** 19, 91; $-\frac{17}{13}$, $\frac{217}{13}$. **910.** $m$, $n$; $\dfrac{m^2-n^2}{2m}$, $\dfrac{n^2-m^2}{2n}$.

**911.** $b$, $a$; $\dfrac{a^2}{b}$, $\dfrac{b^2}{a}$. **912.** $\dfrac{\pm 2}{\sqrt{5}}$, $\dfrac{\pm 1}{\sqrt{5}}$. **913.** $\pm 3$, $\pm 5$. **914.** $\pm 1$, $\mp 2$; $\pm\frac{11}{4}$, $\mp\frac{1}{4}$. **915.** $\pm 4$, $\pm 2$. **916.** 2, 1; 1, 2; $-2$, $-1$; $-1$, $-2$. **917.** 8, 6; 6, 8. **918.** $\pm 3$, $\pm 2$. **919.** $\pm 1$, $\pm 2$. **920.** 5, 3; 3, 5.

**921.** 0, 0; $\dfrac{ab^2 \pm ab\sqrt{2a^2-b^2}}{b^2-a^2}$, $\dfrac{ab^2 \mp ab\sqrt{2a^2-b^2}}{b^2-a^2}$. **922.** 5, 4; 4, 5.

**923.** 0, 0; 4, $-1$. **924.** $\dfrac{\pm a^2}{\sqrt{a^2+b^2}}$, $\dfrac{\pm b^2}{\sqrt{a^2+b^2}}$. **925.** 3, 4; 5, 2.

**926.** 8, $-4$; 28, 56; $-4$, 8; 56, 28. **927.** $a$, $b$; $\dfrac{a+2b}{3}$, $\dfrac{4a-b}{3}$.

**928.** 1, 1; $\dfrac{2}{-1 \mp \sqrt{-3}}$, $\dfrac{-1 \pm \sqrt{-3}}{2}$. **929.** $\pm 2$, $\pm 4$, $\pm 6$.

**Page 300.—930.** $\pm 3$, $\pm 7$, $\mp 2$. **931.** $\pm 9$, $\mp 6$, $\pm 8$. **932.** $\pm 8$, $\pm 7$, $\pm 4$. **933.** $\pm 3$, $\pm 4$, $\pm 5$. **934.** 9, 6. **935.** 7, 4. **936.** 12, 8. **937.** 20 ft., 15 ft. **938.** $3\frac{1}{4}$ ft., $2\frac{1}{2}$ ft. **939.** 480 sq. ft. **940.** 100 rows. **941.** 19 ft., 16 ft. **942.** 3 in., 4 in. **943.** 15 ft., 8 ft. **944.** 10, 12 mi./hr.

**Page 301.—945.** 60 ft., 40 ft. **946.** 16 in., 9 in. **947.** 11,760 sq. yd. **948.** 73. **949.** $\frac{1}{2}(-1+\sqrt{-3})$, $\frac{1}{2}(-1-\sqrt{-3})$. **950.** 248. **951.** 6 da. **952.** 12 in., 6 in. **953.** 7 ft., 4 ft. **954.** 6 mi. **955.** 2 yd.

**Page 302.—956.** 4. **957.** $-333$. **958.** $-26\frac{2}{3}$. **959.** 1, $-2$, $-5 \cdots$. **960.** 280. **961.** $514\frac{3}{4}$. **962.** 108. **963.** 4. **964.** 11,111.

**965.** $-n(-1)^n$. **966.** $-\dfrac{n}{2}$ if $n$ is even, $\dfrac{n+1}{2}$ if $n$ is odd. **967.** $\frac{16}{3}$.

**968.** $\frac{27}{26}$. **969.** $4+3\sqrt{2}$. **970.** $\dfrac{x(x^{10}-1)}{x-1} + \dfrac{y(y^{10}-1)}{y-1}$.

**971.** $\dfrac{x^2(x^{2n}-1)}{x^2-1} + \dfrac{xy(x^ny^n-1)}{xy-1}$. **972.** $36x + \dfrac{y(y^8-1)}{y-1}$. **973.** (a) $\frac{11}{90}$; (b) $\frac{52}{165}$. **974.** $\frac{1}{2}(n^2-3n+4)$. **975.** 0. **976.** 2, 3. **977.** 80.

**Page 303.—978.** 10, 12, 14 $\cdots$. **979.** $\frac{8}{5}$, $\frac{7}{5}$, $\frac{6}{5}$, $\cdots \frac{1}{5}$. **980.** 7 or 30. **981.** 4. **982.** 3. **983.** 5, 11. **984.** $\pm 5$, $\pm 7$, $\pm 9$, $\pm 11$. **985.** 8. **986.** 6, 28, 496, 8128. **987.** $2^{64}-1 = 18,446,744,073,709,551,615$.

**988.** $\frac{1}{14}(2 + 3\sqrt{2})$.   **989.** $6\frac{1}{4}$.   **990.** $\frac{n(n+1)}{2} + \frac{1}{9}\left(1 - \frac{1}{10^n}\right)$.   **991.** $\frac{1}{5}$.

**992.** $2(2 - \sqrt{2})$.   **993.** $6, 18, 54$.

**Page 304.** — **994.** $162, 108, 72, 48$.   **995.** $a = \frac{1}{4}, r = 2$.   **996.** $4, 8, 16$.

**997.** $3, 6, 12, 24$.                    **999.** $(a)\ 2\pi(2 + \sqrt{2}),\ (b)\ 8(1 + \sqrt{2})$.

**1000.** $(a)\ 12(2 + \sqrt{3}),\ (b)\ 4\sqrt{3}$.   **1001.** $(a)\ \frac{1}{1024},\ (b)\ \frac{1}{4^n}$.   **1002.** $6$.

**1003.** $\frac{3\,n^2}{32}\pi$ sq. in.   **1004.** $9$ da.   **1005.** $a^{13} - 13\,a^{12}x + 78\,a^{11}x^2 \cdots -$
$78\,a^2x^{11} + 13\,ax^{12} - x^{13}$.

**Page 305.** — **1006.** $243 - 810\,x^2 + 1080\,x^4 - 720\,x^6 + 240\,x^8 - 32\,x^{10}$.
**1007.** $1 - 14\,y + 84\,y^2 - 280\,y^3 + 560\,y^4 - 672\,y^5 + 448\,y^6 - 128\,y^7$.

**1008.** $x^n + 2\,nx^{n-1}y + 2\,n(n-1)x^{n-2}y^2 + \frac{4\,n(n-1)(n-2)}{3}\,x^{n-3}y^3$.

**1009.** $192,192\,a^6b^6c^8d^8$.          **1010.** $12,870\,a^8b^8$.          **1011.** $70\,a^{\frac{1}{2}}b^{\frac{1}{2}}$.

**1012.** $-\,92,378\,a^{10}b^9$ and $92,378\,a^9b^{10}$.          **1013.** $1716\,a^7x^6$ and $1716\,a^6x^7$.

**1014.** $126\,x^4$.        **1015.** $252\,a^5b^5$.          **1016.** $126\,a^5x^4$ and $-\,126\,a^4x^5$.

**1017.** $\frac{33 \cdot 31 \cdot 29}{2^{20}x^{23}}$.        **1018.** $35$.   **1019.** $12,870\,a^8x^8$.   **1020.** $-\,5^57^4a^{12}$.

**1021.** $3003$.

# ELEMENTARY ALGEBRA

By Arthur Schultze. 12mo. Half leather. xi + 373 pages. $1.10

The treatment of elementary algebra here is simple and practical, without the sacrifice of scientific accuracy and thoroughness. Particular care has been bestowed upon those chapters which in the customary courses offer the greatest difficulties to the beginner, especially Problems and Factoring. The introduction into Problem Work is very much simpler and more natural than the methods given heretofore. In Factoring, comparatively few methods are given, but these few are treated so thoroughly and are illustrated by so many varied examples that the student will be much better prepared for further work, than by the superficial study of a great many cases. The Exercises are very numerous and well graded; there is a sufficient number of easy examples of each kind to enable the weakest students to do some work. A great many examples are taken from geometry, physics, and commercial life, but none of the introduced illustrations is so complex as to require the expenditure of time for the teaching of physics or geometry. To meet the requirements of the College Entrance Examination Board, proportions and graphical methods are introduced into the first year's course, but the work in the latter subject has been so arranged that teachers who wish a shorter course may omit it.

# ADVANCED ALGEBRA

By Arthur Schultze, Ph.D. 12mo. Half leather. xiv + 562 pages. $1.25

The Advanced Algebra is an amplification of the Elementary. All subjects not now required for admission by the College Entrance Examination Board have been omitted from the present volume, save Inequalities, which has been retained to serve as a basis for higher work. The more important subjects which have been omitted from the body of the work — Indeterminate Equations, Logarithms, etc. — have been relegated to the Appendix, so that the book is a thoroughly practical and comprehensive text-book. The author has emphasized Graphical Methods more than is usual in text-books of this grade, and the Summation of Series is here presented in a novel form.

## THE MACMILLAN COMPANY

**PUBLISHERS, 64-66 FIFTH AVENUE, NEW YORK**

# ELEMENTARY ALGEBRA

By Arthur Schultze. 12mo. Half leather. xi + 373 pages. $1.10

The treatment of elementary algebra here is simple and practical, without the sacrifice of scientific accuracy and thoroughness. Particular care has been bestowed upon those chapters which in the customary courses offer the greatest difficulties to the beginner, especially Problems and Factoring. The introduction into Problem Work is very much simpler and more natural than the methods given heretofore. In Factoring, comparatively few methods are given, but these few are treated so thoroughly and are illustrated by so many varied examples that the student will be much better prepared for further work, than by the superficial study of a great many cases. The Exercises are very numerous and well graded; there is a sufficient number of easy examples of each kind to enable the weakest students to do some work. A great many examples are taken from geometry, physics, and commercial life, but none of the introduced illustrations is so complex as to require the expenditure of time for the teaching of physics or geometry. To meet the requirements of the College Entrance Examination Board, proportions and graphical methods are introduced into the first year's course, but the work in the latter subject has been so arranged that teachers who wish a shorter course may omit it.

# ADVANCED ALGEBRA

By Arthur Schultze, Ph.D. 12mo. Half leather. xiv + 562 pages. $1.25

The Advanced Algebra is an amplification of the Elementary. All subjects not now required for admission by the College Entrance Examination Board have been omitted from the present volume, save Inequalities, which has been retained to serve as a basis for higher work. The more important subjects which have been omitted from the body of the work — Indeterminate Equations, Logarithms, etc. — have been relegated to the Appendix, so that the book is a thoroughly practical and comprehensive text-book. The author has emphasized Graphical Methods more than is usual in text-books of this grade, and the Summation of Series is here presented in a novel form.

## THE MACMILLAN COMPANY
PUBLISHERS, 64-66 FIFTH AVENUE, NEW YORK

# PLANE AND SOLID GEOMETRY

By Arthur Schultze and F. L. Sevenoak. 12mo. Half leather. xii + 370 pages. $1.10

## PLANE GEOMETRY

Separate. 12mo. Cloth. xii + 233 pages. 80 cents

This Geometry introduces the student systematically to the solution of geometrical exercises. It provides a course which stimulates him to do original work and, at the same time, guides him in putting forth his efforts to the best advantage.

The Schultze and Sevenoak Geometry is in use in a large number of the leading schools of the country. Attention is invited to the following important features: 1. *Preliminary Propositions* are presented in a simple manner; 2. The numerous and well-graded *Exercises* — more than 1200 in number in the complete book. These are introduced from the beginning; 3. Statements from which *General Principles* may be obtained are inserted in the Exercises, under the heading "Remarks"; 4. *Proofs* that are special cases of general principles obtained from the Exercises are not given in detail. Hints as to the manner of completing the work are inserted; 5. *The Order of Propositions* has a distinct pedagogical value. Propositions easily understood are given first and more difficult ones follow; 6. *The Analysis of Problems and of Theorems* is more concrete and practical than in any other text-book in Geometry; 7. Many proofs are presented in a *simpler and more direct manner* than in most text-books in Geometry; 8. Difficult Propositions are made somewhat easier by applying *simple Notation;* 9. The *Algebraic Solution of Geometrical Exercises* is treated in the Appendix to the Plane Geometry; 10. Pains have been taken to give *Excellent Figures* throughout the book.

## KEY TO THE EXERCISES

In Schultze and Sevenoak's Plane and Solid Geometry. By Arthur Schultze, Ph.D. 12mo. Cloth. 200 pages. $1.10

This key will be helpful to teachers who cannot give sufficient time to the solution of the exercises in the text-book. Most solutions are merely outlines, and no attempt has been made to present these solutions in such form that they can be used as models for class-room work.

---

## THE MACMILLAN COMPANY

PUBLISHERS, 64-66 FIFTH AVENUE, NEW YORK

# The Teaching of Mathematics in Secondary Schools

BY

## ARTHUR SCHULTZE

Formerly Head of the Department of Mathematics in the High School of
Commerce, New York City, and Assistant Professor of
Mathematics in New York University

### *Cloth, 12mo, 370 pages, $1.25*

The author's long and successful experience as a teacher
of mathematics in secondary schools and his careful study of
the subject from the pedagogical point of view, enable him to
speak with unusual authority. "The chief object of the
book,' he says in the preface, "is to contribute towards
making mathematical teaching less informational and more
disciplinary. Most teachers admit that mathematical instruc-
tion derives its importance from the mental training that it
affords, and not from the information that it imparts. But in
spite of these theoretical views, a great deal of mathematical
teaching is still informational. Students still learn demon-
strations instead of learning how to demonstrate."

The treatment is concrete and practical. Typical topics
treated are: the value and the aims of mathematical teaching; causes of the inefficiency of mathematical teaching;
methods of teaching mathematics; the first propositions in
geometry; the original exercise; parallel lines; methods of
attacking problems; the circle; impossible constructions;
applied problems; typical parts of algebra.

---

## THE MACMILLAN COMPANY

### 64-66 Fifth Avenue, New York

CHICAGO     BOSTON     SAN FRANCISCO     DALLAS     ATLANTA

# AMERICAN HISTORY

*For Use in Secondary Schools*

By ROSCOE LEWIS ASHLEY

*Illustrated. Cloth. 12mo. $1.40*

This book is distinguished from a large number of American history text-books in that its main theme is the development of the nation. The author's aim is to keep constantly before the pupil's mind the general *movements* in American history and *their relative value* in the development of our nation. All smaller movements and single events are clearly grouped under these general movements.

An exhaustive system of marginal references, which have been selected with great care and can be found in the average high school library, supply the student with plenty of historical narrative on which to base the general statements and other classifications made in the text.

**Topics, Studies** and **Questions** at the end of each chapter take the place of the individual teacher's lesson plans.

This book is up-to-date not only in its matter and method, but in being fully illustrated with many excellent maps, diagrams, photographs, etc.

"This volume is an excellent example of the newer type of school histories, which put the main stress upon national development rather than upon military campaigns. Maps, diagrams, and a full index are provided. The book deserves the attention of history teachers." — *Journal of Pedagogy.*

---

## THE MACMILLAN COMPANY

**64-66 Fifth Avenue, New York**

**BOSTON**      **CHICAGO**      **ATLANTA**      **SAN FRANCISCO**